ENERGY RESOURCES IN
AN UNCERTAIN FUTURE

ENERGY RESOURCES IN AN UNCERTAIN FUTURE
Coal, Gas, Oil, and Uranium
Supply Forecasting

M.A. ADELMAN
JOHN C. HOUGHTON
GORDON KAUFMAN
MARTIN B. ZIMMERMAN

BALLINGER PUBLISHING COMPANY
Cambridge, Massachusetts
A Subsidiary of Harper & Row, Publishers, Inc.

NOTE

This report was prepared by the Massachusetts Institute of Technology Energy Laboratory as an account of work sponsored by the Electric Power Research Institute, Incorporated (EPRI). Neither EPRI, members of EPRI, the MIT Energy Laboratory, nor any person acting on behalf of either: (a) makes any warranty or representation, express or implied, with respect to the accuracy, completeness, or usefulness of the information contained in this report, or that the use of any information, apparatus, method, or process disclosed in this report may not infringe privately owned rights; or (b) assumes any liabilities with respect to the use of, or for damages resulting from the use of, any information, apparatus, method, or process disclosed in this report.

International Standard Book Number: 0-88410-644-6

Library of Congress Catalog Card Number: 81-10969

Printed in the United States of America

Library of Congress Cataloging in Publication Data

Main entry under title:

Energy resources in an uncertain future.

  Bibliography: p.
  Includes index.
  1. Power resources—United States. 2. Coal—United States.
3. Gas, Natural—United States. 4. Petroleum—United States.
5. Uranium—United States. I. Adelman, Morris Albert.
TJ163.25.U6E528   553.2'0973   81-10969
ISBN 0-88410-644-6   AACR2

ACKNOWLEDGMENTS

Portions of this work were supported by funds from the Electric Power Research Institute, Palo Alto, California, and from the Center for Energy Policy Research, Massachusetts Institute of Technology, Cambridge, Massachusetts. We are also grateful to Peter Heron for superb editing of the manuscript and to Richard Gordon, Jeremy Platt, Daniel Khazzoom, and David Wood for useful critiques of earlier drafts. Henry Jacoby and Loren Cox provided much needed encouragement of this effort. D. Alec Sargent helped with excellent research assistance and Alice Sanderson skillfully typed our many drafts.

84-5449

CONTENTS

LIST OF FIGURES

LIST OF TABLES

EXECUTIVE SUMMARY

This book analyzes the economics of resource and reserve estimation. Current concern about energy problems has focused attention on how we measure available energy resources. One reads that we have an eight-year oil supply and a 500-year coal supply, implying that we must inevitably turn to coal. Uranium is either seen as scarce, implying the necessity of the breeder reactor, or it is seen as plentiful, implying no need for plutonium in reactors. Motivating these statements, which profoundly affect the tenor of our future, is an interpretation of how resources and reserves are estimated. And, as we shall see, the interpretation is more often than not incorrect: Stocks of resources are confused with flows; flows are what economists call supply.

The five parts of this book take an in-depth look at resource and reserve estimation for oil, gas, goal, and uranium. Our goal is not to provide a "good" estimate of what lies beneath the surface of the earth, but to deal with the crucial concepts lying behind resource and reserve estimation. What do geologists measure and how well does this interact with the economists' notion of supply? The fault in estimation is rarely geological but, rather, economic. The concept of abundance is an economic one and we ask what geological information means when interpreted in light of economic reasoning.

Part I deals with the economic theory of natural resources. The unique aspect of mineral economics is the fixity of supply; new resources represent a stock that once exhausted will not be replenished. Part I examines how markets deal with exhaustible resources.

The succeeding parts examine each resource individually. Parts II and III deal with oil and gas. Part II treats the measurement of already discovered deposits.

It examines the concept of proven reserves and the behavior of reserves over time. Part III turns to undiscovered deposits and asks how the size of undiscovered but suspected deposits are estimated. It casts a critical eye upon the biases inherent in many of the estimation methodologies. Part IV examines coal resource and reserve estimation. In the United States an attempt has been made to define an economically meaningful classification of coal reserves and resources. This classification is analyzed in detail along with the more general issue of coal supply. The focus is on already discovered deposits since knowledge of where coal lies is great and new discoveries play a small role in coal supply.

Part V treats uranium. For uranium, as for coal, extraction costs are high and reserve estimation must reckon with cost estimation. And, like oil and gas, the discovered position of the resource base is only a small portion of the resources likely to be exploited over time. Exploration cost is thus significant. Part VI synthesizes Parts II–V by placing resource and reserve estimates in the context of a cumulative cost curve.

PART I

The fixity of mineral supply has important implications for the behavior of economic agents:

1. If costs rise with output, lowering the rate of output will lower the average cost of producing the fixed amount of ore in the deposit. However, given positive interest rates, one would rather have revenue sooner than later. The trade-off then is between expanding production to earn higher revenue today at greater unit expense versus holding back output to produce at lower cost but postponing revenue. In a mine with a rising marginal cost of production, output will be at a lower rate than supply analysis that does not consider fixity of supply would predict.

2. Costs rise for other reasons related to the fixity of supply. First, the faster the output is removed, the more capital in the form of shafts and mine development must be invested. These shafts are useful only for removing the fixed amount of ore. The faster the rate of output, the more capital is used to produce the given deposit and thus the higher the cost.

Second, fixity of supply also refers to the quality of the reserves. Thus, once the more accessible ore is removed or the higher grade ore is removed, production must turn to the less accessible, lower-grade, higher-cost.

3. The last point generalizes to the industry as a whole. Lower cost deposits are exploited first. Easier to develop oil resources are extracted first. Depletion, the movement from low- to high-cost ore, therefore exists both within individual deposits and for the industry as a whole.

4. The cost movement described above will also be reflected in price behavior. Prices rise as costs rise. Thus the owners of a low-cost deposit knows that by

holding back production they can obtain a higher price in the future. Of course, the higher revenue in the future must be discounted at the rate of interest. The mine owners will compare the present value of the future higher price to today's price. In order to coax forth production now, therefore, today's price must be higher than if no future price increases were foreseen. The higher the future run-run-up in price, the higher today's price. In sum, future scarcity is reflected in today's price. Future cost increases and higher prices tomorrow also mean higher prices today. In order to better understand the future behavior of energy prices, we must understand the cost structure of energy deposits.

5. The foregoing theory predicts smooth price increases as we move to higher and higher cost deposits. Mineral prices have a history of constant long-run decline, however. This is because future cost increases have been retarded or reversed by new technology and new discoveries. We learn about new deposits so we need work already discovered deposits less intensively. We also learn how to exploit any given deposit more efficiently. Finally, new technology permits economic exploitation of previously inaccessible deposits.

PART II

Already discovered deposits of oil and gas are the subject of Part II. There are several stages in the production of oil and gas. Exploration leads to discovery of oil and gas in place. Once oil is discovered, investment activity converts the oil or gas in place into proved reserves.

Proved Reserves

Proved reserves are a measure of expected cumulative output from capacity already in place. Without new investment, the rate of production for any oil well will decline over time as pressure in the reservoir declines. The rate at which production declines, called the decline rate, can be approximated as the inverse of the reserve-to-production ratio. The reserve-to-production ratio cannot be used to calculate how long the reserve will last. It is in no sense a life-index of the deposit. It simply measures the rate of decline absent other investment.

The important factor is the behavior of proved reserves over time, and proved reserves are increased by development activity. Development wells enlarge the known area of a reservoir, recorded by the American Petroleum Institute–American Gas Association (API–AGA) and now DOE. Revisions occur as operating expense leads to greater knowledge of the deposit. New reservoirs may also be found in old fields.

The role of discoveries in this process and in the API–AGA statistical representations is midunderstood. Reserves are not discovered. Pools are discovered and their contents gradually converted into reserves. Another error is to treat the API–AGA statistical series "discoveries" as an indication of what has been discovered in any year. A good estimate of what has been discovered comes only through time as development proceeds and knowledge about the deposit increases. Fluctuations in recorded "discoveries" are random, and independent of actual discoveries.

Table 6–5 in Chapter 6 summarizes the behavior of proved reserves from 1946 to 1975. Note that the misconstrued "discoveries" series is never more than a fifth of what is ultimately recovered from a discovery (column 4 compared to column 1). New oil from old fields (column 6) has contributed an increasing share of gross additions to proved reserves. As it was becoming increasingly costly to find new oil in new fields, more effort was devoted to obtaining oil from old fields. Part II also demonstrates similar trends for natural gas reserves.

Probable Reserves

Probable reserves surround proved reserves. They are estimates of what may be produced from the undrilled portions of known reservoirs. They can be viewed as a "stretch factor" on proved reserves. The stretch factor for all fields in the United States in 1945–1977 was 1.92. For the larger fields in the United States (excluding Alaska) the factor was 2.68, indicating that larger fields grow more than smaller fields.

For the period 1967–1977, the stretch factor for crude oil can be divided into two parts: new oil in old fields and a higher recovery percentage of the oil already included. New oil-in-place added 4.4 billion barrels (BB) in the 1967–1972 period. In the 1972–1977 period, new oil-in-place added only 1.85 BB. Improved recovery factors added 3 BB in 1967–1972 and 4 BB in 1972–1977.

For natural gas, recovery factors have been very high (around 75 percent) and the stretch factor is almost entirely new gas.

The Role of Cost

Proved reserves represent past investment. Operating costs, while not trivial, are a small proportion of price. Therefore, estimates of what will be produced from existing capacity will be robust with respect to price fluctuations. Part III estimates that in 1976 average operating costs for oil were approximately $1.30 per barrel. Of course, operating costs were higher at the margin. Therefore, it is conjectured that controls setting maximum prices for oil were enough to discourage production from marginal wells. Examples suggest that this was a significant factor in reducing output and proved reserves.

In order to predict the rate at which probable reserves will be converted into proved reserves we need to know not only operating cost, but also development cost. The data for a comprehensive calculation of development cost are lacking. Again, however, examples show the deleterious effects of price controls on the rate of creation of proved reserves.

PART III

Possible reserves lead, through investment, to proved reserves. But what about future discoveries? Methods for projecting future discoveries of oil and gas are discussed in Part III. Uncertainty is the salient feature of exploration activity. Wildcat wells — wells drilled to discover new deposits — can be viewed as lotteries. A substantial entrance fee (well cost) is paid to participate in the lottery with a high probability of no positive return and a low probability of a very large return. The assignment of probabilities is influenced by the precision of available geologic information, which ranges from very sparse in frontier regions to very rich in mature or intensely explored petroleum basins.

The rate at which exploratory well drilling generates new reserves and additions to already discovered reserves is a function of explorationists' current perceptions of geologic risks and future economics. On the average, larger deposits in a basin are discovered early in the exploration cycle, so the basin's resource base tends to decline more rapidly at first. Depletion of a basin's resource base is accompanied by a decrease in geologic risk. The positive impact of decreased risk on the economic value of exploration is offset by smaller expectations of what may be discovered, and expectations of increasing exploration cost per barrel of oil or MCF of gas discovered are reinforced.

As the geographic area examined expands to include a mix of frontier, partially explored, and mature provinces, these economic and geologic effects are partially masked. Central issues addressed in Part III are as follows: Can credible forecasts be made of the future evolution of amounts discovered, additions to reserves, and production in large geographic regions by use of analytical methods of inference and prediction? Must these methods explicitly incorporate the economic and geologic features of exploration that influence exploration decision-making? Can the consequences of structural changes in fiscal and regulatory regimes be captured by methods that do attempt to mirror principal features of the economics of exploration, but that rely on regionally aggregated exploration data to generate projections of the future? In contrast, how well do disaggregated approaches to forecasting returns to exploration effort perform?

The range of methods used to project exploration futures vary so widely that an exhaustive description and critical interpretation of methodology is beyond the scope of this book. We restrict our attention to examples of the principal approaches as applied in North America:

- Life cycle
- Rate of effort
- Geologic-volumetric

- Subjective probability
- (Play or province) discovery process
- Econometric

A principal conclusion of this critical review of methodology is that improvement in the quality of projections of responses over time of exploration activity and discovery rates to changes in prices, costs, and regulatory regimes await more extensive and more carefully measured data than are currently available.

PART IV

Part IV focuses on coal reserves and resources. In the United States discovery plays a small role in coal supply. Knowledge about where coal lies is orders of magnitude better than for other minerals. The concern of this part is therefore with already-discovered but yet-to-be-developed deposits.

Coal reserves in the United States are classified according to a two-dimensional system. The first dimension treats the physical characteristics of the deposit. The second dimension is the certainty with which the size of the deposit is known. In the physical dimension deposits are divided into depth and thickness categories. Deposits are placed into three categories of seam thickness and several categories of depth. On the certainty dimension, deposits are classified as measured, indicated, or inferred, depending upon how confident the estimator is about the extent and location of the deposit.

In an attempt to delimit the economically recoverable portion of the resource, a concept called the Demonstrated Reserve Base (DRB) has been introduced by the Bureau of Mines and the U.S. Geological Survey (USGS). The Demonstrated Reserve Base consists of the measured and indicated reserves in seams at least 28 inches thick. It is assumed that all the coal in the DRB is available at current costs.

Part IV tests whether the DRB does in fact delimit the portion of coal resources available at current costs. To answer that question, several subsidiary issues are examined. First, is there an important difference between measured and indicated reserves? Second, is seam thickness an important cost-determining variable and is it sufficient to establish the economics of mining a particular seam?

It is shown that cost differences between the measured and indicated reserves are small. In essence, the difference is a small drilling cost. Inferred reserves, however, are geological speculations and subject to wide revision. Thus the separation along the certainty dimension appears reasonable.

Drawing on previous research, it is seen that results are less satisfactory when one analyzes the physical characteristics of the DRB criteria. Seam thickness is an important variable, accounting for 50 percent of the variability in the cost of underground mining. Yet 50 percent of cost is determined by other mining conditions not included in the DRB criteria, such as gas conditions, roof and floor stability, seam pitch, and so forth.

Ignoring these conditions leads to a biased interpretation of the DRB. Claims that the DRB represents coal available at current costs reflects the following logic: "We are currently mining seams as thin as 28 inches. All coal in the DRB is in seams at least 28 inches thick. Therefore, all the coal in the reserve base is available at costs no greater than today's level of cost." This logic neglects the impact of other cost-determining variables. Coal is currently mined at 28 inches only if there are other offsetting influences. The roof might be very stable, the quality of the coal very high, the other mining conditions extremely favorable, and so forth. Thus the typical new mine will be at some seam thickness with a combination of other characteristics that jointly determine cost. The incremental mine is defined as the last mine that must be opened to satisfy demand. The cost of producing coal in this mine is the incremental cost of coal. If seam thickness were the only cost-determining variable, all new mines of a given coal quality (sulfur content) in a given region would have the same thickness. In fact, because other factors are important, there is a dispersion of seam thicknesses in new mines. Part IV presents the data reproduced in Table 18-3 in Chapter 18. The typical seam thickness of new mines is seen to be far removed from the cut-off criteria of the reserve base. Using cost equations estimated in previous work, it is shown that if the 28-inch cut-off were, in fact, the typical seam thickness of new mines, costs would be much higher than today's level of cost. The ratio of cut-off cost to today's cost is shown as column 4 of Table 18-3.

The Bureau of Mines further breaks the reserve base down into coal available through deep mining techniques and coal available through strip mining techniques. The cut-off criteria for strip mining are shown to be closer to the economic cut-off than was the case for deep reserves. (Note that coal excluded from the strip portion is presumed to be available through deep mining techniques. This is not necessarily true since thin coal under shallow overburden might exceed the overburden ratio cut-off yet be too close to the surface to be exploitable by deep methods.) However, evidence is presented that the cut-off criteria for strippable reserves were not systematically applied.

Part IV also discusses what has been omitted from the DRB. The extent to which mining of inferred reserves will slow down the rate of increase in cost is unknown. It is conjectured that the moderating effect will be small. First, by definition, inferred reserves are further away from current mining areas than the reserve base and likely to require more infrastructure investment than exploitation of the reserve base would require. Further, since mining has been conducted extensively in the eastern United States, we expect that inferred reserves have remained inferred because they are of high cost. In the western United States, where large-scale mining is just beginning, this reasoning does not apply. Yet the massive amount of coal in the reserve base is enough to guarantee that any cost increase over time will be moderate.

PART V

The estimation of uranium reserves and resources combines the estimation problems faced by both oil and coal. Extraction costs are high relative to total cost, and therefore information on the production costs of already discovered deposits is important. In this regard the problems are similar to those for coal. Exploration is also an important element in the supply of uranium and in this regard is similar to oil and gas. Part V analyzes both aspects of uranium supply.

The most basic distinction in uranium resource estimation is between ore reserves and potential resources. Resources are further subdivided, in decreasing order of certainty, into probable, possible, and speculative resources.

Reserves represent uranium deposits identified with some certainty. The grade and physical shape are usually delineated by developmental drilling. Tonnages and grades are calculated assuming some mining dilution and recovery levels, but they are not adjusted for mill recovery. A resource is inferred by some process that gives an expectation of ore occurrence. Reserves and resources are further broken down into forward-cost categories.

Forward cost is one of the most poorly understood aspects of uranium resource and reserve estimation. Forward cost measures the expenditures necessary to produce uranium from a given deposit. For already developed deposits this means that only those expenditures that will be spent in the future are included. The costs for shafts already sunk or mills already constructed are not included. For new deposits where expenditures have been small, most of the cost is forward cost. For undiscovered resources, even exploration cost is part of forward cost and the estimator's guess of what it will cost to discover the deposit is calculated.

If one were creating a supply curve from the data described, one class of adjustments would need to be made and a second would help the process. The first adjustment is the most important. The calculation of forward cost ignores a rate of return on capital. Total forward capital costs are divided by total cumulative output to yield the capital cost component of forward cost. This procedure puts no weight on the fact that capital expenditures occur before output begins and total cumulative output is obtained over a period of years. The correct way to deal with these differing time streams of outlays and receipts is to discount future earnings. This is not done and thus forward cost imputes no rate of return to capital. Forward cost therefore seriously underestimates the true cost of exploiting deposits. The bias is worse where there is a large time difference between years of capital expenditures and years of future income.

In the past forward cost has also led to poor predictions of price. In an industry with excess capacity, producers will not shut down so long as revenues cover variable cost. If the industry is to expand, however, price must cover all costs, including new capital costs, so mine owners will see investment as profitable. Thus, even with forward cost in existing mines of $8 per pound of U_3O_8, the price necessary to attract new investment will be significantly higher.

The other adjustments to forward cost are also discussed. It is suggested that because of the difficulties in calculating exploration cost, this category of cost should be separated out in the calculation.

The Grand Junction Office also produces other data related to uranium supply. They currently produce "could-do" scenarios that estimate what could be produced if economics were not a consideration. Thus they provide poor predictions of actual output, especially for longer time horizons.

Part V concludes by discussing how other groups use GJO data. Included is a comparison of the work by both Hans Lansberg and CONAES, who start with the same GJO estimates and arrive at remarkably different conclusions. There is a discussion of trend projection models, which include life-cycle, rate of effort, and econometric models. Crustal abundance models also are described; they are more useful in long-run applications and tend to ignore or subsume economic considerations, while trend projection models are best used in short-run projections and tend to ignore information about the underlying physical resource base. The part ends with a brief discussion of subjective probability models.

1 INTRODUCTION TO ESTIMATION OF RESERVES AND RESOURCES

M. A. Adelman
John C. Houghton

INTRODUCTION

In the United States concern is widespread about the depletion of natural resources. The concern is an outgrowth of the public's increasing awareness in the
late 1960s and 1970s of environmental degradation and population increase. The
environmental crisis and population explosion crisis gave way to the energy
crisis. An oil cartel brought about rapid increases in the price of oil. The price of
uranium increased rapidly too, and a flock of lawsuits alleged existence of a uranium cartel. Issues of strategic minerals have been raised vocally again. Although
this book deals specifically with fuel minerals, the same principles and many of
the same methodologies apply to nonfuel minerals. Public policy is now very
involved with issues of fuel availability. Decisions are being made about quotas,
price regulations, environmental standards, and subsidies, all based on assumptions about primary fuel cost and supply. Measures of reserves and resources are
used to determine the effect of further depletion on scarcity. The aim of this
book is to help resolve these issues of public policy.

This book deals with the calculation of primary energy supplies. *Reserve* and
resource estimates are constructed in a variety of ways for the primary fuels: oil,
gas, coal, and uranium. What inferences about supply can be drawn from these
estimates? What biases and uncertainties do they contain? Are they adequate for
predicting the response of supply to changes in government policy? A volumi-
nous literature exists on society's optimal use of depletable resources. It treats
such issues as determining the appropriate discount rates and determining the
effects of competitive or monopoly pricing. It assumes that the stocks of depletable resources are known, and it addresses the problems of their best use. We,

1

the authors of this book, ask how much is known about stocks. A short answer is: not much. Except in some places in this introduction to the book and in Appendix A, which downgrades the effect of discount rates on rates of production, we will not deal directly with these issues.

Ideally a supply curve built from reserve and resource estimates should answer the question: Under competition, how much of a particular fuel is available at what price? Supply curves for depletable resources are usually considered basically different from supply curves for clothing, chemicals, and other commodities. Inherent in any estimation of the supply price of anything produced from a depletable resource is the notion that what is produced today cannot be produced tomorrow. Because the resource is likely to become more expensive as the stock is depleted, the price tomorrow will likely exceed the price today. So we start this report by restating the nature of supply curves for depletable resources, showing the use one makes of reserve and resource estimates in the process.

1 USER COSTS AND OPTIMAL DEPLETION OF RESERVES

A mineral reserve is defined as a resource that is worth depleting, given the costs and prices expected between the start of investment at present and the end of exploitation. Hence mineral costs are central to any attempt to estimate reserves.

Mineral costs have always been viewed as peculiar by economists. For, in addition to labor and capital, there is the *user cost*, the sacrifice of depleting the reserve today rather than later. Future profits are sacrificed in favor of current profits by a decision that depends on assumptions about future prices and costs compared with present and about the interest rate that mediates between present and future. This chapter briefly sets forth the theory of depletion, or resource exhaustion propounded by Gray and by Hotelling,[1] then the perceived limits of the theory and its amendment during the past twenty-odd years, and finally the relation of the amended theory to the determination of mineral reserves.

THE THEORY OF DEPLETION OF LIMITED RESOURCES

Since every mineral deposit is finite, a unit is produced today at the cost of not producing it tomorrow. Four main results follow from this assumption:

1. In a nonmineral industry, output in the short run will be expanded to the point where the cost of expanding output (marginal cost) is equal to the price. The *marginal cost* of further straining the labor and capital in place must include some allowance for unusual wear and tear: Overloaded people and machines wear out more than proportionately to the rate of use. In other words, savings

are to be gained from an even rate of output. The phenomenon is magnified for a mineral deposit. In any given short period output will *not* be carried to the point where marginal cost equals price. The marginal unit, produced under strain at a very high cost today, could be produced at a lower cost later. The operator will tend to strive for a steady rate of output, at minimum cost, because the unit produced today at high cost will be gone tomorrow. Or, what comes to the same thing, some of the mineral is reserved for future production at a lower output rate and a lower cost. But the future cost savings should be discounted to a present value before the operator can decide how far to hold back output short of the level that would equate short-run marginal cost to price.

2. Turning now from the short run to development investment: The faster a reserve is depleted, the higher the cost, because more capital and labor are being applied to extract the fixed amount. True, there is some minimum efficient scale and some overhead costs, such as the cost of an access road, a warehouse for supplies and parts, and living quarters. These costs are spread wider as the resource is depleted at a higher rate. But at some point size brings no economies, and overhead cannot be spread any more widely. If the intensity of exploitation is to increase, costs per unit of mineral removed must also increase. It may still be rational to dig another vertical shaft where one would do or to drill another well, but only because it is worth accelerating the flow of receipts at the cost of investing more to get them. The rate of interest is crucial to this decision, though only peripheral to current production decision.

3. The more of a given reserve that is removed, the more expensive it becomes to remove the remainder. Solid mineral deposits are not homogeneous, and costs rise with higher overburden. Ore ratios are important. Fluid mineral reservoirs lose pressure as more of the reservoir is removed, and the rate of output declines, unless additional expenditures moderate the rate of decline.

4. The rule of costs rising with higher rates of output and with higher cumulative output holds as well for all deposits taken together, because it is rational conduct to exploit the better before the worse.

The idea of "worse" or higher cost deposits should be taken broadly; oil from shale or heavy-oil deposits, often called bitumens or tars, should be considered as simply high-cost oil, which future higher prices will draw into production. These deposits will then presumably be impounded from *resources* into *reserves.*

If we assume no additions to resources and reserves and no changes in technology, then costs and prices must rise over time. At any moment, for any given deposit and for all deposits taken together, an interaction exists between the current depletion rate and the future higher price. Whether our viewpoint is that of the owner of a deposit, who wishes to maximize profits, or that of a whole society, to produce more today is to sacrifice a higher future sales value and to incur a higher present and future cost.

The value, net of all costs, of the mineral unit to be produced in the present must be compared with that net value at all points of time in the future. The

Figure 1-1. Price over Time, with Corresponding Demand Function.

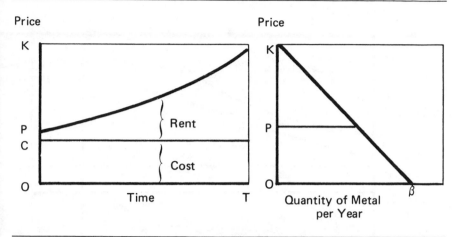

Source: Orris C. Herfindahl, "Depletion and Economic Theory," in *Extractive Resources and Taxation*, ed. Mason Gaffney, © University of Wisconsin Press, Madison, Wisconsin, 1967, titled "Price over Time, with Corresponding Demand Function," p. 67. Reprinted with permission.

rate of interest is the mediator, or balancing element. Figure 1-1 illustrates the process of establishing the value. For simplicity, the figure starts with constant costs over time. The demand curve is linear and holds in every period. Above some price K, the demand is zero. The rent (the difference between P and C) must increase at the interest rate. Producing one unit should be equally attractive each year. Postponing output one year costs the interest that could have been earned in that year. Assume now an industry that is out of equilibrium, say if the cost of mining drops. It will move toward a new equilibrium system, as shown in Figure 1-2. In either case, the resource is "exhausted" exactly at time T when demand goes to zero at price K. There is more mineral left to mine, but it is so depleted that the cost of extraction is greater than the price at which demand is effectively zero.

The increase in the whole industry's cost over time, as the total of reserves is depleted, can be stated as follows. Let x = cumulative output and c = cost per unit. Then the increase in cost over time is

$$dc/dt = (dc/dx)(dx/dt) \, . \qquad (1-1)$$

If p is the price net of cost, i.e., "rent," then in a competitive industry, p will rise over time: $rp = (dp/dt) + (dc/dt)$;

$$\frac{dp/dt}{p} = r - \frac{dc/dt}{p} \qquad (1-2)$$

Figure 1-2. The Effect of Lower Cost from C_0 to C_1.

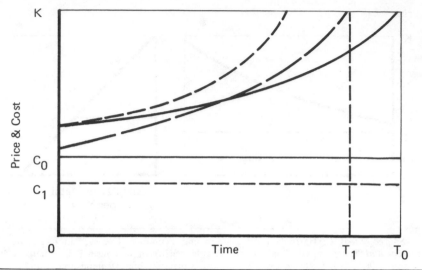

Source: Orris C. Herfindahl, "Depletion and Economic Theory," in *Extractive Resources and Taxation*, ed. Mason Gaffney, © University of Wisconsin Press, Madison, Wisconsin, 1967, titled "The effect of lower cost from C_0 to C_1," p. 69. Reprinted with permission.

The percentage increase in rent in any time period is equal to the rate of interest less the increase in cost, as a percentage of rent.

It makes sense to hold reserves rather than to produce and sell them if and only if the net value ("rent" or "royalty") will increase at least as fast as the rate at which future costs and benefits are discounted. The rate of discounting is therefore of central importance. Consider a mineral owner initially in an optimal position, now confronted with a changed interest rate. A lower rate of interest would make future receipts more attractive compared to present receipts. Higher user cost would force down current output; prices would be higher. The rate of increase of price would henceforth be lower than the previous long-term trend. The optimum profile of production and consumption would tilt clockwise. Less mineral would be used up today, thus rationing consumption to more essential and higher value uses and postponing the day when the last unit is consumed and the mineral is displaced by higher cost and less suitable alternatives.

Consider next a change in expectations about future prices. Suppose they are expected to rise more quickly than previously assumed, causing the net value to rise faster than the rate of interest. If it does, the present discounted user cost will be greater today, just as it would be with a lower interest rate. The effect is the same: It will pay to withdraw minerals from current extraction to hold for the future. This raises the price today until the rate of increase in the "royalty" declines to the rate of interest, whereupon the system returns to equilibrium. Production and consumption have been delayed from present to future.

IMPLICATIONS OF THE THEORY OF EXHAUSTION

As will be shown, there are severe limitations to the theory of exhaustion. Nobody would deny its basic importance in showing how a market economy makes the best use of limited mineral resources, however. (A centrally planned economy can simulate the process.) Coming shortages cast their shadows before they actually occur, by way of the discounting mechanism. If the holders of mineral bodies act in their own interests, the economy is in no danger of either hitting a brick wall of sudden shortage or going over the cliff. The limited stock is rationed out over time until the last unit is used up, at the highest price at which anybody wants any of the mineral in view of whatever alternatives exist.

The rationing process does not need perfect information or perfect markets to operate. On the contrary, perceived data on resources and costs (that is, data on reserves), demands, prices, and interest rates form an interacting system to put everyone on notice as to the market consensus. There is an incentive to adapt, but also an incentive to profit by better knowledge. The market becomes a forum for sharing information and concepts, changed perceptions of the data, and with them changed perceptions of mineral prices and rate of use.

The theory of exhaustion has begotten a vast literature on the "right" discount rate. Are private discount rates any indication of the rate at which society should discount? What value judgments on the welfare of present versus future populations might be implicit in a rate? How much of a limited patrimony does any generation have the "right" to use, and how much is it under a "duty" to save for the use of posterity?

Much of this debate is pointless, because it is confused on the nature of mineral reserves and assigns too much importance to the rate of interest. But even crude data may be enough to rule out false questions and invite true ones. Thus, if it is assumed that "proved and probable" reserves of crude oil will have been exhausted by 2020 A.D., having forced the world in the interim to build an industry supplying synthetic oil at a cost of, say, $50 per barrel (in 1980 prices), that determines the user cost of removing a barrel today. Discounting $50 down to 1980 at 6.0 percent real rate of discount implies a price of about $5 per barrel. Lower discount rates would imply higher prices. But no plausible discount rate, nor any current or expected scarcity, could explain a crude oil price around $30. One has to look elsewhere; for example, to see whether perhaps there was monopoly control.[2]

REFORMULATION OF THE THEORY
OF EXHAUSTIBLE RESOURCES

A scientific theory that is internally consistent and seems to fit the facts may yet lead to predictions and results that contradict other data. If it does, something has been left out. The Gray–Hotelling theory set forth the economic response to limited mineral reserves: increasing scarcity and higher prices over time. It was also taken as a paradigm that minerals prices *did in fact* inevitably rise over time. Nothing is more untrue. The current deafening clamor over "limits to growth" and with it the certainty that mineral prices must inevitably rise, has disregarded what has long been known: (1) There are dramatic increases in some mineral reserves; (2) there is no tendency for royalties in the long run to rise at the rate of interest; and (3) there is no tendency for mineral prices to rise at all. Indeed, if anything there has been a tendency for the prices of "nonrenewable" raw materials to decline while the prices of "renewable" materials, like timber, persist in rising.[3] Moreover, if user cost were a really important fact, some owners of reserves would often be observed to hold output short of what would maximize profit. Yet nobody can come up with such examples. Some recently alleged examples, Saudi Arabia and other Middle East nations, are inconsistent with even wildly implausible discount rates and future prices. A simple and consistent explanation is that they restrain output to maintain prices. A counter example is Algeria, a country that, with relatively high-cost reserves, would benefit most by postponing output. Yet Algeria plans to deplete practically all its oil reserves over a twenty-five-year period; Algeria has similar plans for gas, but its exploitation has been delayed by marketing problems.[4] This is logical conduct in a producer with a small market share who disregards cartel discipline and feels free to use all the capacity that can be created.

The modern theory of resource depletion was formulated in a series of papers and books.[5] Minerals are limited, but no one knows where those limits are. The economic asset "reserves" is only the visible part of a wide spectrum. It is also the end product of an investment process. From vague guesses about the incidence of a mineral in the crust of the earth we go to estimates of what may exist in some areas on the basis of its known geology, then to estimates of what exists on the basis of exploration and actual discoveries, then to much more precise estimates based on continued finding and development. Reserves move from possible to plausible to probable to proved and are then extracted, according to a scheme that might be called the five equalities at the margin.

In any given mineral deposit and for all deposits taken together, there is a constant movement toward equating price with marginal cost in several stages:

1. The current operating margin, or rate of production, which is governed by the proportion of the reserve already depleted.

2. The intensive development margin, which includes investment costs for the already known deposits and is governed by the trade-off between rising investment requirements and quicker realization of revenue.

3. The extensive development margin, where exploitation is begun of known but previously unecomic deposits.

4. The exploration margin, where a search for new deposits is conducted. The expected cost per unit is highly uncertain, and the certain costs of many failures must be balanced against the chance of finding something worth finding, something with total marginal costs no higher than at margins 1-3.

5. The technology margin, which interacts with the first four.

The Gray-Hotelling theory, it is apparent, is a special case because it covers only stages 1-3, setting 4 and 5 to zero. Yet the last two are, in the long run, far more important.

If the five functions are aggregated into one supply curve, the true paradigm can be stated as follows: At any given moment, mankind is unwillingly crawling up the leftward-moving supply curve toward higher minerals costs and *also* pushing the curve over to the right toward lower costs (see Figure 1-3). The outcome is indeterminate. So far the human race has won almost every round and made minerals increasingly cheap. Hence the "pure theory of exhaustion," which deals with optimal use of dwindling stock, is at most a special or transitory case and should not be built into any scheme of calculating reserves.

It does not follow that lower or stable prices are to be expected for every mineral at every point in time or that most mineral costs will continue forever to be pushed down. For example, metallic ores have grown cheaper as the refining industries have learned to use lower and lower grades. Improved technology has unlocked ever larger resources, because the amount of metal (perhaps) is inversely proportional to the grade. This relation does not hold for oil,[6] but it does hold for natural gas, in a rather exaggerated form. Total proved and probable reserves of conventional natural gas in the United States, plus past production, are somewhat less than 1,000 trillion cubic feet (TCF). The amount of gas dissolved in pressurized water in deep formations is estimated at about 3,000 TCF in a portion of Louisiana alone (see Part II of this book).

Figure 1-3.   Effect of Expanding Estimates of Supply.

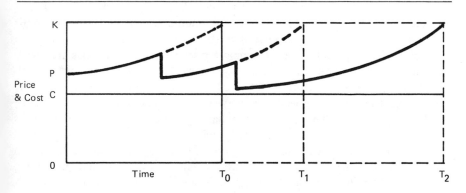

The processing of lower and lower grades of ore may necessitate turning over such large quantities of dirt as to create environmental costs that dwarf labor and capital costs, however.[7] Similarly there is as yet no assurance that *any* perceptible fraction of gas dissolved in water will ever be recovered, since huge amounts of brine would need to be brought to the surface, then reinjected once the methane had been extracted.

THE UNDERLYING CONCEPT OF RESERVES

This book, it is hoped, will be a bridge between the calculation of reserves and a somewhat wider economic theory of exhaustible resources. As shown thus far, the idea of reserves as a fixed stock is wide of the mark. In trying to count the amounts of mineral as they flow from the large undefined bucket of possible or potential to the smaller bucket of probable to the small, precisely defined bucket of proved, the aim is to generate a perpetual inventory of the holdings in these various categories. Without much more information, these reserve numbers do not reveal the cost of using up what is known to exist, nor how quickly it will be used up.

THE RATE OF INTEREST AND THE OPTIMUM DEPLETION RATE

Replacing the theory of exhaustion by the five marginal equalities set forth above allows a fresh look at the place of the discount rate in the exploitation of minerals. In Appendix A the case of the optimum depletion rate of a known deposit of oil or gas is treated. It is easier to handle mathematically than other minerals because the rate of depletion of the deposit can be taken to approximate the rate of decline of output per unit of time. However, the general idea is easily stated.

There would be widespread assent to the following proposition: The higher the discount rate, the faster the optimal depletion of a mineral deposit, since the value of a future unit decreases relative to a present unit. But the conclusion does *not* necessarily follow. By common consent, mineral industries are highly capital intensive. Development may require massive installations, such as offshore oil production platforms. But in a larger sense, they are capital intensive because of the usual long lead time between the start of expenditures and the start of production. The ratio of accumulated payments to a year's revenues is commonly high. The higher the interest rate, the higher the cost of mineral development. Thus a higher interest rate would, all else being equal, slow down the rate of investment, hence slow down the rate of depletion, and raise prices, not lower them. Because the discount rate pushes in both directions it is equivocal with respect to the optimum depletion rate. The net effect of a change in the interest rate is unlikely to be strong, although this has not been worked out with any precision.

The usual formulation, "Higher discount rate means faster depletion," assumes that the only capital in question is the mineral reserve itself and that the only cost is user cost. But user cost is only one particular kind of capital cost. User cost may be and nearly always has been small or imperceptible because of the amount of the existing reserves and the flow of increments to reserves over time. Even when nonnegligible, though, user cost may be small relative to the costs of labor and capital needed to find and develop the mineral.

The controversy over the economics or ethics of comparing the welfare of one generation with that of its successors is even more complex than supposed. Our successors may be poorer with respect to some mineral or richer. To exploit it more quickly may not deprive them of much mineral, but it may deprive them of the resources that a better use of the capital would have accumulated for them. These issues cannot be resolved by reference to user cost and the rate of interest, because interest cuts both ways and because user cost in any case may be negligible. In fact, it usually is. But general principles do not decide concrete cases, and what we really need to know are the actual sizes of the stocks and flows of the particular mineral. To elucidate the process of gaining that knowledge is the purpose of this book.

REFERENCES

1. L. C. Gray, "Rent under the Assumption of Exhaustibility," *Quarterly Journal of Economics* 28 (1914): 466–89. Harold Hotelling, "The Economics of Exhaustible Resources," *Journal of Political Economy* 39 (1931): 137–75.
2. M. A. Adelman, "The World Oil Cartel: Scarcity, Economics, and Politics," *Quarterly Review of Economics and Business* 16 (1976): 1–18.
3. Neal Potter and Francis Christy, Jr., *Trends in Natural Resource Commodities* (Washington, D.C.: Resources for the Future, 1962). Robert S. Manthy, *Natural Resource Commodities – a Century of Statistics* (Washington, D.C.: Resources for the Future, 1978).
4. Bechtel Corporation, *The Hydrocarbon Development Plan of Algeria* (San Francisco, August 1977).
5. Harold J. Barnett and Chandler Morse, *Scarcity and Growth* (Baltimore: The Johns Hopkins University Press, 1963). Richard L Gordon, "A Reinterpretation of the Pure Theory of Exhaustion," *Journal of Political Economy* 75 (1967): 274–86. Orris C. Herfindahl, "Depletion and Economic Theory," in *Extractive Resources and Taxation.* ed. Mason Gaffney (Madison: University of Wisconsin Press, 1967). Anthony C. Scott, "The Theory of the Mine under Condition of Certainty," in Gaffney, ed., *Extractive Resources.*
6. Leigh C. Price, in "Deep Pressured Waters: An Alternative Energy Source?" *Oil and Gas Journal* (December 15, 1978) suggests that the relation does

hold for oil— with the same qualifications as for gas. However, this remains at best an untried hypothesis.

7. This is minimized by H. E. Goeller and Alvin M. Weinberg in "The Age of Substitutability," *Science* 191 (1976): 683–89, reprinted in *American Economic Review* 68 (1978): 1–11. They point out that coal mining produces much more spoil than all nonenergy minerals. But this may not be true in the future if nonenergy mineral consumption grows at historic rates or if there is massive development of oil shale and of the uranium content of the Chattanooga shale.

2 RESERVES VERSUS RESOURCES

Mineral resource terminology is particularly slippery. Perhaps the most authoritative work is by Schanz.[1] The emphasis in the present part of this book is not so much on sorting out different terms but on explaining in some detail the process that brings about the differences. When people distinguish between *reserves* and *resources, reserves* are generally meant to include material that has already been identified, whereas *resources* refer to material that is inferred by some method. A more precise definition for each particular fuel appears in each part of this book.

THE RESOURCE BASE

The resource base for any particular mineral is larger than the known reserves for two primary reasons. First, some deposits are identified that are of too low grade for economical exploitation at current prices. Second, some relatively unexplored areas undoubtedly contain deposits economic to exploit at current prices. McKelvey cleared away some of the confusion regarding these two concepts in the late 1950s by developing what is now known as the McKelvey diagram. Figure 2-1 is an example. He developed the box to separate these two extensions for the resource estimates along orthogonal axes. Originally intended as an aid in defining terminology, the diagram has since been used to categorize actual estimates of minerals; for example, the Grand Junction Office (GJO) of the U.S. Department of Energy publishes uranium reserve and resource estimates as shown in Table 2-1. Uranium is unique in that the economic axis is actually measured in dollar values.

Figure 2-1. Adaptation by U.S. Department of Interior of McKelvey Diagram.

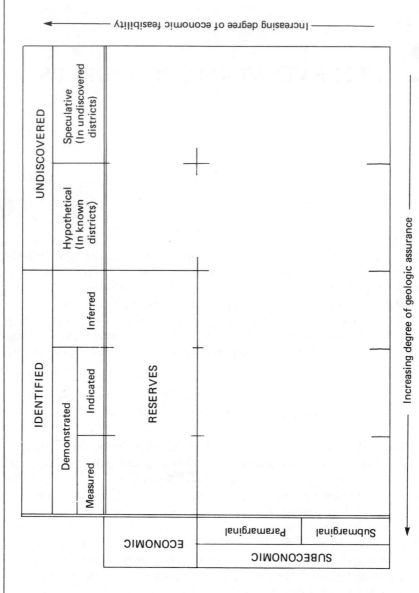

Source: John J. Schanz, Jr., *Resource Terminology: An Examination of Concepts and Terms and Recommendations for Improvement*, EPRI Report 206 (Electric Power Research Institute and Resources for the Future, August 1975), p. 9. Reprinted with permission.

Table 2-1.   Estimates of Uranium Reserves and Resources
in a McKelvey Diagram Format, as of January 1, 1977.

$/lb U_3O_8 Cost Category[a]	Tons U_3O_8			
	Reserves	Potential		
		Probable	Possible	Speculative
$10	250,000	275,000	115,000	100,000
$10-15 increment	160,000	310,000	375,000	90,000
$15	410,000	585,000	490,000	190,000
$15-30 increment	270,000	505,000	630,000	290,000
$30	680,000	1,090,000	1,120,000	480,000
$30-50 increment	160,000	280,000	300,000	60,000
$50	840,000	1,370,000	1,420,000	540,000

a. The category is that of *forward cost*. For its limitation, see Part V, on uranium.

Source: Grand Junction Office of the U.S. Department of Energy, *Uranium Industry Seminar*, GJO-108(77) (Washington, D.C.: 1977).

RESERVES

Reserves at any moment are by definition a stock, a fixed amount. But over time, both reserves and resources form part of a flow. Figure 2-1 stops the flow. It has two dimensions: cost per unit and degree of certainty. But uncertainty is an element of cost. A best area estimate of X tons, which time may revise to zero or to something larger, is no foundation for investment in production, processing, or use. It is therefore of value less—or cost greater—than an estimated X tons within narrow limits of error, so in theory at least both dimensions could be assimilated into a single one: a cumulative cost curve, embodying more and more cost and risk, going from the best proved reserves toward the worst ultimate resources. More specifically, over time additional investment will subject the classes in the diagram to endless change. Investment partly in hardware but partly also in highly skilled labor is forever finding new habitats of primary energy. More exploration and research will push some prospects and even some mineral bodies farther into the unpromising lower right-hand corner of low probability and high cost and finally off the diagram: to know them is to damm them.[2] Other prospects are pushed—they do not migrate on their own—toward the desirable upper left-hand corner of high probability and low cost, where additional investment will ready them for production.

Proved versus Probable Reserves

Thus every estimate of reserves, in any sense, is essentially a forecast of investment outputs and the return to investment. *Proved* reserves accurately forecast the amounts to be produced in the near term, mostly from developed mines and reservoirs. *Probable* reserves are a forecast, usually without specification of time, of future increments to proved reserves. The judgments of companies and governments about adequacy and cost are guided by their notions of how time-consuming and costly will be the flows from unknown to known, from less developed to fully developed.

The notion of resources as a flow is important enough to reiterate in a different way. The McKelvey box diagram is a snapshot of a process that operates through time. One can think of the columns of the box (for example reserves, probable reserves, possible resources, and speculative resources) as different tubs, each containing water. As water is emptied at a fairly constant rate from the nearest tub (reserves), it is being replaced in spurts (exploration and development) by water from the second tub (probable reserves). And although the farther tubs cannot be seen as well (uncertainty is higher), essentially the same thing is happening. The water emptying out of closer tubs is being replaced irregularly by tubs farther down the line. This cannot go on forever. As water is drawn from the tubs there is some depletion, which makes it harder (costlier) to transfer. But water is also in some sense manufactured through evolving technology and then added to one or more of the tubs.

The important point is that a McKelvey diagram represents only the level of water inside each tub. It says nothing about the changes in the levels, the dynamics of the process. The dynamics are composed in part of exploration, development, and production models, a crucial link in the whole system. Estimating each of the levels and clarifying the semantics of which tub gets called what is only a small part of the story. To predict the level of the nearest tub ten years from now, the present level is not the only important piece of information. Perhaps even more significant is the estimation of the rate at which water pours into the tub.

Different types of estimates of resources are based on different assumptions concerning these characteristics. For example, crustal abundance methods, defined in Part V, skip most assumptions about technology and prices, whereas econometric models skip many geological issues.

REFERENCES

1. The terminology used here has been adapted from John J. Schanz, Jr., *Resource Terminology: An Examination of Concepts and Terms and Recommendations for Improvement*, EPRI Report 336 (Electric Power Research Institute and Resources for the Future, August 1975), p. 9.

2. For example, the west coast of Greenland has been consigned to the outer darkness (see *Oil and Gas Journal Newsletter* (July 10, 1978): p. 4). Yet one wonders whether this is any more final than the dismissal of the plain on the western shores of the Persian Gulf in the 1920s and early 1930s.

3 ECONOMIC ISSUES FOR THE MINING INDUSTRY

Several economic concepts are applied to the classification scheme of the McKelvey diagram (see Figure 2-1) to determine the behavior of cost as it depends on the rate of output and the amount of past cumulative output. Chapter 1's very summary treatment of the net value of the mineral unit to be produced is elaborated here.

PRICE AND MARGINAL COST

In competitive long-run equilibrium, all firms equate price and marginal cost. A firm whose marginal cost is just equal to minimum average cost and to price is in equilibrium, earning a rate of return just large enough to keep it in business but not large enough to induce expansion or new entry. Firms with above-minimal returns—that is, earning tents—cannot expand if additional output would put marginal cost above price.

In a mineral industry, the latter constraint is an obvious physical fact: mines and firms are not identical. Mines with thicker seams in better locations earn a high rate of return at prices that are just high enough to keep the less favored mines in business. The highest cost mine can be considered the "incremental mine," just breaking even by producing at minimum average cost, which is the long-run marginal cost to itself and to the industry. If at any moment it were known how the incremental mine would change as output cumulated over time, the behavior of long-run marginal cost would be known. Ideal reserve data would show the rate at which the better deposits would be exhausted and how the nature of the deposits exploited would change. The economic interpretation

17

of this geological information translates the deterioration of geological conditions into a rate of cost increase.

SUPPLY CURVES

Each mine has a supply curve, representing output q as a function of price p, as shown in Figure 3-1. In the short run, with no new investment possible, the supply curve depends only on short-run variable costs; it neglects sunk capital costs, and it represents the output each year as a function of an expected price that is stable over time. (The supply curves do not reflect the response to a sudden change in price. Such a supply curve is imaginable, but it would be very complicated, perhaps not even monotonically rising.) The mine manager knows the price and has invested to optimize the size of the mine and mill. The supply curve represents the range of choice for each year's output. By horizontally adding the firms' supply curves shown in Figure 3-1 one arrives at Figure 3-2, an industry supply curve. It represents a steady state, in which capital has been fixed, but the prices are long term relative to operation decisions. A supply curve of the same form could be drawn for a longer time period, allowing for fresh investment. The problem would be to modify a previously optimal depletion rate and corresponding investment in the light of new data, or to choose the optimal depletion rate for a new mine. It would be an exercise in the kind of choice discussed in Chapter 1 and in Appendix A.

The common notions of breakeven points, shutdown points, and marginal firms can be applied to both types of supply curves. Only the marginal mine produces at a point where marginal cost equals average cost, in the short run or the long run as the case may be. The rest of the mines are earning a rent; that is, there is a greater than minimal rate of return. For the short run the minimal return is zero, and above-normal would mean covering some fixed charge or even earning a profit. For the longer run it would mean earning an above-normal return on investment in the mine itself and covering some exploration expense or even earning a profit on exploration—up to the best mines earning large profits in the whole finding-developing producing investment.

The supply curve is a function of cumulative production. To predict a short-run or medium-run supply curve for the year fifteen years from the present would require postulating very carefully the pattern of exploitation between now and then. One would need to know the future resource base, as increased by interim discoveries and improved technology in finding and exploiting the new mix of low-cost and high-cost sources. Both the exploration-development and the production stages take relatively long. Hence the mixture of resources in each of the phases at any particular time is a function of the history of market conditions and expectations. All of the following determinants of a supply curve are in fact functions of time paths of supply, demand, price, and

Figure 3-1. Supply Curve for Two Mines.

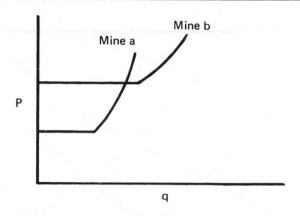

Figure 3-2. Industry Supply Curve.

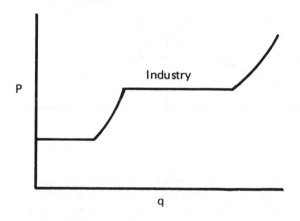

expectations: exploration, development, technology research and development, production, and even such regulatory conditions as depletion allowances and availability of public lands. Figure 3-3 shows a simplified example. Suppose Q_0 is a supply curve for the present, and D_0 is the present demand. Then production is set at q_0 and the associated price is p_0. For the next period, assuming no new discoveries or technological improvements, the supply Q_0 has been shifted leftward by the amount q_0 to a new supply curve Q_1. If the demand curve in the second period is D_1, then the third period's supply curve is Q_2. Over the periods from zero to 3, output expands mildly, while supply price nearly triples. In the fourth period, the price jumps, quantity shrinks, and the industry ends.

Figure 3-3.   Industry Supply Curves as a Function of Cumulative
Production and Industry Demand Curves as a Function of Time.

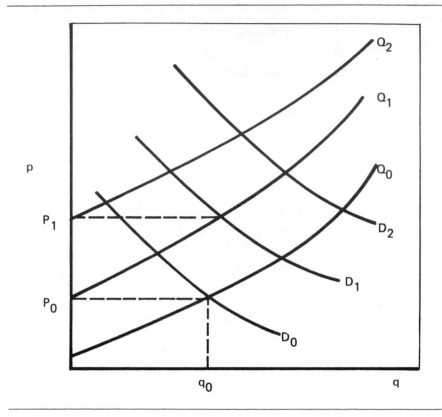

But the knowledge of rising prices has a feedback effect on supply. There is a
benefit to producers in producing less today, shifting some output from present
to future. The present value of the benefit (user cost) is the sum of lower production costs today (movement down the curve) plus the higher future price,
discounted to present values. Thus in Figure 3-3, the "true" supply curves are
all shifted leftward. Prices are higher at first and lower later. (Compare Figure
1-2, p. 6.) This is the traditional effect of reckoning with user cost, over and
above operating cost and development cost, as part of the supply curve. But the
process is incomplete. Exploration and improved technology will push the supply curve rightward.

Figure 3-4 shows a family of curves for given exploration intensities E_1,
E_2, and E_3. Depletion and intensity of exploration are only two determinants of
the path industry takes during the period in question. The price of the mineral,
forecasts of its price, changes in cost, effectiveness of exploration, and so on,
all affect this path.

Figure 3-4.   Industry Supply Curves as a Function of Cumulative Exploration.

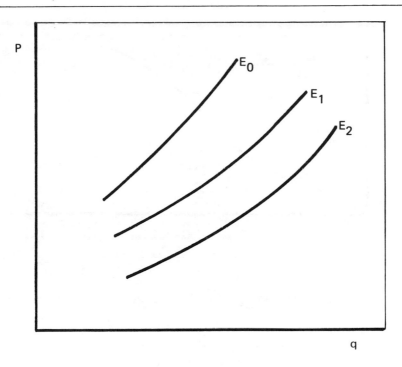

So far we have been describing supply curves as price p (or cost), as a function of the level of production q. One can also think of a cost curve as a function of cumulative production Q as in Figure 3-5. One might assume some backstop technology for mining uranium, say Chattanooga Shale or seawater, at a price p. (A backstop technology is a source of nearly unlimited resources at a price higher than that prevailing currently. Mining uranium from the low-grade Chattanooga Shale would be expensive, but the formation contains a very large total amount of uranium.) The depletion of uranium deposits would be portrayed as a monotonically increasing curve that finally approaches or intersects the horizontal supply curve at p. To construct the curve, strong assumptions would be needed about rates of extraction, price scenarios, competition among firms, and so on.

A cost curve can aid policymaking. Let us say coal-fired power plants are to be compared with nuclear plants. The cost of the fuel for the marginal nuclear plant is an average price corresponding to the interval $Q_0 + Q$, where Q_0 is the cumulative supply of uranium needed to supply lifetime requirements of every

Figure 3-5.   Price as a Function of Cumulative Production.

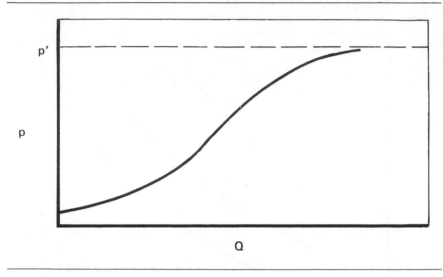

existing or planned plant and Q is the lifetime requirement for the marginal plant.

EXPLORATION MODELS

A simple example of an exploration model is the assumption of a specific relation between effort and yield. For a particular price, say $30 per pound, the model predicts that uranium deposits will be found, on the average, at the rate of a certain number of pounds per foot drilled. Often the predicted yield per unit effort is assumed to decline in a certain way with cumulative discoveries. Effort/yield models evade the need for explicit estimates of extent; assumptions about the resource base are inherent in the model but are usually unstated. An estimate of the resource base itself sometimes can be made by integrating under the effort/yield curve, but it is only a by-product of the underlying model. It may thereby reveal assumptions not made explicit or perhaps not even realized by the model-builder. In models such as these exploration costs are merely another expense incurred in the production of uranium.

A simple effort/yield relation is unrealistic, however. Fuel and nonfuel minerals are typically found in bunches, known variously as districts, basins, and plays. When a new district is located, the yield is high at first because the larger deposits are found first; then the yield usually tapers off. The decline is a highly nonlinear process, which typical effort/yield models fail to capture. The more complicated Kaufman–Barouch model of the discovery process will be discussed more fully in Part III, on undiscovered oil and gas. Using the first few discoveries in any oil play, Kaufman and Barouch predict the size and quantity of future discoveries in the same play. Given assumptions about the probabilities of success and failure for any particular drill hole, exploration costs and expected yields can be built into supply curves for that district.

The basic idea of sharply diminishing returns to additional effort has also been applied to very wide areas, even the whole world.[1] The particular assumptions are not revealed; they probably cannot even be made explicit. But geologists have long regarded the whole earth as their oyster and the various published estimates as a way of capturing their subjective but important guesses. Exploration is a way of tempering or even reversing the otherwise inevitable increases in development costs from known deposits—the fourth margin discussed in Chapter 1. It is perhaps the most important and the least known.

REFERENCE

1.   *World Energy Conference, The Limits of World Oil Supply*, Pierre Instánbul, 1978.

SUMMARY

Resource appraisals are conducted for many reasons. These estimates of availability are data used to make public policy decisions. With the federal government increasingly involved in decisionmaking about energy issues, much research that was previously private (for example, which area to explore next) is becoming public (for example, how fast breeder reactors should be introduced and how much domestic uranium stock should be guaranteed as supply for foreign countries). Resource availability models, then, should be evaluated in terms of their contributions to strategic decisions as well as their sensitivity to regulatory decisions, such as environmental legislation. The remainder of this book will show that many mineral supply models are of value to strategic questions only in a general way. For example, crustal abundance models add information only to questions concerning very long-run availability, and they do not contribute to questions regarding the efficiency of price incentives. However, the more detailed and the more specific a model, such as the one used in the resource estimates of the Grand Junction Office of the U.S. Department of Energy (see Table 2-2), the more reliable it is in analyzing alternative policy scenarios.

OIL AND GAS Estimation of Discovered Reserves

M.A. Adelman

INTRODUCTION

A 1978 publication of the World Energy Conference estimated "ultimate resources" of oil at about 7,500 billion barrels (BB), which is at least ten times the current figure for "proved and probable reserves of conventional oil."[1] Although both numbers are valid, they do not measure the same thing. *Proved reserves* (excluding probable) is a more precise measure of shelf inventory. *Ultimate resources* is a very imprecise estimate of the outer limits of shelf inventory, plus warehouse inventory, plus what is still in the quarry, plus what is thought to be "out there." It includes discovered, plus to-be-discovered oil, plus as-yet unexploited sources (shales, heavy oil).

The next two chapters explore the area between proved reserves and ultimate resources. Within the scheme of definition and classification (see Figure 2-1 in Part I), the existing numbers will be explored to show their meaning and interrelations. Some of these data may be of interest for their own sake, as are diminishing returns for U.S. exploration since 1950, for example. Others will, upon examination, show contradictions and gaps in knowledge. The primary interest here is with principles rather than facts, however.

The principles go beyond the consistency of definitions. *Reserves* at any moment are by definition a stock, a fixed amount. But over time, *reserves* as well as *resources* form part of a flow, as explained in Part I.

Proved reserves being nothing more nor less than cumulative output from producing installations in place, they are measured by cumulating current producing capacity, whose decline rate over time is approximated by the depletion rate (see Appendix B). In the past, probable reserves of oil in the United States at any moment have been much larger than proved reserves, two to three times as large. The rate is much lower for gas because improved recovery is not important. Some such proportions probably hold in other countries with younger industries, but not in this country, where the return to discovery investment is probably less than one-tenth that of the period just after World War II, , and the return to development investment is also lower and diminishing.

For any given owner or for the whole producing system, a larger optimal plan could be drawn up, including the time and investment needed to convert probable into proved reserves or even to convert potential (undiscovered) into probable reserves using some kind of certainty equivalent. Of the infinite number of possible time profiles of conversion and ultimate production, the most desirable would be the profile with maximum present value. For every given change in price, there would be a different maximum and a different production response. This would be the true long-run supply function.

The scarcity of information in the public domain makes calculations of this function difficult. Some private companies (or governments that do not make information public) are doing some calculations of probable reserves or at least could do so. For the present though we must be content to show what kind of information exists on proved and probable reserves, leaving to Chapter 4 the understanding of reserves yet to be discovered.

REFERENCE

1. Summarized in *Petroleum Economist* (March 1978): 86.

4 PROVED RESERVES
The API-AGA-CPA System

Between 1946 and 1978, with a notable expansion in 1966, the American Petroleum Institute (API) and the American Gas Association (AGA) (and the Canadian Petroleum Association [CPA])[1] each year estimated proved reserves through a national committee (Tables 4-1, 4-2). The committee brought together the estimates made for about fourteen large regions subdivided into many smaller ones. In the later years the regional subcommittees comprised about 160 API members and 110 AGA members. Estimates of proved reserves were made for individual reservoirs, and individual reservoir records were maintained by the district subcommittees. The underlying data were kept strictly confidential; only the district reserve estimates were transmitted to the national committees, and beginning in 1974 the API to the annual publication added a table on the 100 largest oil fields in the United States. This estimation procedure ceased in 1978 and has been only partly and insufficiently replaced by government estimates.

The subcommittee structure was the result of the great decentralization of the American petroleum industry, where a given field or reservoir may be operated by several or many firms. The subcommittee structure permitted estimating without violating antitrust laws, since the subcommittee member was not permitted to convey the information to anyone else in his or her company. Members were drawn from local operators, a highly variable mixture of large and small firms, for a company that is negligible in the national picture may be large in a small area. Other members were drawn from state agencies or oil consulting firms.

Table 4-1. Annual Estimates of Proved Crude Oil Reserves in the United States, 1946–1977, Thousand Barrels of 42 U.S. Gallons.

(1) Year	(2) Proved Reserves at Beginning of Year	(3) Revisions	(4) Extensions	(5) New-Field Discoveries	(6) New Reservoir Discoveries in Old Fields	(7) Total of Discoveries, Revisions, and Extensions	(8) Production	(9) Proved Reserves at End of Year	(10) Net Change from Previous Year[a]
1946	19,941,846	1,254,705	1,158,923	n.a.	244,434	2,658,062	1,726,348	20,873,560	931,714
1947	20,873,560	749,278	1,269,862	n.a.	445,430	2,464,570	1,850,445	21,487,685	614,125
1948	21,487,685	1,958,853	1,439,873	269,438	127,043	3,795,207	2,002,448	23,280,444	1,792,759
1949	23,280,444	603,566	1,693,862	544,319	346,098	3,187,845	1,818,800	24,649,489	1,369,045
1950	24,649,489	663,378	1,334,391	407,739	157,177	2,562,685	1,943,776	25,268,398	618,909
1951	25,268,398	1,776,110	2,248,588	205,959	183,297	4,413,954	2,214,321	27,468,031	2,199,633
1952	27,468,031	743,729	1,509,131	280,066	216,362	2,749,288	2,256,765	27,960,554	492,523
1953	27,960,554	1,264,832	1,439,618	344,053	247,627	3,296,130	2,311,856	28,944,828	984,274
1954	28,944,828	537,788	1,749,443	307,625	278,181	2,873,037	2,257,119	29,560,746	615,918
1955	29,560,746	696,114	1,697,653	219,824	257,133	2,870,724	2,419,300	30,012,170	451,424
1956	30,012,170	804,803	1,702,311	234,727	232,495	2,974,336	2,551,857	30,434,649	422,479
1957	30,434,649	465,421	1,543,182	207,437	208,760	2,424,800	2,559,044	30,300,405	(134,244)
1958	30,300,405	954,605	1,338,908	151,210	163,519	2,608,242	2,372,730	30,535,917	235,512
1959	30,535,917	1,518,678	1,778,705	165,695	203,667	3,666,745	2,483,315	31,719,347	1,183,430
1960	31,719,347	787,934	1,323,538	141,296	112,560	2,365,328	2,471,464	31,613,211	(106,136)
1961	31,613,211	1,087,092	1,209,101	107,423	253,951	2,657,567	2,512,273	31,758,505	145,294
1962	31,758,505	759,053	1,041,257	92,488	288,098	2,180,896	2,550,178	31,389,223	(369,282)
1963	31,389,223	966,051	858,168	96,732	253,159	2,174,110	2,593,343	30,969,990	(419,233)
1964	30,969,990	899,292	1,419,182	126,682	219,611	2,664,767	2,644,247	30,990,510	20,520
1965	30,990,510	1,783,231	792,901	237,335	234,612	3,048,079	2,686,198	31,352,391	361,881
1966	31,352,391	1,839,307	814,249	160,384	150,038	2,963,978	3,864,242	31,452,127	99,736
1967	31,452,127	1,900,969	716,467	125,105	219,581	2,962,122	3,037,579	31,376,670	(75,457)
1968	31,376,670	1,320,109	776,780	166,291	191,455	2,454,635	3,124,188	30,707,117	(669,553)
1969	30,707,117	1,258,142	614,710	96,435	150,749	2,120,036	3,195,291	29,631,862	(1,075,255)

1970	29,631,862	2,088,927	631,354	9,852,512	116,125	12,688,918	3,319,445	39,001,335	9,369,473
1971	39,001,335	1,600,426	560,596	91,469	65,241	2,317,732	3,256,110	38,062,957	(938,378)
1972	38,062,957	820,107	459,311	123,210	155,220	1,557,848	3,281,397	36,339,408	(1,723,549)
1973	36,339,408	1,551,777	390,141	116,097	87,816	2,145,831	3,185,400	35,299,839	(1,039,569)
1974	35,299,839	1,310,929	368,918	226,163	87,563	1,993,573	3,043,456	34,249,956	(1,049,883)
1975	34,249,956	677,271	340,128	173,177	127,887	1,318,463	2,886,292	32,682,127	(1,567,829)
1976	32,682,127	488,659	466,279	67,842	62,511	1,085,291	2,825,252	30,942,166	(1,739,961)
1977	30,942,166	769,678	365,007	159,950	109,145	1,403,780	2,859,544	29,486,402	(1,455,764)
1978	29,486,402	707,190	366,589	199,994	73,483	1,347,256	3,029,898	27,803,760	(1,682,642)

a. Negative numbers are shown in parentheses.

Source: API-AGA-CPA Reserves of Crude Oil, Natural Gas Liquids, and Natural Gas in the United States as of December 31 (Washington, D.C.: API-AGA, various years). Reprinted with permission.

Table 4–2.   Annual Estimates[a] of Proved Natural Gas and Natural Gas Liquids Reserves in the United States, 1945–1977.

Year	Proved Natural Gas, Million Cubic Feet, 14.73 psia at 60°F				Natural Gas Liquids, Thousand Barrels of 42 U.S. Gallons		
	Nonassociated	Associated-Dissolved	Underground Storage	Total	Nonassociated	Associated-Dissolved	Total
1945	110,113,066	36,873,657	c	146,986,723	c	c	c
1946	115,807,949	43,895,864	c	159,703,813	1,929,926	1,233,293	3,163,219
1950	129,919,009	54,325,898	339,838	184,584,745	2,372,189	1,895,474	4,267,663
1955	151,229,351	69,892,358	1,360,835	222,482,544	2,619,926	2,818,639	5,438,565
1960	185,291,523	74,862,658	2,172,145	262,326,326	3,686,986	3,129,073	6,816,059
1961	190,669,393	73,272,560	2,331,689	266,273,642	3,852,152	3,196,944	7,049,096
1962	198,687,335	71,100,603	2,490,920	272,278,858	4,237,659	3,073,858	7,311,517
1963	201,219,649	72,186,931	2,744,653	276,151,233	4,571,636	3,102,342	7,673,978
1964	207,122,360	71,189,331	2,939,763	281,251,454	4,791,833	2,954,799	7,746,632
1965	213,315,274	70,063,403	3,090,246	286,468,923	5,040,024	2,983,510	8,023,534
1966	217,426,169	68,681,867	3,224,769	289,332,805	5,229,261	3,099,705	8,328,966
1967	221,751,275	67,870,256	3,376,172	292,907,703	5,575,956	3,038,275	8,614,231
1968	220,990,299	62,864,813	3,494,740	287,349,852	5,693,001	2,905,107	8,598,108
1969	211,873,282	59,633,644	3,601,909	275,108,835	5,416,898	2,726,276	8,143,174
1970	204,098,552	82,643,929	4,003,927	290,746,408	5,110,939	2,592,002	7,702,941
1971	195,953,617	87,537,773	4,314,228	278,805,618	4,867,070	2,437,157	7,304,227
1972	186,072,643	75,541,412	4,470,791	266,084,846	4,572,721	2,213,838	6,786,559
1973	172,245,938	73,587,760	4,116,509	249,950,207	4,124,031	2,330,676	6,454,707
1974	162,192,222	71,002,190	3,938,085	237,132,497	4,109,128	2,241,321	6,350,449
1975	156,785,551	67,173,979	4,240,646	228,200,176	4,040,128	2,227,702	6,267,830
1976	148,637,755	63,335,223	4,053,096	216,026,074	3,857,728	2,544,239	6,401,967
1977	142,460,202	61,917,650	4,500,026	208,877,878	3,653,747	2,340,618	5,994,365
1978	138,991,322	56,661,626	4,648,759	200,301,707	3,726,022	2,199,830	5,925,852

a. Year-end reserves.
b. Includes offshore reserves.
c. Not estimated.

ENGINEERING AND ECONOMIC CONCEPT
OF PROVED RESERVES

With this sketch of the estimating machinery, let us turn to what was being estimated: that which "geological and engineering data demonstrate with reasonable certainty to be recoverable in future years from known reservoirs under existing economic and operating conditions,"[2] that is, under current prices and costs.

The definition of proved reserves is not stretched much by stating that it is the forecast of production from existing installations, with zero additional investment. This becomes clear in examination of the division of reservoir areas between "(1) that portion delineated by drilling and defined by gas-oil or oil-water contacts, if any; and (2) the immediately adjoining portions not yet drilled but which can be reasonably judged as economically productive on the basis of available geological and engineering data."[3]

Because the area is delimited by drilling, which is the major part of production investment, the knowledge is not available until the money is mostly spent. Furthermore, "oil which can be produced economically through application of improved recovery techniques (such as fluid injection) is included in the proved classification when successful testing by a pilot project, or the operation of an installed program, provides support for the engineering analysis on which the project or program was based."[4] Even known productive reservoirs expected to respond to the improved techniques are not included in proved reserves "when an improved recovery technique has been installed but its effect cannot yet be fully evaluated."[5] These less certain estimates, along with estimated production from known techniques not yet installed, are classified as "indicated additional reserves," not proved reserves.

A development plan exists for each reservoir, consisting of the number of wells to be drilled and the estimated production from these wells. Production per day or per year will decline over time due to falling reservoir pressure and/or increased water production from wells at the moving boundary between oil and encroaching water. Proved reserves are the area under the decline curve, the summation of all planned production from existing wells, surface facilities, and knowledge.

The undrilled reserves must be a small and often negligible addition, since it is well established that the amount to be recovered from a given reservoir is independent of the number of wells to be drilled into it.[6] A single well might (and a few would) completely drain an oil or gas reservoir, if one were content to wait until doomsday to get it all out. The number of wells to be drilled is an economic decision, trading off the additional investment in wells against the quicker conversion of the asset oil into the asset money. The optimal depletion rate maximizes the present value of the flow of receipts from sale of the oil or gas.

Production decline for any reservoir may be approximated by a constant percentage rate. If Q is current annual production and R is proved reserves, then it is approximately correct that $R = Q/a$, or $a = Q/R$; that is, the decline rate is the same as the depletion rate. This is explicitly assumed in a recent report of the National Petroleum Council, whose membership and staff overlaps greatly with the API and AGA, in forecasting production from current proved reserves.[7]

The ratio of reserves to production, R/Q, is in no sense a life index, nor is it a guide to how long the reservoir will produce. That depends basically on the relation between price and current (variable) operating cost, a subject on which more will be said below. But the reciprocal, Q/R, is an indication of how fast production is likely to decline—that is, the rate at which reserves can be produced. The optimal rate varies considerably among reservoirs, depending on how much incremental investment is needed for an incremental barrel of capacity. Within limits, the relation may be linear, but at some point serious "well interference" develops and greater capacity is increasingly dearly bought by disproportionately high investment. Or too high a rate of production decreases ultimate recovery, usually rather abruptly, to where present value is much decreased.

Production/reserves ratios among forty-three producing areas in the United States varied in 1977 from 9.0 percent (California, San Joaquin Basin) to 24.6 percent (Texas, District 5), with a mean of 13.5. Individual reservoirs of course vary more widely, from as low as 5 to as high as 30 percent per year. The national average production/reserve ratio for natural gas is much lower, 10.8 percent.[8] This is partly because of large seasonal swings in consumption, which gear capacity to the peak rather than to the annual average.

One additional item of evidence confirms, with qualifications, the concept of proved reserves as the forecast of production from developed reservoirs. The API table of the 100 largest oil fields lists only one as undeveloped; it is Hondo, off the California coast, with 94 million barrels of proved reserves. In two other fields "development is just being completed." All others have been producing.[9]

LESSONS OF THE 1971 NATURAL GAS "AUDIT"

In 1973 a valuable experiment was performed when a task force of the Federal Power Commission National Gas Survey made an independent estimate of proved reserves.[10] The special task force was free to examine proprietary company information, and members operated under strong legal sanctions against unauthorized disclosure. According to a scheme devised by commission employees and other government personnel, all large fields were sampled with certainty, while a stratified sample was drawn for all other fields. The survey's estimates of natural gas reserves were 9 percent lower than the AGA estimates for 1971. The difference is barely within the 95 percent confidence limit of 10 percent. It is consistent with the following hypothesis: The original estimating teams, had they been reassembled and asked to redo the task with the hindsight of two years of operating experience, would have downgraded their original estimates.

A 9 percent revision of 280 trillion cubic feet (TCF) is 25 TCF. Some large fraction of this down-revision must actually have been entered in subsequent years, for from 1971 to 1976 cumulative net natural gas revisions were *minus* 7.9 TCF. Revisions are normally positive, and much routine upgrading must have been done in 1971–1976 period. Hence the gross downgrading that took place during that six-year period must have been roughly twice as great, say about 15 TCF. It is unfortunate that upward and downward gas revisions are not tabulated separately. The effect of a small relative adjustment in a very large reserve total was to decrease drastically the year-to-year *gross increments* to reserves. The validity of reserves added in both 1971–1976 and to a much lesser degree to some earlier years is thereby damaged.

REFERENCES

1. In addition to sources already listed the preceding chapters of this, the account in this chapter is based on a number of sources: Morris Muskat, "The Proved Crude Oil Reserves of the United States," *Journal of Petroleum Technology* (September 1963): 915–21. Wallace F. Lovejoy and Paul T. Homan, *Methods of Estimating Reserves of Crude Oil, Natural Gas, and Natural Gas Liquids* (Washington, D.C.: Resources for the Future, 1965). American Petroleum Institute, *Organization and Definitions for the Estimation of Reserves and Productive Capacity of Crude Oil*, Technical Report No. 2, Washington, D.C., 1970. American Petroleum Institute–American Gas Association–Canadian Petroleum Association, *Reserves of Crude Oil, Natural Gas Liquids, and Natural Gas in the United States and Canada as of December 31*, (Washington, D.C.: API–AGA, annually, with slight title changes, since 1946), cited as API–AGA–CPA, *Reserves*. Also, conversations with Ralph W. Garrett and Morris Muskat and my experience as a public member of the API Coordinating Committee on Statistics from 1966 to 1975.
2. API–AGA–CPA, *Reserves*, 1978, p. 14.
3. Ibid.
4. Ibid.
5. Ibid.
6. Rupert C. Craze, "Development Plans for Oil Reservoirs," in *Petroleum Production Handbook*, ed. Thomas C. Frick (New York: McGraw-Hill, 1962), pp. 33-14 to 33-20.
7. National Petroleum Council, *Enahnced Oil Recovery* (Washington, D.C.: NPC, December 1976), p. 216, Table 47; also see Appendix 1, "Note on Decline Curves." The problem is important and needs more attention than given here.
8. API–AGA–CPA, *Reserves*, 1977, pp. 22, 114.
9. Ibid., Table VIII. There is no corresponding table for natural gas.
10. Federal Power Commission National Gas Survey, *National Gas Reserves Study*, Washington, D.C., FPC rev. September 1973, p. 3. M. A. Adelman was a member of the survey panel responsible for this reserve study.

5 CANADA'S "NONCONVENTIONAL RESERVES"

Canadian reserve statistics are comparable to American statistics in most respects except that they include estimates of "nonconventional reserves." Two plants extracting oil from Athabasca oil sands, formerly called "tar sands," are enumerated. "Synthetic crude oil reserves associated with each plant are calculated on the basis of the plant's rated output capacity over a 25-year period, the 25 years being indicative of a reasonable economic life for the facilities. New projects are recognized 3 years prior to scheduled startup date of each such project."[1] This is consistent with the concept of proved reserves as cumulative planned production from facilities already or nearly in place.

In July 1978 Petro-Canada's manager of synthetic crude and minerals said that Canada would, by 1984–1986, have two more plants on stream to produce from the bitumen (heavy oil) deposits at Cold Lake, Alberta.[2] Their combined production would be about 275,000 barrels daily, and the plants would have a twenty-five-year life. It would be consistent, once construction started, to credit 2.5 billion barrels to probable reserves, and then, three years before completion, to proved reserves.

Toward the end of 1977 Albertan bitumen changed from a geological fact to an economic asset; they were potential reserves, soon to become probable reserves. The Canadian Department of Energy, Mines, and Resources estimated production costs at C$7–C$11 per barrel (including capital costs, but not allowing for income taxes[3]). Then Imperial Oil, an Exxon subsidiary, proposed in November 1977 to invest *its own money* in a recovery project, believed to be economic at "world prices." The Canadian regulated price was then slated to reach C$12.75 at the beginning of 1978. In 1977 prices, Imperial's investment was C$2.5 billion, which—at 15 percent return before taxes on the total investment and the same after taxes on the equity portion—would indicate an annual capital charge of C$380 million. Added to annual operating expenses of $188 million (in 1977 prices), this would indicate total costs of C$568 annually, or C$11.83 per barrel.[4] At higher oil prices, this would seem like a sure winner; *but* the greater the rent, the more incentive to Federal and provincial governments to make claims totalling over 100 percent to produce a lasting deadlock.

This cost can probably be lowered, in time, by the addition of a third "train" at modest incremental investment, not to speak of the benefits from the learning process and improved technology. Moreover, companies are rarely accused of understating their costs when trying to prove they need a given price. Waiving any claims to precision, there is probably general agreement that the Imperial and other projects are viable at 1978 world prices, provided they are taxed at ordinary corporate rates, or at least on net profits rather than gross receipts. Therefore, since 1977, it has been up to the provincial and federal governments in Canada to decide how much bitumen is to be unlocked. The decision in late 1980 seems to be no nearer. The amounts in question are obviously large. Imperial had previously estimated the amounts underlying all its Cold Lake leases at 44 billion barrels of producible bitumen,[5] which would imply 33 to 40 billion barrels of recoverable oil. According to the *Application to Alberta Energy Resources Conservation Board* total deposits in the Cold Lake region contain about 164 billion barrels, and Imperial estimates Canada's total heavy-oil resources at nearly 1,000 billion barrels, including the (most expensive) Athabasca tar sands with about 300 million barrels.

Thus, proved reserves of bitumen-derived oil *may* increase from the current 1.5 billion by a factor of as much as 100, depending on the speed of development versus the speed of depletion. The supply curve would probably start as low as C$4.75 (1978 prices) for the lighter or Lloydminster-type heavy crude. This would reflect a capital coefficient of about 40 percent of the Cold Lake proposal; operating costs would be correspondingly lower, since the upgrading process is much less severe.[6] Current technology could extract about 5–10 percent of the Lloydminster oil now in place; more sophisticated (and more expensive) technology could raise this to 40 percent. As for the bitumen itself, the supply curve, wherever it starts, is probably rather flat over a long range. First, the deposits are very large and thick, with overburden between 1,000 and 2,000 feet. Second, perhaps most of the cost is attributable to building a plant that will convert the heated bitumen, brought to the surface by injected steam, into a refinable crude oil.

The Imperial project comes at the end of a decade of laboratory and field research, including three small pilot plants costing a total of $C40 million. Because not much of this research and development work has been done in Venezuela, the heavy oil and bitumen deposits there, which are estimated as three times larger than the Canadian, cannot be called even potential reserves at this time. When enough is known about the costs, the government of Venezuela will need to make decisions that in some respects are simpler than Canadian decisions. The comparison of prices with costs is uncomplicated by division of profits. The owners—the government—need not fear that by asking too much or too little per unit, they will get too little in sum. But the decision is more complex because Venezuela, as a cartel member, must consider the effect of any additional production on the world market, and will not go forward unless conventional oil reserves rapidly disappear, as seems now to be the case.

REFERENCES

1. American Petroleum Institute–American Gas Association–Canadian Petroleum Association, *Reserves of Crude Oil, Natural Gas Liquids, and Natural Gas in the United States and Canada, as of December 31, 1977* (Washington, D.C.: API–AGA), p. 263.
2. *IGT Highlights* July 31, 1978): 2.
3. *Oil and Gas Journal* January 30, 1978): 88.
4. *Application to Alberta Energy Resources Conservation Board* (AERCB). 11 November 1977. Calculations and conclusions are mine.
5. *Wall Street Journal* November 14, 1977): 14.
6. *Petroleum Economist* (August 1978): 344.

6 THE GROWTH OF PROVED RESERVES IN THE UNITED STATES

The process by which proved reserves are increased at one end while being depleted at the other is the movement of inventory from the poorly lit warehouse of probable reserves onto the well-lit shelves of proved reserves, from where it is sold and disappears.

TYPE OF WELLS

Reserves are added by drilling, and wells have for years been classified by the Lahee system, devised by the geologist Frederic H. Lahee (Figure 6-1). The 1978 statistics, with the percentage of dry holes an indicator of increasing risk, are as follows[1]:

Type of Well	Percentage Dry Holes
Development	22.7
Exploratory:	
Shallower pool	19.2
Outposts	58.3
Deeper pool tests	65.0
New-pool wildcats	61.8
New-field wildcats (approximate 1966–1970 average)	98.3

Each type of well, if successful, adds to proved reserves. Development wells may enlarge the area of a known reservoir, and additional reserves thus credited are the "extensions" of Tables 6-1 and 6-2. As reservoirs become better known

41

Figure 6-1.   AAPG and API Classification of Wells.

OBJECTIVE OF DRILLING			INITIAL CLASSIFICATION φ WHEN DRILLING IS STARTED	FINAL CLASSIFICATION AFTER COMPLETION OR ABANDONMENT	
				SUCCESSFUL ● ○ ■	UNSUCCESSFUL ⊕
Drilling for a new field on a structure or in an environment never before productive			1. NEW-FIELD WILDCAT	NEW FIELD DISCOVERY WILDCAT	DRY NEW-FIELD WILDCAT
Drilling for a new pool on a structure or in a geological environment already productive	NEW POOL TESTS	Drilling outside limits of a proved area of pool	2. NEW-POOL (PAY) WILDCAT	NEW-POOL DISCOVERY WELLS *(Sometimes extension wells)* — New-Pool Discovery Wildcat *(Sometimes an extension well)*	DRY NEW-POOL WILDCAT
		Drilling inside limits of proved area of pool — For a new pool below deepest proven pool	3. DEEPER POOL (PAY) TEST	Deeper Pool Discovery Well	DRY NEW-POOL TESTS — DRY DEEPER POOL TEST
		For a new pool above deepest proven pool	4. SHALLOWER POOL (PAY) TEST	Shallower Pool Discovery Well	DRY SHALLOWER POOL TEST
Drilling for long extension of a partly developed pool			5. OUTPOST or EXTENSION TEST	EXTENSION WELL *(Sometimes a new-pool discovery well)*	DRY OUTPOST OR DRY EXTENSION TEST
Drilling to exploit or develop a hydrocarbon accumulation discovered by previous drilling			6. DEVELOPMENT WELL	DEVELOPMENT WELL	DRY DEVELOPMENT WELL

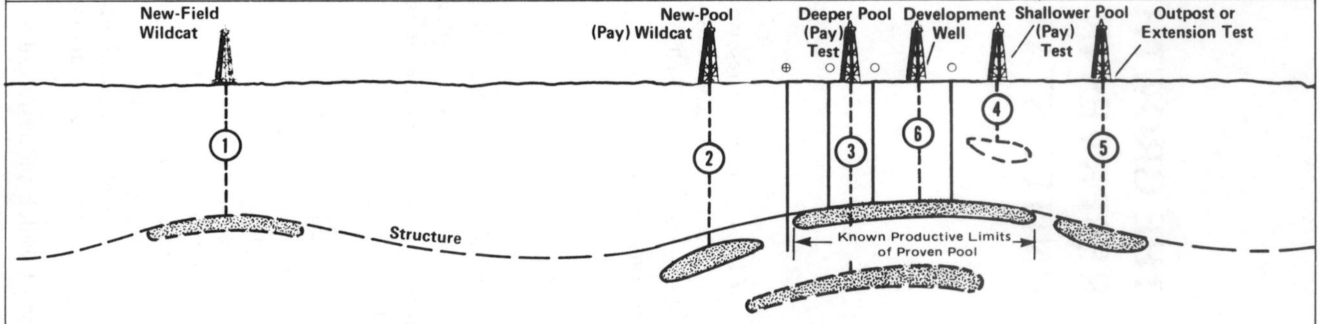

New-Field Wildcat ①

New-Pool (Pay) Wildcat ②

Deeper Pool (Pay) Test ③   Development Well ⑥   Shallower Pool (Pay) Test ④   Outpost or Extension Test ⑤

⊕ ● ○

Known Productive Limits of Proven Pool

Structure

LAHEE CLASSIFICATION OF WELLS, AS APPLIED BY CSD

Source: The American Association of Petroleum Geologists, "North American Drilling Activity," *AAPG Bulletin* (Annual Drilling Issue) 63, no. 8 (1979) Figure 8, titled "AAPG and API Classification of Wells," p. 1202. © American Association of Petroleum Geologists, Tulsa, Oklahoma.

Table 6-1. Crude Oil Reserves in the United States,[a] by New and Old Fields, 1946-1975.

Period	(1) New Field "Discoveries"	(2) Exploratory Wells (000s) Total	(3) Exploratory Wells (000s) NFW[b]	(4) Recovery (1976 Evaluation) from Fields Discovered in Period	(5) Gross Additions to Proved Reserves	(6) New Oil from Old Fields (5)-(4)	(7) Ratios of Columns (1)/(4)	(8) Ratios of Columns (4)/(3)	(9) Ratios of Columns (6)/(5)
1946-1950	1.8	42.7	20.6	12.7	14.8	2.1	.142	.616	.142
1951-1955	1.4	65.5	35.3	8.8	16.2	7.4	.161	.246	.457
1956-1960	0.9	69.0	31.0	7.1	14.1	7.0	.130	.222	.496
1961-1965	0.7	52.7	33.1	4.3	12.8	8.5	.163	.130	.664
1966-1970	0.6	45.5	27.8	3.6	13.8	10.4	.176	.122	.754
1971-1975	0.7	39.7	26.4	2.1	8.3	6.2	.467	.057	.747
Total	6.1	315.1	174.2	38.6	80.0	41.6	.163	.215	.520

a. Excluding Alaska.

b. New-field wildcats.

Sources: Cols. (1), (4), (5) from API-AGA-CPA, *Reserves of Crude Oil, Natural Gas Liquids, and Natural Gas in the United States as of December 31, 1978*, (Washington, D.C.: API-AGA, 1976), Tables II and III. Columns (2), (3) from *AAPG Bulletin* "North American Drilling," issue (various years).

Table 6–2. Natural Gas Reserves in the United States,[a] by New and Old Fields, 1946–1975, Trillion Cubic Feet.

Period	(1) "Discoveries,"[b]	(2) Exploratory Wells (000s) Total	(3) Exploratory Wells (000s) NFW	(4) Recovery (1978 Evaluation) from Fields Discovered in Period	(5) Gross Additions to Proved Reserves	(6) New Gas from Old Fields (5)–(4)	(7) Ratios of Columns (1)/(4)	(8) Ratios of Columns (4)/(3)	(9) Ratios of Columns (6)/(5)
1946–1950	18.4	42.7	20.6	67.8	81.8	14.0	.27	3.3	.17
1951–1955	26.0	65.5	35.5	66.9	82.1	15.2	.39	1.9	.18
1956–1960	32.5	69.0	41.0	74.2	98.1	23.9	.44	2.4	.24
1961–1965	32.2	52.7	33.1	50.0	96.4	46.4	.64	1.5	.48
1966–1970	23.5 (11.0)[c]	45.5	27.8	27.8	70.3	42.5	.84 (.40)[d]	1.0	.60
1971–1975	21.5 (9.3)	39.7	26.4	30.2	45.4	15.2	.71 (.31)	1.1	.33
Total	154.1 (129.4)	315.1	174.2	316.9	474.1	157.2	.49 (.41)	1.8	.33

a. Excluding Alaska.

b. Numbers represent "total discoveries," that is, new-field discoveries plus new pools in discovered fields.

c. Numbers in parentheses represent new fields alone.

d. Numbers in parentheses represent calculations using the new-field values from column (1).

Sources: Columns (1), (4), (5) from API-AGA-CPA, *Reserves*, 1978 issue, Tables II and III. Columns (2), (3) from *AAPG Bulletin* "North American Drilling Statistics," issue (various years).

through operating experience, changes up or down in their reserves are the "revisions." Upward and downward revisions in estimates of oil were and perhaps still are approximately equal in number, but upward revisions are almost invariably larger because "individual upward revisions are more concentrated among the larger reservoirs which experience has begun to show often turn out to be better with respect to reserves and of longer life than they were judged to be in their earlier years."[2] New reservoirs may be found in old fields shallower, deeper, or removed horizontally. Finally, there are wells drilled to find new fields.

REPORTED "DISCOVERIES"

Discoveries is a much abused term. One often reads of new reserves "discovered" that are in fact recent additions to proved reserves in known pools. Much more misleading is the implication that petroleum producers need to *discover* some amount to maintain or expand output. They need new reserves, but nobody ever found a reserve. Oilmen find reservoirs, the contents of which they gradually develop into reserves.

Another error is in treating the reserves-added in any given year by API-AGA "discoveries" as though they were an approximation, or at least an indication of, what has in fact been discovered. Year-to-year fluctuations in recorded "discoveries" are largely independent of the year-to-year variation in what has actually been newly found. In fact, oil and gas operators drill development wells into an assortment of reservoirs, depending on convenience and cost. New reservoirs are typically overrepresented, of course, because they tend to be flush production at lower cost. But as column (5) of Table 6-1 shows, from 1946 to 1975 approximately 80 billion barrels were added to proved reserves of oil (excluding Alaska), of which about 39 billion barrels represented newly found oil fields. But the aggregate of annual oil "discoveries," in column (1), over the period amounted to only 6 billion. (Corresponding natural gas numbers were 474, 317, and 157 trillion cubic feet.[3]) The reserves contained in the new fields discovered unfolded by development in subsequent years were first published in the 1961 National Petroleum Council Report,[4] and since 1966 have appeared in the annual API-AGA-CPA *Reserves* volumes. It is surprising that after 1961 there was any econometric work utilizing the series "discoveries" of oil and gas. In fact, there was much.

Table 6-1 summarizes the respective contributions of new and old fields to American crude oil supply over thirty years. Discoveries in column 1 is a misleadingly low total, as explained. Except for 1971-1975, the total is never as much as one-fifth of what is actually discovered in a given year, and probably 1971-1975 only confirms the rule of thumb that it takes six years to evaluate an oil or gas field. It must be said, however, that with the average size of new fields decreasing, the time lag should become much shorter.

Column (4) records the reserves from fields discovered in those years, as estimated in 1978, with the benefit of operating experience, development drilling, exploration in and around the original reservoirs, and improvement in recovery methods. Column (4) is the real *discovery* record. Newly discovered oil has been diminishing since at least 1950, despite increased effort, shown in columns (2) and (3). The payoff to exploratory effort, column (8), dropped by 60 percent from 1946-1950 to 1951-1955, declining more gradually thereafter.

SOURCES OF NEW RESERVES IN OLD FIELDS

By subtracting the newly discovered oil, column (4), from the gross additions to reserves in each period, column (5), an indicator of the contribution of new oil from old fields can be derived. The resulting figure in column (6) is imprecise, however, because some of the new oil from 1946-1950 discoveries, for example, was not impounded into reserves until many years later. It is as though column (5) represented cumulative transfers of cash into an account for outpayment while column (4) represented a mixture of cash and short-term securities received by the corporation. With this qualification, column (6) is an estimate of how much new oil was made available from old fields. A sharp turnabout is visible around 1950. As it was becoming increasingly costly to find new oil in new fields, more effort was devoted to obtaining oil from old fields. The absolute amount tripled from the first to the second period, and continued to increase through 1970, when it in turn confronted sharply rising marginal costs. By 1970 new oil from old fields was providing over four-fifths of new proved reserves, column (9).

Table 6-2 shows a similar but less dramatic change for natural gas. The contribution of new gas from old fields rose from less than 20 percent before 1955 to over 50 percent in 1960-1970, falling sharply again in 1971-1975. For reasons to be stated later, new gas from old fields is not as important as new oil from old fields.

In both tables, column (8), the amount of new hydrocarbon per new-field wildcat well, is of ordinal significance only. That is, it is not known how many such wells ought to be considered as aimed at gas or at oil. The best that can be done is to divide discoveries of each hydrocarbon by the total of all such wells. There are sharply decreasing returns to exploration during thirty years. In oil, the decline in discoveries is continuous and very steep; in gas, not nearly as much.

NEW OIL OR GAS: AFTER DISCOVERY

The API report on definitions for estimation of oil reserves[5] outlines some typical sequences in the discovery and development process. A new reservoir has been found. If the porosity (liquid per cubic foot of sands) and permeability (rate of fluid flow through the sands) are low and interrupted, 10-20 acres around the well is assumed as the area for reserves calculation. Given better conditions of thickness, the area is assumed to be 20-40 acres. A thicker but steeply plunging section is allotted the same 20-40 acres, whereas if the dip is gentle the discovery is allotted 60-100 acres. Such judgments will be made on the spot.

Another case is one with a dry hole followed by a success, which between them furnish enough information for an estimate of a proved drilled area of 40 acres or of a proved undrilled area of 170 acres. But doubt exists as to the cross section. A fault seals one end of the reservoir and water seals the other, but between the low of the proved oil and the high of the proved water there is a gap; then the intervening space is "prospective," not proved.

Still another case has three successful wells in sequence, which expand the "proved drilled" area from 40 to 100 acres; the "proved undrilled" area rises from 50 to 425 acres. An additional well hits salt water, outlining the pool, but it also finds a new pool, which is assigned 15-20 acres. This oversimplified sequence, moreover, gives no hint of what is being discovered or about the driving mechanism for the oil—gas, water, or gravity—which may strongly affect the estimates of both oil-in-place and of the fraction that can be recovered. This process of development cum fringe exploration accounts for the great bulk of oil production investment and additions to reserves.

REFERENCES

1.   "North American Drilling Activity in 1978," *AAPG Bulletin* 63, no. 8 (August 1979): Tables 1, 3, 16, 17. Note that only a minor fraction of the "successful" new-field wildcats find commercial fields, but this evaluation takes years to make. Hence, statistics are only published with a six-year lag; e.g. for 1970 in the volume covering 1976 developments.
2.   Morris Muskat, "The Proved Crude Oil Reserves of the United States," *Journal of Petroleum Technology* (September 1963): 919.
3.   In this case, discoveries include "total discoveries" (see Table 4-1).
4.   National Petroleum Council, *Report of the National Petroleum Council Committee on Proved Petroleum and Natural Gas Reserves and Availability*, Washington, D.C., NPC, 15 May 1961.
5.   American Petroleum Institute, *Organization and Definitions for the Estimation of Reserves and Productive Capacity of Crude Oil*, Technical Report No. 2, Washington, D.C.: 1970), pp. 38-42.

7 THE RELATION OF PROBABLE RESERVES TO PROVED RESERVES

Thus far the topic has been oil reservoirs (or pools). A *reservoir* can be rigorously defined as a closed hydrodynamic system with precise limits in which fluid pressure (gas, oil, water) is balanced by the strength of the containment materials. Changes in pressure at any one place produce changes in pressure everywhere else. By contrast, *field* is a loose term meaning an assemblage of adjacent or overlapping reservoirs that are inside the ill-defined boundaries of a geological structure. Adjacent fields have some geological event or cataclysm in common, and may be called a *trend.* Two or more trends may make up a *basin.* The oil or gas *play* is an investment in gathering knowledge about the limits of the first reservoir, the presence of neighboring reservoirs, and their aggregation into fields, basins, and trends.

Table 7-1 shows that surrounding the proved reserves are probable reserves, which consist of estimates of what may be produced from the undrilled portions of inown reservoirs, from the new horizons in those reservoirs, and from adjacent pools. Most of these pools exist in *structures*, deformations of the rock strata in areas about which a great deal is known. The number of pools and the likelihood and amount of hydrocarbon content are estimated by analogy and extrapolation.

The ratio of these probable reserves to proved reserves must be fairly high. In the United States, for example, during 1946-1965 the production/reserve ratio stayed, within narrow limits, around 8 percent. In 1978 it was up to 13.7 percent.[1] Thus an operator planning to maintain constant output despite a 14 percent decline rate must plan ahead ten years with 1.745 times the amount

Table 7-1. Classification of Petroleum Reserves.

Energy Source	Degree of Proof	Development Status	Producing Status
Primary Reserves	*Proved*	*Developed*	*Producing*
These reserves recoverable commercially at current prices and costs by conventional methods and equipment as a result of natural energy inherent in the reservoir	Primary reserves that have been proved to a high degree of probability by production from the reservoir at a commercial rate of flow or in certain cases by successful testing in conjunction with favorable complete core-analysis data or reliable quantitative interpretation of log data	Proved reserves recoverable through existing wells	Developed reserves to be produced by existing wells from completion lateral(s) open to production
			Nonproducing
			Developed reserves behind the casing or in certain cases at minor depths below the producing zones, which will be produced by existing wells
		Undeveloped	
		Proved reserves from undeveloped spacing units in a given reservoir that are so close and so related to the developed units that there is every reasonable probability that they will produce when drilled	
	Probable		
	Primary reserves that have not been proved by production at a commercial rate of flow, but being based on limited evidence of commercially producible oil or gas within the geological limits of a reservoir above a known or inferred water table are susceptible to being proved by additional drilling and testing		
	Possible		
	Primary reserves that may exist but where available data will not support a higher classification		

Secondary Reserves

Those reserves recoverable commercially at current prices and costs, in addition to the primary reserves as a result of supplementing by artificial means the natural energy inherent in the reservoir, sometimes accompanied by a significant change in the physical characteristics of reservoir fluids

Proved

Secondary reserves that have been proved to a high degree of probability by a successful pilot operation or by satisfactory performance of full-scale secondary operations in the same reservoir or in certain cases a similar nearby reservoir producing from the same formation

Developed

Proved reserves recoverable through existing wells from a reservoir where successful secondary operations are in progress

Producing

Developed reserves to be produced by existing wells in that portion of a reservoir subjected to full-scale secondary operation

Nonproducing

Developed reserves to be produced by existing wells upon enlargement of the existing secondary operations

Undeveloped

Proved reserves that will be produced upon the installation and operations of a secondary recovery project and or by drilling of additional wells

Probable

Secondary reserves that are thought to exist in a reservoir by virtue of past production performance or core, log, or reservoir data, but where the reservoir itself has not been subjected to successful secondary operations

Possible

Secondary reserves from reservoirs that appear to be suited for secondary operations but where available data will not support high classification

Source: Jan J. Arps, "Estimation of Primary Oil and Gas Reserves," in *Petroleum Production Handbook*, eds. T. C. Frick and R. W. Taylor, © McGraw-Hill Book Company, New York, 1962. Reprinted with permission.

Table 7-2.   Canada: Proved and Probable Hydrocarbon Reserves.

	1963	1976
Crude oil, BB		
Proved	4.48	6.65
Probable	2.21	1.18
$P + P'$	6.69	7.83
Natural gas liquids, BB		
Proved	0.70	1.59
Probable	0.12	0.16
$P + P'$	0.82	1.75
Natural gas, TCF, 1964		
Proved	21.2	57.0
Probable	6.2	13.5
$P + P'$	37.4	71.5

Source: American Petroleum Institute–American Gas Association–Canadian Petroleum Association, *Reserves of Crude Oil, Natural Gas Liquids, and Natural Gas in the United States as of December 31, 1978* (Washington, D.C.: API–AGA, 1964, 1978), CPA Section, Tables I and II. Reprinted with permission.

of the proved reserves.[a] Small wonder that two reasonable persons would differ 75 percent when estimating the operator's "reserves." The evaluation of known structures and pools, their contents known not with reasonable certainty but only with reasonable probability, would well repay immediate systematic estimation.

Some efforts are being made to fill the gap between estimates of proved and probable. The Canadian Petroleum Association defines *probable reserves* as follows: "Probable reserves are a realistic assessment of the reserves that will be recovered from known oil or gas fields based on the estimated ultimate size an and reservoir characteristics of such fields."[2] The term will be used here to denote amounts in known fields not now in proved reserves; the term "proved plus probable" $(P + P')$ is their sum.

Table 7-2 shows that Canadian probable crude oil reserves have diminished since 1963, the first year estimated, absolutely and even more relatively; natural gas liquids have increased absolutely but not relatively; probable and proved natural gas reserves have doubled.

In the United States, the American Gas Association devolved the estimate of nonproved reserves to a Potential Gas Committee, whose work it supports, monitored by the Potential Gas Agency, a group at the Colorado School of Mines. They define probable reserves as "resulting from the growth of existing fields" trough extensions and new pools. Table 7-3 shows that probable reserves in the lower forty-eight states diminished by over one-third, from 300 trillion cubic

a. That is, over ten years the cumulative output will be $(.137 \times 5.44) = .745$ of the existing reserves, and new reserves must be provided to replace them.

Table 7-3.   United States: Probable Reserves of Natural Gas,
End of 1976, Trillion Cubic Feet.

| | Lower 48 States | | | Alaska |
Year	Probable Reserves	Increment	Gross Increments to Proved Reserves	Probable Reserves
1966	300	–	–	n.a.
1968	238	–62	+35	22
1970	218	–20	+20	39
1972	212	– 6	+19	54
1976	192	–20	+34	23
		–108	+108	

Sources: Potential Gas Committee, *Potential Supply of Natural Gas in the United States*, School of Mines, Golden, Colo., 1976, p. 16; gross increments to proved reserves, from API–AGA–CPA, *Reserves*, various years. Reprinted with permission.

feet (TCF) in 1966 to 192 TCF in 1976. Probable reserves drained into proved reserves (the exact coincidence of 108 TCF, is of course, accidental). The drainage was not offset by exploration and development. The years 1973–1976 inclusive was a time of relatively low, though rapidly increasing, exploratory drilling. The effect of higher drilling rates remains to be seen.

Probable reserves in Alaska diminished by much more than might be expected in a new area. Much reserves seemed likely, but further evaluation was very disappointing.

PROBABLE RESERVES AS THE STRETCH
FACTOR ON PROVED RESERVES

It is evident that the flow from probable reserves into proved reserves is of the first importance. Because of the long decline in American discovery efforts, the pressure to stretch has been greatest in this country. Hence, experience in the United States anticipates what will happen elsewhere.

Tables 7–4 and 7–5 show two samples of the process of transition from probable reserves (not estimated nor tabulated) into proved reserves. If PR = proved reserves, PBR = probable reserves, UR = ultimate recovery, O = original year, and T = final year, then $PBR_O = UR_T - (UR_O + PR_O)$, and the stretch factor for reserves is S.

$$S = \frac{PR_O + (UR_T - UR_O)}{PR_O} = \frac{PR_O + UR_O((UR_T/UR_O) - 1)}{PR_O} \quad (7\text{-}1)$$

If an experienced observer in 1945 anticipated that ultimate recovery in 1977 (UR_T) from those fields known in 1945 would be about 98 BB, then 1945 prob-

Table 7-4. Proved Reserves, Ultimate Recovery, and Intervening
Production and Reserve Additions in Large Fields, 1957-1977,
Million Barrels.

End of 1957	Already produced.	15,364	Ultimate recovery,
	Remaining reserves,	9,900	25,264
1958-1977	less production,	-16,686	
	plus reserves added,	+16,638	
End of 1977	equals remaining reserves.	9,852	Ultimate recovery,
	Already produced.	32,050	41,902

Sources: For 1977, API-AGA-CPI, *Reserves*, 1978, Table VIII. For 1957, *Oil and Gas Journal* (January 27, 1958): 163-68. Proved reserves for those two years for the total United States were 30.4 billion and 19.7 billion.

able reserves were 30 BB (= 98 - 48 - 20), and the stretch factor on proved reserves in the base year 1945 would be equal to

$$\frac{20 + 48\ (98/45\ (-1))}{20} = 3.5\ .\tag{7-2}$$

Similarly, in 1972, with proved reserves at 28 billion barrels, another 5.4 billion were due to be added in the next five years in fields known in 1972. The importance of unestimated probable reserves in oil supply is also seen by a comparison in Table 7-4 of sixty-six large oil fields over twenty years. The list was obtained by comparing the 1977 100 largest, minus Alaskan fields, with any of the same name that could be found in the *Oil and Gas Journal* listing of large fields twenty years earlier. Some fields are missing from the earlier tabulation simply because no data were available; some few because they originally were not large enough to be counted and have subsequently grown. Hence the sample is one of large mature fields, which in 1957 and 1977 amounted to about one-third and one-half of all U.S. reserves, excluding Alaska.[3] Using formula (7-1), the stretch is estimated to be 2.68, as compared with a thirty-year stretch of 3.5 for all fields in 1945-1977.

All these estimates exclude the North Slope of Alaska. There and in the lower forty-eight states one of the things we would most wish to know is the volume of probable reserves in known fields that can be added into proved reserves in the near future. There is nothing automatic about the process. The amount depends on what is in the ground, on the cost of developing it, and on expected prices: The price-reserve increment curve is the heart of the problem.

A little can be gleaned by a careful examination of Tables 7-5 and 7-6. As might be expected, the early years after discovery are usually the most rewarding, since there is most to be learned then. This aside, fields discovered before 1920 are most consistent in stretching. The 1931-1940 group outdoes them before the middle 1960s, then slumps.

Table 7-5. Increased Ultimate Recovery of Oil in Old Fields, United States,[a] Billion Barrels, and Growth in Percent per Year.

Year of Estimate: Proved Reserves, BB:	1945 19.9	Growth to 1960, %	1960 20.5	Growth to 1967, %	1967 31.4	Growth to 1972, %	1972 28.2	Growth to 1977, %	1977 22.0	Growth to Total, %
Period of discovery										
Pre-1920	14.6	1.22	17.4	2.72	21.1	2.43	23.8	2.31	26.7	1.89
1920–1930	17.9	2.36	25.5	1.80	28.9	0.61	29.8	0.00	29.3	1.53
1931–1940	15.9	2.82	24.1	1.90	27.5	1.28	29.3	0.10	30.8	2.09
1941–1950			18.5	1.55	20.6	0.68	21.3	1.02	22.4	1.13
(1951–1954)			(5.9)	(2.26)	(6.9)	(1.02)	(7.3)	(0.08)	(7.3)	(1.25)
1951–1960					13.7	1.97	15.1	0.97	15.9	1.47
(1961–1967)							(4.4)	(1.54)	(4.7)	(1.54)
1961–1970							6.1	-1.35	5.7	-1.35
Total	48.5	—	91.37	—	111.8	—	125.38	—	130.68	

a. Excluding Alaska.

t. Numbers in parentheses are excluded from totals to avoid double counting.

Sources: 1945 and 1960 data from National Petroleum Council, *Report of the National Petroleum Council – Committee on Proved Petroleum and Natural Gas Reserves and Availability* (Washington, D.C.: NPC, 1961). Other years, API-AGA-CPA, *Reserves*. Both sources are drawn from the same basic data, which were published regularly in 1966–78.

Note: Both sources are drawn from the same basic data, which were published regularly in 1966–78.

Table 7-6. Increased Ultimate Recovery of Natural Gas in Old Fields, United States,[a] Trillion Cubic Feet, and Growth in Percent per Year.

Year of Estimate:	1960	Growth to 1967, %	1967	Growth to 1972, %	1972	Growth to 1977, %	1977	Growth to Total, %
Period of discovery								
Pre-1920	75.7	1.74	85.4	0.58	89.9	0.76	91.3	1.11
1920–1930	88.1	1.05	94.8	2.38	95.4	1.46	102.6	0.90
1931–1940	110.1	2.30	129.1	0.28	130.8	-0.59	127.1	0.85
1941–1950	94.9	3.22	118.5	0.47	121.3	1.03	127.7	1.76
(1951–1954)	(50.1)	(0.45)	(52.7)	(0.97)	(54.2)	(0.80)	(56.4)	(0.70)
1951–1960			135.9	1.08	143.5	-1.41	140.6	0.34
(1961–1967)					(67.1)	-1.28	(62.8)	(-1.28)
1961–1970							78.7	—
Total	419.1	—	563.7	—	646.1	—	668.0	—

a. Excluding Alaska.

b. Numbers in parentheses are excluded from totals to avoid double counting.

Sources: 1945 and 1960 data from National Petroleum Council, Report. Other years, API–AGA–CPA, Reserves.

Note: Both sources refer to the same basic data, which were published regularly in 1966–78.

The poor showing in 1972-1977, after the price explosion, is to some extent explained by price ceilings. Although this regulation undoubtedly distorted the reserve-adding effort, it can hardly be said to have offset the large 1972-1977 increase in development wells and new-pool wells that furnish the new reserves in old fields. Putting this alongside the generally poorer showing as one goes from older to newer fields (pre-1920 to 1960-1970) it is hard to avoid the conclusion that diminishing returns do finally overcome the stretch factor. If we think of every barrel of proved reserves as carrying a complement of probable reserves equal to the difference between present and future ultimate recovery, that complement, if we could estimate it, would be shrinking.

For the period 1967-1977, the stretch factor for crude oil can be divided into two parts: new oil added in the old fields and a higher recovery percentage of the oil already included. Tables 7-7 and 7-8 show that new oil in place furnished 4.4 billion barrels in 1967-1972 but only 1.85 billion barrels in 1972-1977, despite the slightly larger base.

However, the growth of oil in place is understated. New pools discovered in old fields are not credited to the discovery year of the old field if the new pools "are themselves geologically significant and were discovered through application of a new exploration concept." For such new pools, "the assigned discovery years are the ones in which they were actually discovered," following "geological and exploratory judgments best developed by the experts in the local Subcommittees."[4] This procedure is biased because it understates the contribution of new knowledge in an old area. In any case, newfound oil in old fields provided most of the additional reserves in 1967-1972, but improved recovery was more important in the next five years, first 3.0, then 4.0 billion barrels. These large amounts constitute only a tiny fraction of oil in place, but the apparent decline (see Chapter 8) in spending on "improved recovery programs" and in service wells drilled may not bode well for contributions here.

The future lies with unconventional or, as they are usually known, "enhanced oil recovery" methods. Estimates of available reserves vary widely, as might be expected. Two studies had access to a very large data base of 245 individual reservoirs in Texas, Louisiana, and California, and used very similar methods, but results differed considerably (see Table 7-9).

The National Petroleum Council (NPC) task force estimated lower efficiencies and higher unit costs in the new processes, thereby ruling out many reservoirs included by the Mathematica group. Perhaps the NPC estimates are a better estimate of what can actually be done now, and the Mathematica numbers are a proxy for what later technology might do. The analysis of small areas for probable reserves should have been done years ago, to discover how much stretch there is in existing reserves. It still remains to be done.

Natural gas (Table 7-6) requires little comment. "Stretch" consists almost entirely of new gas, since the recovery factor is high (modal value around 75 percent) and is rarely amenable to pressure maintenance or fluid injection. A brief

Table 7-7. Contributions of Improved Recovery Factor (*RF*) and of New Oil in Place (*OIP*) to Increased Ultimate Recovery (*UR*), 1967–1972, Billion Barrels.

	RF, %		OIP, BB		UR, BB		Contribution of:	
Year of Estimate:	*1967*	*1972*	*1967*	*1972*	*1967*	*1972*	*RF*	*OIP*
Period of discovery								
Pre-1920	24.9	25.7	84.7	92.4	21.1	23.8	.68	1.92
1920–1930	34.0	34.7	84.9	86.0	28.9	29.8	.59	.37
1931–1940	34.3	36.6	80.2	80.0	27.5	29.3	1.84	-.07
1941–1950	29.1	29.7	70.9	71.7	20.6	21.3	.43	.23
1951–1960	28.4	28.4	48.2	53.2	13.7	15.1	0	1.40
Total	30.3	31.1	368.9	383.3	111.8	119.3	2.95	4.36

Source: API–AGA–CPA, *Reserves*, Table III, various years. Reprinted with permission.

Table 7-8. Contributions of Improved Recovery Factor (*RF*) and of New Oil in Place (*OIP*) to Increased Ultimate Recovery (*UR*), 1972–1977, Billion Barrels.

	RF, %		OIP		UR		Contribution of:	
Year of Estimate:	*1972*	*1977*	*1972*	*1977*	*1972*	*1977*	*RF*	*OIP*
Period of discovery								
Pre-1920	25.7	27.1	92.4	98.3	23.8	26.7	1.30	1.50
1920–1930	34.7	36.0	86.0	81.2	29.8	29.3	1.10	-1.66
1931–1940	36.6	37.5	80.0	82.2	29.3	30.8	.72	.80
1941–1950	29.7	30.4	71.7	73.6	21.3	22.4	.50	.56
1951–1960	28.4	29.0	53.2	54.8	15.1	15.9	.32	.45
1961–1965	28.0	28.5	11.5	12.2	3.21	3.5	.06	.20
Total	31.0	32.0	394.8	402.3	122.5	128.5	4.0	1.85

Source: API–AGA–CPA, *Reserves*, Table III, various years. Reprinted with permission.

Table 7-9. Additional Reserves, Billion Barrels.

	National Petroleum Council		Mathematica, Inc.
Oil Price (1976 $/barrel)	*U.S.*	*3 States*	*3 States*
5	2	2	7
10	7	7	28
15	13	12	37
20	20.5	18	—
25	24	19	—

Source: National Petroleum Council, *Enhanced Oil Recovery* (Washington, D.C.: NPC, December 1976), pp. 57–61 and Appendix G.

comparison may be made between the factors suggested by Table 7-6 with those of the Potential Gas Committee (Table 7-3). For example, ultimate recovery in fields known in 1967 was stretched by 1977 from 564 to 589 TCF (668.0 less 78.7 equals 589.3). Anyone able to estimate this in 1967 from existing data would have seen that the 300 TCF probable reserve was much too high.

REFERENCES

1. AGI-AGA-CPA, *Reserves*, Table II, dividing production by average reserves for the year.
2. API-AGA-CPA, *Reserves*, 1978, p. 263
3. The source for 1977 is API-AGA-CPA, *Reserves*, 1978, Table VIII. For 1957 *Oil and Gas Journal* (January 27, 1958): 163-68. Proved reserves for those two years for the total United States were 30.4 billion and 19.7 billion barrels.
4. API-AGA-CPA, *Reserves*, 1978, p. 18.

8 THE ROLE OF COSTS AND PRICES IN ASCERTAINING PROVED AND PROBABLE RESERVES

Proved reserves are the fruit of some exploration and much development investment; probable reserves are the fruit of much exploration and some development. In both cases, the investment is not made unless a return is expected. Reservoirs being highly varied in many factors that affect their cost, they should be arrayed from lowest to highest to form a supply curve, such that successively higher prices would mean larger and larger reserves.

As concerns proved reserves, the estimator can plead ignorance and not be too badly hurt, for when the time comes to estimate reserves, the investment has already been committed. The estimate of oil or gas to be produced is robust, because the variable cost per unit tends to be very low in relation to price. If all producing units are arrayed in order of variable cost, then fairly substantial changes in price will not make much difference in the production out of proved reserves.

VARIABLE COSTS

Information on variable costs is sadly lacking. The *Joint Association Survey*, Part II (now superseded by the U.S. Bureau of the Census *Annual Survey of Oil and Gas*) compiled total operating costs but did not even divide between oil and gas. Table 8-1 uses a rough division made by the National Petroleum Council (NPC) to show national averages per barrel and per million cubic feet (MCF). The U.S. Bureau of Mines has for some years been tabulating producing wells by production rates, but there are no corresponding cost numbers. Table 8-1 suf-

Table 8–1. Oil and Gas Operating Costs, Selected Years 1971–1981.

Year	(1) Total Operating Costs, $ Millions	(2) Oil Wells, (000s)	(3) Gas Wells (000s)	(4) Oil Production,[a] MMB	(5) Oil Operating Costs, % of Total Operating Costs[b]	(6) Oil Operating Costs, Cents/ Barrel	(7) Gas Production, Trillion Cubic Feet[c]	(8) Gas Operating Costs, Cents/ Thousand Cubic Feet
1971	2,969	517	117	3,522	77	65	17.1	4.0
1973	3,277	500	123	3,457	75	71	17.8	4.6
1974	4,092	494	127	3,298	75	93	17.1	6.0
1976	5,372	503	138	3,099	74	128	15.6	9.0
1978	9,995	516	154	3,030	67	221	15.7	21.0
1981	22,388	519	190	3,112	67	484	16.2	45.6

a. Oil production includes natural gas liquids produced through oil wells (associated and dissolved).

b. Operating costs are calculated as follows:

In the National Petroleum Council's *U.S. Energy Outlook – Oil and Gas Availability* (Washington, D.C.: NPC, 1973, p. 617) gas wells are estimated to cost 1.33 times oil well operating costs. Therefore,

$$\frac{\text{Total operating costs}}{\text{oil wells} + (1.33 \times \text{gas wells})} = \text{operating costs per oil well.}$$

(Operating costs per oil well) $\times$ (oil wells) = total cost for oil wells.

$$\frac{(\text{Total cost for oil wells})}{(\text{total operating costs})} = \text{oil operating costs as a percentage of total operating costs.}$$

$$\frac{(\text{Total cost for oil wells})}{(\text{oil production})} = \text{costs per barrel of oil.}$$

c. Gas production includes only gas produced through gas wells (nonassociated). By omission of gas produced from oil wells, and liquids produced from oil wells, and liquids produced from gas wells, this overstates somewhat the cost per barrel and per mcf.

Sources: For column (1), data for 1971–1976 are from *1975 Joint Association Survey*, Section II, March 1977, API, with estimate from Tables
1, 3a and 3d, which includes both oil and gas (taxes and royalties excluded); 1976 data and later, *Annual Survey of Oil and Gas*, U.S. Department
of Commerce, Bureau of the Census, MA–13K. For columns (2) and (3), *World Oil* (February 15, various years). For columns (4) and (7) API–
AGA–CPA, *Reserves of Crude Oil, Natural Gas Liquids, and Natural Gas in the United States as of December 31* (Washington, D.C.: API-AGA,
various years) and Energy Information Administration, *U.S. . . . Reserves* (Washington, 1982).

fices for the proposition that operating costs are usually low and that it would hence take a radical change to affect reserves. Nevertheless, as will be seen below, operating costs are far from negligible even for proved reserves.

In the process that converts probable into proved reserves, both operating and development costs are of concern; the latter are predominantly investment requirements per unit of output. Here too, as will be seen, very little in the public domain permits the drawing of anything resembling a supply curve. The Joint Association Survey has for years made good estimates of drilling expenditures for oil, gas, and dry holes separately, by states and by well depths, onshore and offshore separately. We could simplify to a tolerable degree by assuming depth to be the only systematic source of cost variations (all else being random), and arraying wells by drilling cost. This array might also be compared with output per well by states or subdivisions thereof to make a rough approximation a supply curve, but this would of course involve the assumption—now almost certainly wrong in this country—that the *incremental* output per well will be close to current *average* output.

Thus the serious study of reserve creation lacks the essential ingredient of data, which, did we know them, would undoubtedly expose gaps in analytic equipment, for science proceeds by an endless shuttle between concepts and facts. However, we can at least apply a correct method and some fragments of data, to see a little way into the failure of reserves to grow in the United States after the price explosion in 1973-74.

LACK OF GROWTH OF RESERVES

The failure was shown in Tables 4-1 and 4-2. The annual "discoveries" are of very little importance. However, the annual AAPG estimates[1] are good, if imprecise, evidence that the volume of newly discovered oil and gas has been very small. To some unknown extent, this may be only a time lag; to some extent, it means sharply diminishing returns.

The failure to obtain large volumes of new oil and gas in old fields is not to be explained by time lags, because the conversion from *provable* (the original meaning of "probable") to *proved* is not a lengthy process. Some of it can be done in no more time than is required to decide to keep producing from wells when prices have risen to where sales revenues now suffice to cover the operating expenses. This would show up as "revisions," within the next year or two at most. The other increases do require drilling and appraisal. But more drilling has thus far given us less reserves. In 1974-1977 inclusive, the number of development and exploration wells drilled (65,000) exceeded the number drilled in 1969-1973, but the additional reserves proved in 1974-1977 were 5.8 billion barrels; in 1969-1973, 10.8 billion barrels.

Lacking data to construct a supply curve but knowing that the curve exists helps us to see a little way into the darkness. A preliminary note: Oil "revisions" upward have declined from 2.2 billion in 1973 to 1.4 billion in 1978, while "revisions" downward have shown no trend.[a] This suggests that something unfavorable has been happening to reserves *already* proved. The reversal after 1979 bears this out.

EFFECTS OF PRICE CONTROLS AND PER-BARREL TAXES

Let us now pick four points on the unknown short-run supply curve.[b] In Table 8-2 four reservoirs are considered. Reservoir A was marginal in 1973, with operating costs just equal to the 1973 average price less taxes and royalties. B was economic because it paid no 1973 income tax (benefiting from percentage depletion). C had operating costs just one-half of bare break-even; D, one-fourth of break-even.

In 1977, all these reservoirs would be considered "old" or "lower tier" oil, with an average price of $5.19 for the year.[2] After taxes and royalties of about 20 percent, net price was about $4.15. Operating costs have increased for two reasons: first, the rise in factor prices, the result of general inflation, plus the surge of demand against supply limited in the short run. Second, there is the continued decline of output per well. I approximate the first factor by the IPAA series and the second by the rate of depletion, which in 1973-1977 was 12.4 percent per year.

Taking both forces into account, operating costs increased by a factor of 2.68.[c] Wells with costs less than half of break-even in 1973 would have been unprofitable in 1977 under the "lower tier" ceiling. The rise in costs by way of depletion and decline is overstated for some reservoirs and understated for others. Wells sand up, clog up, and need cleaning by acidizing, fracturing, and workovers (partial redrilling). There is a trade-off: Incurring higher operating costs per barrel through excessive decline versus making additional variable and/or investment outlays. A zero-expenditure decline rate is estimated at about 23 percent.

a. Upward revisions, in billions of barrels, 1972-1978: 1.6, 2.2, 1.9, 1.4, 1.4, 1.4, 1.4. Downward revisions: 0.8, 0.7, 0.6, 0.7, 1.0, 0.6, 0.7.
The EIA series: upward 1977-81: 1.5, 2.8, 2.4, 2.9, 2.2; downward: 1.1, 1.4, 2.0, 1.0, 0.9.
b. This section was written before decontrol of oil prices. However, because the "windfall profits tax" is an excise tax on production, it amounts to the perpetuation of price controls at a higher level. Costs have also increased above 1978 levels. Hence the analysis has been left unchanged, since the theory is no less valid, while the particular number must in any case be quickly obsolete.
c. Note that the average for all wells did not increase as much (Table 8-1, col. [6]) since the mixture is always "freshened up" by new wells, which have not been subject to decline.

Table 8-2. Four Hypothetical Reservoirs, 1973-1977.

	1973				1977			
	A	B	C	D	A	B	C	D
(1) Price	3.89	3.89	3.89	3.89	5.19	5.19	5.19	5.19
(2) Price less 20 percent for royalties and taxes	3.11	3.11	3.11	3.11	4.15	4.15	4.15	4.15
(3) Operating cost	3.11	2.33	1.56	.78	8.32	6.24	4.18	2.08
(4) Depletion allowed	0.0	.86	.86	.86	.75	.75	.75	.75
(5) Taxable income (2) - (3) - (4)	0.0	0.0	.69	1.47	0	0	0	1.32
(6) Income tax (5 × 0.5)	0.0	0.0	.35	.74	0	0	0	.66
(7) Cash flow (2) - (3) - (6)	0.0	0.68	1.20	1.60	0	0	0	1.41

Sources:

(1) *Monthly Energy Review* (June 1978): 57.

(2) *Joint Association Survey*, Section II, adding to 15 percent royalties plus a 5 percent allowance for state taxes (Table I), which in 1975 were $1,818 million, or 5.6 percent of total revenues increased by 15 percent, i.e., 1,818/[27,252].85 = .056.

(3) Assumed for 1973; factor of increase = 1.654 × 1.1244 = 2.68. For source, see Table 8-4, "Price Index."

(4) Percentage depletion is 22 percent of price in 1973; cost depletion 5 percent of $15.02 net (one-half the gross) fixed assets per barrel of production of liquids as calculated from U.S. Bureau of the Census, *Annual Survey of Oil and Gas 1976* (ASOG), Table 3, i.e., $41,010/[2,348 + 270 + 112] = $15.02.

Table 8-3. Loss of Reserves in Two Hypothetical Reservoirs, 1973-1977, Barrels.

Year	Reservoir	Ratio, Current Cost to Price (= break-even cost b)	Years of Production Remaining $(1.13^{-t} = b)$	1973 Reserves = $(1 - 1.13^{-t})/.13$	Cumulative Production 1974-1976
1973	C	1/2	5.7	3.8	2.4
	D	1/4	11.3	5.8	2.4
				9.6	4.8
1977	C	1	0	0	—
	D	1/2	5.7	2.3	—
				2.3	

1973 reserves less *1974-1976 output* less *1977 reserves* equals *lost reserves.*

| 9.6 | 4.8 | 2.3 | 2.5 |

Under the old price-cost regime, at the end of 1973, reservoir A would have shut down; by the end of 1977 B would also have stopped. Turning to C and D (Table 8-3), their end-1973 reserves amounted to 9.6 barrels for every barrel they were then producing. By the end of 1977, they had produced 4.8 barrels between them, C was about to shut down, and D still had 2.3 barrels to go. Thus 2.5 barrels would be unproduced, lost reserves. To explain how these figures were derived in Table 8-3: At end-1973, the C operating cost was one-half of net price. Then at 13 percent decline rate, it would operate for 5.7 years before cost rose to equal price and the well would be shut down; that is, $1.13^{-5.7} = 0.5$. In those 5.7 years, 3.8 barrels would be produced.

We cannot begin to translate Table 8-3 into any calculation of reduced "lower tier" oil production, because the actual distribution of reservoirs by operating costs is not known. But plainly a large amount of "old oil" was worth producing in 1973 and even more worth producing in 1977. Its marginal social cost was a small fraction of the price of a barrel of imported oil that they would displace (around $14). Yet it was not worth the attention of private producers in 1977. This is largely but not wholly due to price controls. Not for the first nor the one-hundredth time have excise taxes and royalties (private excise taxes) been shown to cause economic waste, suppressing worthwhile production.

A little can also be said about the inducement to invest in order to convert probable into proved reserves at these levels of cost. Tables 8-4 and 8-5 contain a rough calculation of the average 1976 development investment needed to install the capacity to produce one barrel per year (or $6,072 per barrel daily).[3] Let us assume that four new wells (like A, B, C, and D) are available, each with productivity of one barrel; operating costs are at the 1973 levels, adjusted for a factor cost increase of 65.4 percent. The initial net cash flow available from well A is negative, that of the other three are $0.30, $1.16, and $1.80, respectively, for a return on investment of 1.8, 4.8, and 10.8 percent. Since revenues are subject to a 13 percent decline, none of the wells is worth drilling.[d]

If, however, these reservoirs are classified as *new* or *upper tier* oil, which is entitled to $11 per barrel, the respective returns will be 13, 17, 21, and 25 percent. C and D are worth drilling.

This brings us to the distinction between *new* or *upper tier* oil and *old* or *lower tier* oil. Oil from every producing property in operation in 1972 is old oil and receives the lower tier price for production equal to its 1972 production, which is designated as the base production control level (BPCL). If the operator is able to overcome the decline rate and actually increase production above BPCL, the excess is considered new oil and receives the upper tier price, which is

---

d. As a check: Gross fixed assets divided by total 1976 production is $15.02 per annual barrel (Table 8-2, note 4); investment per barrel of incremental capacity at $16.65 is 11 percent more. The incremental barrel should probably be higher relative to the average; however, given the high decline rate, most capacity is fairly recent. The two estimates are altogether independent and hence confirm each other.

Table 8-4.   United States Development Expenditures.

Year	Wells Drilled Development	Wells Drilled Service	Price Index	Joint Association Develop-ment	Joint Association Improved Recovery	U.S. Census Bureau Develop-ment	U.S. Census Bureau Improved Recovery
				Expenditures, $ Millions			
1967	23.4	1.4					
1968	21.8	1.3	75.2	2,333	222		
1969	22.5	1.5	77.8	2,559	303		
1970	20.4	1.2	82.5	2,631	285	Not in	
1971	18.9	1.4	90.9	2,671	323	existence (n.i.e.)	
1972	18.9	1.4	96.2	3,093	310		
1973	19.1	1.0	100.0	3,255	276	3,039	n.i.e.
1974	23.1	1.1	119.8	4,476	399	4,413	n.i.e.
1975	28.0	1.5	139.4	6,985	556	6,423	n.i.e.
1976	30.5	1.5	152.6			7,735	378
1977	35.0	1.3	165.4			8,830	461
1978	36.4	1.4	184.3	Series		10,560	693
1979	39.0	1.6	203.6	discontinued		11,793	528
1980	48.9	1.6	236.0			16,164	941
1981	63.4	1.9	276.7			20,256	1,458

Sources:
Wells – API, *Quarterly Review of Drilling Statistics* (respective years).
Price Index – IPAA, *Report of the Cost Study Committee*, various years.
Expenditures – *Joint Association Survey*, Part II, respective years; U.S. Bureau of the Census *Annual Survey of Oil and Gas*, MA–13K, various years. Transfer payments excluded.

Table 8-5.   Investment per Barrel Oil-Producing Capacity, 1976-81
(*Expenditures in $ millions*).

Year	(1) Devel. Expend.	(2) Oil Fraction	(3) Oil Devel. Expend. ([1] × [2])	(4) Change in Production (millions of barrels)	(5) Production Decline Made Good	(6) Gross Capacity Increment ([4] + [5])	(7) Investment per Barrel (current $)
			I. Excluding Alaska				
1976	7,735	0.52	4,023	−61	307	246	16.39
1977	8,830	0.53	4,690	−71	415	344	13.63
			II. Including Alaska				
1978	10,560	0.44	4,670	113	289	402	11.63
1979	11,793	0.44	5,162	−62	293	231	22.35
1980	16,164	0.47	7,518	90	297	386	19.45
1981	20,256	0.50	10,094	−25	296	271	37.24

Sources. (1), (2), (4): U.S. Bureau of the Census, *Annual Survey of Oil and Gas*, various years. (5): Decline rate assumed equal to depletion rate, from API–AGA–CPA, *Reserves*, for 1976–77; later years, from EIA, *U.S. Reserves*.

about $11 nationally (varying somewhat among regions). But for every barrel of new oil, one barrel of old oil is considered released oil, eligible to be sold as new oil. Thus the net price for each barrel of "new oil" is about $17 (= $11 + $11 - $5). In reservoirs with a naturally low decline rate, it may be worthwhile to invest in new plus released oil; in others, it may not. The rules scatter bonanzas and hardships with no relation to costs and benefits.

Investment to bring production above BPCL would be detectable in the expenditures on "improved recovery programs." Table 8-4 shows that in real terms (1973 prices) this expenditure averaged $341 million during 1968-1972. It dropped to $276 million in 1973 and rose to $398 in 1975. The 1976 figure is only $229 million, but we cannot be sure that the Census Bureau's estimate is strictly comparable with that of the JAS. As a check, we can compare the number of service wells drilled, which dropped in 1973 and 1974 but then resumed at about the same level as during 1967-1972. In real terms, there was obviously no boom in fluid injection operations and other types of improved recovery until the decontrol boom of 1980-81.

In conclusion, there must have been substantial loss of reserves due to the control of crude oil prices, but we cannot estimate them because we know almost nothing about the distribution of operating and investment costs, by reservoirs or even by small areas, for either oil or gas. The boom in drilling has accelerated production, but it has not increased proved reserves—the amount of oil to be produced through existing installations—and there is no indication that it has increased probable reserves.

REFERENCES

1.   Published annually by the American Association of Petroleum Geologists as the "North American Drilling in 19xx [previous year]" issue of the *AAPG Bulletin.*
2.   U.S. Department of Energy, *Monthly Energy Review* (June 1978): 56.
3.   *Oil and Gas Journal* (December 5, 1977): 70 reports a study by four companies where these types of outlays reduce the decline from 23 to 6 percent.

9 PROVED AND PROBABLE RESERVES OUTSIDE THE UNITED STATES AND CANADA

For many years *Oil and Gas Journal* and *World Oil* have published estimates of "reserves" for all producing countries, including the Communist bloc and other closed societies. For the United States and Canada, they reprint the API–AGA–CPA data discussed in the preceding chapters, and some countries' official estimates (sometimes with a grain of salt.)

But the most important source of most foreign estimates are the estimates of private companies and industry personnel. The social function of the oil trade press is to do informally what the API–AGA meetings do by set rules: to pool experience and opinions while preserving anonymity. The press can thereby bypass government or company over- or underestimates for bargaining purposes. It is a byword that the best source of information about reserves in Country X are oil company people in adjoining Country Y. At its best the combination of geological continuity, professional curiosity, and contacts bred of isolation from the local community produce an independent estimate that an experienced editor may prefer to, or use to correct, official numbers.

The chief weakness of the trade press estimates is the lack of agreed, known standards. They take what they can get. One person's "proved" may be another person's "proved-plus" with little notion of what the "plus" may be. The CIA for years estimated "proved and probable reserves," but with little explanation. Table 9-1 presents the two remaining sources. They have come close together; one hesitates to call this a good or bad sign.

In 1971, a task force of the National Petroleum Council polled some American oil companies operating outside the United States, asking for their estimates of 1970 *proved reserves* following the API definition, as a percentage of the *Oil*

Table 9-1. World Proved Reserves of Crude Oil (Billion Barrels)
and Natural Gas (Trillion Cubic Feet), 1982.

	Oil and Gas Journal		World Oil	
	Oil	Gas	Oil	Gas
North America	37	301	35	291
South/Central America	78	186	76	183
Western Europe	23	157	18	161
Eastern Europe	66	1,245	87	1,404
North Africa	36	177	41	175
South/Central Africa	22	13	14	55
Middle East	369	770	351	731
Asia, Far East	24	82	23	169
Oceania	16	94	16	35
Total	670	3,024	661	3,204

Sources:
Oil and Gas Journal (December 27, 1982): 78-79.
World Oil (August 15, 1982: 46. Excludes natural gas liquids.
Central Intelligence Agency, International Energy Biweekly Statistical Review (January 25, 1978): 4. United States and Canada include Arctic gas deposits and natural gas liquids; Iran includes "recent" discoveries.

and Gas Journal reserves. The range of estimates (shown in Table 9-2) is wide, except for certain countries.

Possibly as a result of this comparison, in 1975 the Oil and Gas Journal changed its definitions somewhat and now designates its numbers "proved reserves." In 1975 and 1976 it stated: "All reserves indicate proved reserves recoverable with today's technology and prices, and exclude probable and possible reserves." The definition for the Soviet Union is less stringent (see Appendix C).

It is impossible to say just what difference this has made. There seems to have been some downgrading in 1975 from 1974 and no upgrading from 1976 to 1977. And a comparison of the Oil and Gas Journal figures with the CIA's shows the latter to be consistently higher, as they should be. Of course, the two series are not fully independent.

For several years, there was an important variant of the Oil and Gas Journal that came closer to the concept of proved reserves. The Journal publishes an International Petroleum Encyclopedia, which lists the most important large fields, with reserve estimates. The chances are good that a field by field estimate will be more accurate than a global country estimate, for three reasons. First, the combined error of the sum of many individual observations should be less than the error of a single observation or estimate. Second, the sum of separate field

Table 9-2. Oil Company Estimates of Proved and Probable Reserves (API Concept) as a Percentage of Reserves Published by Oil and Gas Journal, 1970.

Area	Proved Reserves	Proved Plus Probable Reserves
Latin America	.97 to .99	
Europe	.97 to .98	Not
Africa	.50 to .73	available
Middle East	.67 to .80	
Total	.66 to .81	.88 to .97
Rough Point Estimate	About .75	About .95

Source: National Petroleum Council, *National Petroleum Council Committee on U.S. Energy Outlook: An Interim Report. An Interim Appraisal by the Oil Supply Task Group*, (Washington, D.C.: NPC, 1972), pp. 21-24. Reprinted with permission.

listings makes it more likely that the estimator is not allowing for probable reserves. Third, they permit the observer to eliminate fields that should not be counted. These are of two types, best shown by example.

In Abu Dhabi, in 1976, four large fields produced 486 million barrels in 1976 and were credited with reserves of 4,323 million, a depletion rate of 11.2 percent. Other fields produced 94 million barrels. If we are to credit Abu Dhabi with a total of 29 billion barrels of proved reserves, we would need to suppose a depletion rate in the small fields of (94/24,677), or 0.38 percent. This is unlikely. If we assume the same depletion rate for the smaller fields, then dividing 94 by 112, we estimate their reserves at 839 million, and total Abu Dhabi proved reserves at (4,323 + 839) or 5,162 million. The other 24 billion are probable reserves. The error is unknown but is at least an order of magnitude less.

At the other extreme is the oldest field in the Persian Gulf, Masjid-i-Suleiman in Iran. In 1976, it was credited with reserves of 1,510 million barrels, but produced only 4 million, for a depletion rate of 0.26 percent. Plainly only a negligible fraction of the 1,500 million barrels will ever be recovered, barring some change in technology. Hence the oil underground is not inventory and should not be counted in proved reserves.

Adjustments of this kind are needed to put the reserve data to some use. Proved reserves are significant because they state the amount of oil ready and connected for production. Reserves multiplied by the target steady-state depletion rate yield a good forecast of supply under current cost and price conditions; $Q = aR$.

Once the proved reserves are nailed down, probable reserves can be approached: How soon, and at what cost, can they be made into proved reserves and productive capacity? If the estimator's interest lies in the oil market and in forces determining capacity, the first step is therefore to get good estimates of proved reserves, probable reserves, and investment needed to convert probable to proved. To-be-discovered reserves are proved reserves at one additional remove of discovery time and cost.

The insistence on a three-stage process, from undiscovered to probable to proved, is needed for protection from the mindless addition of all three, to produce a hash of meaningless arithmetic. Worse yet is to compare proved reserves in one place with proved-plus-probable-plus-discovered in another. One of the worst (surely not the last, unfortunately): "Iraq's reserves may run as high as 130 billion barrels—a staggering total that approaches the 150-billion-barrel reserves of Saudi Arabia."[1] In fact, the proved reserves of Iraq are estimated at 34 billion by the *Oil and Gas Journal*, and the CIA "proved and probable" reserves were 35 billion barrels.[2] A figure comparable to the Iraq 130 (supposing it to be a respectable guess) would be upwards of 300 for Saudi Arabia. But more important, the Saudi proved and probable reserves (discussed below) imply something about how much that country can produce within a given number of years from oil that has been located and is or can be developed into producing capacity. The estimate for Iraq means nothing of the kind.

REVIEW OF SELECTED RESERVE ESTIMATES

For some countries it is possible to discern something of how proved and probable reserve estimates are made, and how they change over time. Though here and there the data may be of current interest, note again that the purpose throughout is with method rather than substance.

Venezuela

The Ministry of Energy and Mines (formerly the Ministry of Mines and Hydrocarbons) has long maintained a series of proved reserves very similar to that in North America. In the ten years prior to 1974, "revisions" fluctuated widely, from a high of 1.1 billion in 1964 to a low of 145 million in 1969. In 1974, following the price explosion, revisions were 5.5 billion, which represents 11.1 percent of 1974 ultimate recovery, that is, cumulated production plus current proved reserves. Probably most of the revision represents reservoirs that can be kept producing longer, and a little represents previously submarginal pools that are now worth exploiting, since the number of exploratory and development wells decreased in 1974-1978.[3]

The *Oil and Gas Journal* increased its Venezuela estimate from 15.3 billion barrels at end-1976 to 21.5 bb at end-1982; adding 4.7 bb cumulative production, this indicates a 71 percent addition to proved reserves in six years. Since there were no large discoveries announced, it is evident that Venezuela achieved a solid success in converting a large stock of probable into proved reserves.

Mexico

The petroleum industry of Mexico[4] appeared in 1971 to be winding down. Production had fallen short of consumption, and the prospect was for ever-increasing imports, at a real and foreign-exchange cost that appeared large then, however modest it may look today. But Pemex geologists had already wished to probe the hypothesis of Francisco Viniegra of a huge atoll-like chain of structures, going offshore around the Yucatan peninsula and extending overland to the northwest over several hundreds of miles. The old Poza Rica field, at the far northwestern end of the supposed chain, gave promise of success. With no threat of a local shortage, the hypothesis might not have seemed worth trying; but exploratory wells were completed in Chiapas–Tabasco state in 1972, and it was at once apparent that the find was the best in many years. But that was not saying much; its importance seemed only local. In 1974 the press began to carry tidings of great things and the world industry credited them with too much too soon. The *Oil and Gas Journal* "oil" estimates for Mexico rose to 14 billion barrels in 1974, down to 7 in 1976, back up to 16 in 1978. There was much talk then, as there is now, of huge reserves being hidden for political reasons. More likely Mexican oilmen, like their colleagues everywhere, made their best guess, then probed for new pools and new fields' horizontal and vertical limits, and found these hard to reach. Productive areas were gradually determined to be 6 to 15 times as thick as in Poza Rica, the oil-water contact was hard to find and in many fields it has not yet been found, and as wells began to be exploited, there appeared to be much more natural gas. Three good fields in one area ran into each other to be recognized as one giant field, and so on.

As shown in Table 9–3, by early 1978 it was possible to give a better accounting of Mexican reserves than those of any other country, because Pemex had put into the public domain much information that is usually company-confidential. Most important, in two of the documents one can make an explicit connection between current production and proved reserves, and between near-term production and probable reserves, in the new producing area. Proved reserves as of end-1977 were the expected aggregate production of reservoirs already producing, plus reservoirs to be producing in 1978, whose output would peak in 1979 and decline thereafter by 7 percent per year. *Probable* reserves refer to "cuencas productoras conocidas" ('known productive basins'—terminology being no more precise in Spanish than in English) with production slated to begin in 1979 but not to peak until 1996, then declining at the rate of about 11 percent per year.

These proved and probable reserves are (end-1982) in about twenty-seven known fields onshore, four offshore. Scattered around them are about 100 structures onshore, and forty offshore that are thought to be prospective; the "potential" reserves are an estimate of (1) how many of those structures are likely to prove productive, multiplied by (2) their average contents. Since the volume of

Table 9-3. Mexican Proved, Probable, and Potential Reserves.[a]

I. 1976–78

	Proved					Probable			Potential	
		Mid-1977								
	End-1976	Developed	Undeveloped	End-1977	End-1978	Mid-1977	End-1977	End-1978	End-1977	End-1978
Oil (including condensate), BB	7	6	2	10	26	22	n.a.	29	n.a.	n.a.
Gas, TCF	19	12	7	28	70	41	n.a.	79	n.a.	n.a.
Gas (oil equivalent), BB	4	2	1	5	14	8	n.a.	45	n.a.	n.a.
Total, BB	11	9	3	16	40	30	31.1	45	73	115

II. End-1982

	Proved	Probable	Potential
Oil (including condensate), BB	57	56	69
Gas, TCF	(90)	(120)	(145)
Gas (oil equivalent), BB	15	24	70
Total, BB	72	80	250

a. Probable reserves denote known reserves awaiting development. Potential = undrilled structures in productive areas, estimated by analogy. Detail may not add to total because of rounding.

Sources: Petroleos Mexicanos, *Memoria De Labores,* various years; *Generalidades del Proyecto de Construccion del Gasoducto Cactus–Reynosa,* pp. 27, 19, Tables II, IV. *Oil and Gas Journal Newsletter* (11 September 1978): 4. *Petroleum Intelligence Weekly* (8 January 1979). Chicontepec reserves excluded entirely.

the structures to be drilled is many times larger than those already drilled, it is obvious that the average hydrocarbon content per unit of volume is estimated as drastically lower than in the structures already drilled. This prudence is, of course, the only right way to make such guesses, although many of the larger structures offshore are larger, simpler, and hence more prospective than those onshore.

Beyond the areas already known as productive, many more structures have been identified—perhaps an additional 100 offshore. The Ixchel-1 well offshore is slated for drilling at a distance of 200 kilometers from any productive reservoir. If it turns out to be productive, the intervening structures can be assigned some probability of hydrocarbon occurrence, and the *potential reserve* number would need to be increased, perhaps several times. How much of the "atoll" will be explored, and how soon, cannot be said. It is not clear that proved reserves as of end-1978 were defined as strictly as the year earlier. If not, we might guess that proved reserves of oil and gas were somewhere around 16 and 8 billion barrels respectively, and probable reserves 32 and 18 BB. Most striking was the transfer of nearly 30 BB from potential into probable reserves, representing large offshore discoveries later in the year.

An infinite number of oil and gas production curves over time can be drawn, their limitation being that the total area under the curve not exceed the total of reserves and that the slope of the curve—the rate of increase of production in any given time span—not be impossibly or uneconomically high. Obviously Mexico has the capability to become a major producer. Without much doubt, proved reserves can be retained (in round numbers) at a steady-state 70 BB (including oil-equivalent of gas), which, depleted at a low 5 percent per year (characteristic of a quickly growing industry) would mean about 9.5 million barrels daily (see Table 9-4). This would exceed the 1978 capacity of Iran, which had about equal "proved reserves" in the *Oil and Gas Journal* reckoning, but whose maximum economic depletion rate seemed much less than what is feasible in Mexico, because marginal costs turn up too soon.[5] It now seems likely that reserves to be discovered will make possible a higher proved reserves and capacity, but this will take additional time and a political decision.

Mexico is the one country where we can look at a snapshot of proved, probable, and potential reserves, with the detailed rationale for each. In a real sense, all three numbers are *probable*. But the margin for error around proved reserves, on a production schedule from existing wells, is small; that around reserves scheduled out of undrilled installations not yet in place is wider; that around appraisals of untested structures in the productive area is much wider; and that around unknown structures in the "atoll" is very much wider.

The much-discussed Chicontepec "field" is actually a set of low-grade shallow reservoirs almost bordering on tar sands. The amount of oil in place is very large, perhaps 100 BB, but almost irrelevant.

Table 9–4. Projection of Flow, Mexican Reserves and Production, Billion Barrels of Oil Plus Oil-Equivalent Gas.[a]

	Reserves		Additions to Reserves Per Year, 1972–1977	Flow Variant A[b]			Flow Variant B[c]		
	End-1972	End-1977		Additions to Reserves Cumulative, 1978–1987	Cumulative Production, 1978–1987	Reserves, 1987	Additions to Reserves Cumulative, 1978–1987	Cumulative Production, 1978–1987	Reserves, 1987
Proved	3	16	2.6	+41	8	49	+67	12	71
Probable	3	31	5.6	{ −41 +73	0	63	{ −67 +73	0	37
Potential	6	73	13.0	−73	0	0	−73	0	0
		120		00	8	112	00	12	108

a. Given a constraint of 5 percent depletion rate, no more discoveries, and time needed to convert potential to probable proved, many time paths might be followed; only two are shown. Which is optimal can be decided only with reference to the various constraints, but the aim is a steady-state relation of proved reserves and production.

b. Five percent depletion, 2.45 billion barrels yearly (BBY) or 6.7 million barrels daily (mbd) for 26 years, after which 5 percent decline. Output (1977, 0.5 BBY) and reserves, both increasing 10 percent per year.

c. Five percent depletion, 3.55 BBY or 8.7 mbd for 10.4 years, after which 5 percent decline. Output (1977, 0.5 BBY) and reserves, both increasing 20 percent per year.

As of end-1982, oil proved reserves were one third of Saudi reserves, 55 compared with 165. Proved plus probable were 96 compared with 250. (As explained below, Saudi proved plus probable are about 250 billion.) Adding Mexico gas to oil makes sense when next to a huge gas market; the Mexican proved plus probable is some 35 percent below the Saudi total. There is no publicly available estimate of Saudi potential.

Saudi Arabia

Table 9-5 shows how the *Oil and Gas Journal* estimate lies between Aramco's "proved" and "proved" plus "probable." There are major problems at the border, as information grows ever more thin. For example, as of the end of 1975 the 16 operating fields had reserves of 103.5 BB. Another 16 fields were not producing; some of them were credited with as little as 1 million, the total with 4.3 billion barrels.[6] (By the end of 1979, the number of inactive fields had grown to thirty, but with no indication of their content.[7]) This is almost surely a fraction, perhaps a minor fraction, of what those fields could produce if operating; but they are not going into operation soon, so they should not be counted as proved reserves. But in general, the difference—not a discrepancy—between the two sources is reconciled by the statement of the deputy petroleum minister that "Aramco's 'proved' represents reserves that could be recovered without development drilling in known fields."[8] In general, the government estimate of proved reserves adds on an allowance "for oil discoveries not yet delineated."[9]

Saudi productive capacity has been variously estimated in recent years between 9 and 12 million barrels daily (MBD). In mid-1978, expansion to 16 MBD

Table 9-5. Saudi Arabia Reserve Estimates, Billion Barrels.

	Oil and Gas Journal (Saudi government)	Aramco	
	Proved	Proved	Probable
1981	164.6	116.7	60.5
1980	165.0	113.5	65.2
1979	163.4	113.5	64.4
1978	165.7	113.3	64.4
1977	150.0	110.4	67.3
1976	144.6[a]	110.2	67.3
1975[b]	148.0	107.8	68.0
1974	164.5	103.2	69.3
1973	132.0	96.9	67.6

a. In 1976, the Saudi Arabian government first reported 110 BB, later corrected to 144.6 BB.

b. Starting in 1975, *Oil and Gas Journal* defined *proved reserves* in the "Worldwide Oil" issue. This suggested a closer correspondence to the *API-AGA* definition.

Sources: Oil and Gas Journal "Worldwide Oil" annual issues; Aramco *Annual Reports.*

(5.84 BB per year) was projected, but nothing near it is expected any more. The stated reason is "conservation," which in the United States used to mean restriction of output to support prices above the competitive level. At 1974-1978 prices (and, a fortiori, at 1980 prices), the market is not there. Given stagnant production, there is no need for more inventory. Growth will come overwhelmingly from two old fields.[10] In 1973, a careful Aramer assessment was "that our true reserves were 245 billion based on the method that is commonly accepted for determining these figures."[11] They could sustain more than 20 mbd production.[12] There will be very little growth in Saudi Arabian proved reserves in the immediate future, unless the current drilling program aims to delineate the Ghawar colossus fully, or to develop some of the thirty undeveloped fields mentioned earlier.

Some informal guesses of Saudi Arabian "potential reserves" are in the air— up to 350 BB—but one cannot tell whether these are estimates of prospective undrilled structures, an intuition that there must be a lot more, an allowance for much higher recovery factors, or something else.[13] Since Saudi production will not reach the planned 16 MBD capacity for years, if ever, the measurement and estimation of potential or even probable reserves is on a far back burner.

The amount of additional reserves that could be added in known Middle East fields without any discoveries was estimated for 1969 at 205 BB, compared with proved reserves of 345 BB, that is, the stretch factor is 1.59.[14] A much more knowledgeable estimate of probable reserves was made for 1975, when proved reserves were 368 BB. The amount that could be added in known fields was estimated at between 300 and 550 BB, a stretch of 1.82 to 2.49.[15]

REFERENCES

1. J. P. Smith, "Iraq Sits atop Huge Reserves," *The Washington Post* (August 7, 1978).
2. "World Wide Oil," *Oil and Gas Journal* (December 27, 1977). CIA, *International Energy Statistical Review*, 1978-1980.
3. Ministerio de Energia y Minas, Republica de Venezuela, *Petroleo y Otros Datos Estadisticos* (Caracas), various years, tables entitled "Pozos Completados, por Tipo," and "Reservas Probadas de Petroleo Crudo."
4. This section is based on a number of sources. Petroleos Mexicanos (Pemex), *Memoria de Labores*, various issues 1970-1977. *Generalidades Del Proyecto de Construccion del Gasoducto Cactus Reynosa*, July 1977. *Informe del Director General de Petroleos Mexicanos*, annual, March, various issues 1975-78. *Comparecencia del Sr. Ing. Jorge Diaz Serrano . . . ante el H. Congress de la Union*, October 1977. *Potencial Actual y Futuro de la Industria Petrolera en Mexico*, March 1978. Martin Nava Garcia, "Correlacion de Registros Geofisicos de Pozos Perforados en el Area Tabasco-Chiapas y la Plataforma Continental de Campeche y Yucatan," *Energeticos*, Commision de Energeticos, Secretaria del Patrimonio, Mexico, enero

1978. *AAPG Bulletin* "Foreign Developments" issues, various years. *International Petroleum Encyclopedia*, 1975-1978. *Oil and Gas Journal*, "World Wide Oil," 1972-77. And last but not least, conversations with Juan Eibenshutz, Ignacio Leon, Arthur A. Meyerhoff, Jaime Perez Olazo, Jose Santiago Acevedo, and Leopoldo Solis.

5. Maureen S. Crandall, "The Economics of Iranian Oil," MIT Energy Laboratory Working Paper 75-003WP, Massachusetts Institute of Technology, Cambridge, Mass., March 1975.

6. *AAPG Bulletin* 60 (October 1976): 1,869, 1,882.

7. *AAPG Bulletin* 63 (October 1979).

8. *Oil and Gas Journal* (February 14, 1977): 62.

9. *Oil and Gas Journal* (June 26, 1979): 84-86.

10. David Mansfield, "Saudi Arabia: Uncertainty Hits Forward Planning," *Petroleum Economist* (August 1978): 333.

11. Hearings before the subcommittee on multinational corporations of the Committee on Foreign Relations. U.S. Senate, 93d Congress, 2d session, Part 7 (Washington 1974), p. 539.

12. *Ibid.*, pp. 519, 453.

13. Sheik Yamani is quoted in *International Petroleum Encyclopedia*. "There is as much oil again in Saudi Arabia as has already been found" (1977, p. 9).

14. M.A. Adelman, *The World Petroleum Market* (Baltimore, Md.: The Johns Hopkins University Press, 1972), p. 71.

15. Z.R. Beydoun and H.V. Dunnington, *The Petroleum Geology and Resources of the Middle East* (London: Scientific Press, 1975), p. 84.

SUMMARY

In this part, an attempt has been made to sketch a sequence. Hydrocarbon reservoirs are discovered and their contents first loosely, then more precisely estimated as they begin to become part of a programmed flow. Hydrocarbons are always being produced out of an ever-dwindling inventory of proved reserves that is always replenished from probable reserves, even as the probables are replenished from the stock of prospects.

The rate at which reserves flow from one stage to another depends on investment and operating costs, and on prices. Reserves can thus be priced or costed into or out of existence; the economist's basic object is to see what difference a price change makes for the hydrocarbon flow. Our knowledge is painfully scant. It is no surpirse that there have been so many disappointments in reserve estimation. Some workers used nondata ("discoveries"), a fragment masquerading as an observation. Others, without realizing it, assumed that probable reserves existed within the cost range of proved reserves; unfortunately, this was unfounded. For the United States, diminishing returns to exploration go back nearly thirty years in oil, over twenty years in natural gas, and the domestic industry has long maintained its working inventory by running the tailings through the mill, so to speak. New oil in old fields and slightly better recovery factors each contributed about half. The rest of the world has a much larger stock of probable reserves, but much of it will remain uncounted for lack of a market. Better knowledge of proved and probable reserves would give us a valuable handle on near-term developments outside the United States, which we now largely lack.

Moreover, the data in the public domain are shrinking. As foreign oil companies are expelled, less is published, and the old unofficial forum has dwindled.

OIL AND GAS
Estimation of
Undiscovered
Resources

Gordon Kaufman

INTRODUCTION

A forecast of the amount of petroleum remaining to be discovered in a large region is a forecast of an uncertain quantity several orders of magnitude more uncertain than a forecast of what is ultimately recoverable with current technology from discovered deposits. This is a commonly accepted assertion, and yet until very recently no forecast of undiscovered petroleum for large regions such as the onshore United States explicitly incorporated and reported measures of uncertainty. Uncertainty is uncomfortable, and coping with it in an orderly, logical way is not easy. Forecasts of undiscovered petroleum expressed as single numbers or point forecasts forecasts reflect a natural human tendency to avoid uncertainty, a trait reinforced by our culture and our educational system. Winkler[1] expresses this idea forcefully by linking together three passages dealing with uncertainty as a cognitive concept:

> The psychological reduction or omission of uncertainty is in itself a useful cognitive simplification mechanism. The notion that events are uncertain is both uncomfortable and complicating. Indeed, even in the supposedly "rational" world of business, there is evidence that businessmen are averse to admitting uncertainty.[2]
>
> Above all, the educational advantage of training people – possibly beginning in early childhood– to assay the strengths of their own opinions and to meet risk with judgment seems estimable. The usual tests and language habits of our culture tend to promote confusion between certainty and belief. They encourage both the vice of acting and speaking as though we were certain when we are only fairly sure and that of acting and speaking as though the opinions we do have were worthless when they are not very strong.[3]

83

Table III-1. Forecasts of Ultimately Recoverable Oil in the United States,[a] Billion Barrels.

Forecaster	Year	Estimate of Ultimate Recoverable	Estimate of Amount in Place
Day	1909	10–24.5	...
Shaw	1919	8–10	30–50
White	1920	11	...
USGS	1922	15	...
Pratt, Weeks and Stebinger	1942–1950	100–110	...
Pogue and Hill	1956	165	...
Pratt	1956	180	...
Hubbert	1956	150–200	...
U.S. Department of the Interior	1956	315	...
Hill	1957	250	...
U.S. Department of the Interior	1960	400–600	2,000–3,000
Netchert	1960	235–250	...
Zapp	1961	590	...
Hubbert	1962	165–175	...
Resources for the Future	1962	250+	500
Hendricks	1965	400	1,600
Hendricks and Schweinfurth	1966	500	2,000
Moore	1966	420	...
Eliot and Linden	1968	430	...
Schweinfurth	1969	530	2,700
Arps	1970	165	...
Hubbert	1970	200	...
Moore	1971	224–353	587
AAPG/NPG	1970–1971	443	...
AAAS	1973	280 years ultimate potential	33 × 10^{19} joules per year
USGS	1975	218	...
Exxon	1976	285	340

a. Excluding Alaska as of January 1975, 122 billion barrels have been produced.

Source: T.A. Bryant; "Policy Implications of Oil and Natural Gas Reserve Estimates," M.S. thesis, Massachusetts Institute of Technology, Cambridge, Mass., May 1976.

Our culture does not encourage explicit representation of uncertainty or its manifestations in other than extreme forms. The notion of risk and probability inherent in business does not seem to be handled very effectively in our educational system as a whole.[4]

Prior to 1975 most forecasts of aggregate amounts of undiscovered oil and gas in U.S. territories were point forecasts, although in some cases a range of possible values were reported. Meaningful comparisons are difficult because of variations from forecast to forecast in the methods used and in the definitions of the quantities being forecast. At one extreme are forecasts generated by using a deterministic model fit to historical data to project the future; at the other extreme are forecasts based on expert opinion alone.

Table III-2. Forecasts of Ultimately Recoverable Gas in the United States.[a]

Forecaster	Year	Estimate of Ultimate Recoverable, TCF
Miller	1958	1,250
Weeks	1958	1,200
Averett	1961	2,000
Zapp	1961	2,650
Hubbert	1962	1,000[b]
Resources for the Future	1962	1,200
Hendricks	1965	2,000[c]
Hubbert	1967	900-2,000[b]
Potential Gas Committee	1967	1,290[d]
American Association of Petroleum Geologists	1970	1,543
Potential Gas Committee	1971	2,197
	1970	1,518
American Association of Petroleum Geologists	1973	110
COMRATE	1975	1,380
Exxon	1976	1,112

a. As of January 1975, 477 trillion cubic feet (TCF) have been produced.

b. Excluding Alaska.

c. Excluding submarginal gas of 2,000 TCF.

d. Includes 690 potential, excludes Alaska.

Source: Bryant, "Policy Implications of Oil and Natural Gas Reserve Estimates."

As a prelude to detailed discussion of methods used to forecast undiscovered petroleum it is worth reviewing briefly the evolution over time of forecasts of ultimately recoverable oil and gas in the United States. It is the most intensively explored country in the world, accounting for the greater part of all exploratory drilling done throughout the world since petroleum exploration began. Forecasts of aggregate United States petroleum resources stretch back to the late 1800s.

Most widely quoted forecasts are in units of ultimately producible reserves based on implicitly or explicitly stated assumptions about what fraction of oil or gas in place can and will be produced in the future. Amounts in place that remain to be discovered can be deduced in some but not all such cases. Tables III-1 and III-2 are chronologically ordered lists of often quoted estimates of ultimately recoverable reserves of oil in billions of barrels. The principal feature of estimates of ultimately recoverable oil in the coterminous United States is that from 1909 until the early 1960s the estimates tended to increase with time from a low of 10-25 billion barrels in the early part of the century to a high of 400-600 billion barrels in 1960. Subsequent estimates turn downward (if the 1973 American Association for the Advancement of Science (AAAs) estimate of 280 years of potential is ignored), and most recent forecasts are of the order of 200-300 billion barrels ultimately recoverable. Natural gas estimates have a similar history.

Why there is so much variation over time is a long story. Pieces of the plot are the subject of Part III. We start with a preview of possible forms of a forecast of undiscovered petroleum.

WHAT IS A FORECAST OF UNDISCOVERED PETROLEUM?

Forecasts of amounts of undiscovered oil and natural gas vary greatly in form and content. Each is generated by one or more logically distinct forecasting methods and is responsive to a *forecasting question* dictating in part how the forecast is made. The flavor is suggested by an omnibus question such as the following:

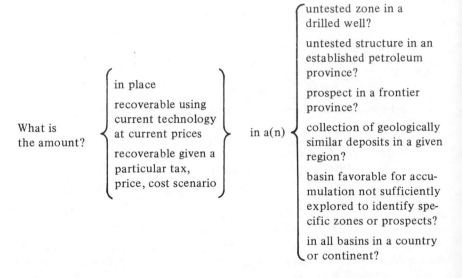

Often a particular time horizon is imposed and the forecasting question is elaborated by asking for the time rate of amounts discovered or alternatively the rate of amounts discovered per unit of exploratory effort.

This omnibus question covers enormous ground, at one extreme requiring rich, detailed geological and geophysical information (sufficient to identify structures, prospects, and untested zones) to the other extreme of little or no information; from focus on a specific geographical location to focus on really enormous regions; and from measurement of in-place amounts (at least conceptually independent of economic considerations) to measurement of amounts dependent on the interaction of in-place petroleum with exploration and production. The answer to the question of amount is a statement about properties of a resource base and is *not* a projection of future supply from this base. The question clarifies what is possible, and if it is properly posed, the answer can be used in the construction of a family or collection of projections describing the range of possibilities for future supply at one or more points in time. Each member of this family accords with one or possibly several paths of evolution of geological, geophysical, and engineering technologies, of alternative energy technologies, and of the economic and political structure within which these technologies evolve on the way to the "long run."

If we possessed a telephone line to the Lord, He called and told us precisely where each undiscovered deposit of petroleum in the world lies and then described its salient attributes to us, we would be able to reduce drastically the size of the set of realizable supply functions, one of which will obtain in the long run. It would still be a set with many elements. But a principal source of uncertainty would be resolved. This is, of course, a fantasy.

Resolution of uncertainty about the resource base is achieved incrementally and at nonnegligible cost. If uncertainty cannot be resolved into certainty, given the current state of knowledge, degrees of uncertainty about the world's petroleum resource base can at least be represented in an orderly way. How this is done depends critically on the current state of knowledge about the petroleum resource base, about substitute or backstop technologies, and about the political and economic environment.

There is a conceptual directionality to geologic and geophysical data: Highly disaggregated drilling and deposit attribute data can be recast into less finely partitioned taxonomies, but the converse is not generally true. There is no simple mechanism for refining highly aggregated geologic and geophysical data into their original constituents. The character of available data for a particular region, basin, or sedimentary unit determines what options are available for a formal quantitative analysis of future supply.

It is obvious that the amount and type of objective data—data generated by predrilling exploration activities and by drilling—available for inference about model parameters and for prediction of supply from new discoveries in a petroleum basin, are strongly conditioned by the intensity with which the basin has been explored. What is not so obvious is how the character of the models used to generate supply forecasts should change, if at all, as more and more predrilling and drilling information accumulates. At one extreme are mature basins where reconaissance survey information is available but with sparse or nonexistent data. At the other extreme are frontier basins where an enormous quantity of drilling data is available and detailed reservoir engineering studies of a large number of fields can be done. In between is a continuum of partially explored basins. A traditional paradigm for spanning this crude classification of information has been to apply geologic-volumetric methods to frontier basins and to generate estimates of aggregate amounts of undiscovered in-place or recoverable oil and gas in the basin, restricting application of models requiring a finer partitioning of geologic and economic attributes to basins that are closer to "mature" than to "frontier." Unfortunately, this paradigm falls short in that forecasts of supply from frontier basins are then at best only crudely responsive to changes in economic and policy parameters that influence supply.

Perhaps the most challenging strategic research problem underlying methods for forecasting petroleum supplies from new discoveries is to design a sequence of logically compatible models of deposition and discovery that span the information spectrum from frontier to mature.

It is the rule rather than the exception that within a single basin stratigraphic units are sometimes intensely drilled or "mature" and others are unexplored or "frontier," so this problem arises at the basinal level as well as at a higher level of geographical aggregation. In order for a model of discovery and supply from a mature region or time-stratigraphic unit to be logically compatible with a model for a relatively unexplored unit or region, the two models must share a common conceptual core. One important component of such a core is the concept of *deposit size distribution.*

In the search for petroleum, oilmen drill *prospects*, geologic anomalies known or inferred to contain producible hydrocarbons; they develop *deposits.* Hence prospects and deposits are economic and technological decisionmaking units. The supply over time of petroleum from a petroleum basin or region is a flow composed of a sum of flows from individual deposits, and the shape of this flow is determined by both economic and technological decisions made about *individual* prospects and deposits. Ideally then, a model of petroleum supply that pretends to incorporate industry behavior would somehow reflect the impact of prices, costs, technology, and the regulatory regime on individual deposits and prospects; that is, it would be a composite of highly disaggregated micromodels, one for each prospect and one for each deposit. Such a model of supply is at one extreme on a qualitative scale of level of aggregation. This highly disaggregated approach to projecting supply requires much geological engineering and economic information.

The converse, highly aggregated models of petroleum supply from a region, models built from a perspective that holds that the total flow over time from that region is a "time series" of values adequately modelable using traditional econometric and/or time series models for the aggregate, may fail to capture essential features of industry behavior and often project future supply poorly as a consequence.

The data-intensive character of a completely disaggregated model can be partially sidestepped and its inherent conceptual advantages kept by modeling properties of *collections* or *populations* of descriptively similar deposits located in a geological region. If the relative frequency or probability distribution of deposit sizes for such a population is specified, along with the number of deposits in the population, then the impact of changes in economic and technological factors on the subset of deposits in the population which are economic to produce can be computed in a conceptually straightforward way.

Specifically, suppose that there are N descriptively similar deposits in a given basin whose frequency distribution of sizes, measured in barrels of oil equivalent (BOE) is as shown in Figure III-1. The hatched portion of each vertical bar

Figure III-1.   Discovered and Undiscovered Deposit Size Distributions.

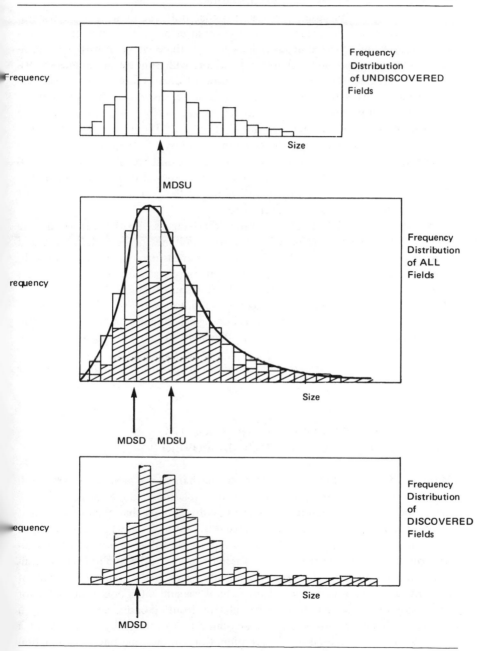

represents the fraction of deposits in the size interval spanned by the bar that has already been discovered. The unhatched portion of the bar is the fraction of deposits in that size interval not yet discovered.

For illustrative purposes, suppose also that both the number of deposits in each size class and the predevelopment costs in effort and investment in discovery in as yet undiscovered deposits are known with certainty. A particular specification of economic and technological factors will truncate the frequency distribution, dividing the frequency distribution of deposit sizes into *four* classes: discovered deposits that are economic to produce, discovered deposits that are uneconomic to produce, undiscovered deposits sufficiently large to be economic to explore, and undiscovered deposits that are not. The economic and technological factors that determine the economic viability of a deposit can be said to determine a minimum deposit size profitable for discovered deposits (MDSD) and a minimum deposit size profitable for undiscovered deposits (MDSU). The difference between MDSD and MDSU is the added cost of "front-end" effort and investment required to discover a deposit.

As economic and technological variables change so do the MDSD and the MDSU. The impact of changes on the total volumes of both discovered and undiscovered petroleum producible is directly measurable and follows directly from a (micro)economic evaluation of each individual deposit. If, however, a deposit size distribution is *not* given and the resource base is characterized by a statement of total volumes of undiscovered and discovered petroleum alone, the impact of changes in economic and technological variables on producible petroleum must be treated in an ad hoc way. The logic of individual deposit economics is lost. For this reason, *the concept of deposit size distribution is an essential component of models of petroleum supply designed to reflect industry behavior in a logical way.*

MODELS FOR PROJECTING AMOUNTS
OF HYDROCARBON TO BE DISCOVERED

Most of this part of the book centers on the character of *models*, mathematical representations of observable quantities and relations among them, that have ben proposed for projecting amounts of hydrocarbons that exploration might reveal. There are many, and they are very different in form, which presents both opportunities and difficulties. The set of approaches that appeals to an individual consumer of resource projections is narrowed by the application the consumer has in mind; for example, a regionally aggregated econometric model of field market supply and demand for natural gas and oil is not at all useful for predicting future supply from a speculative frontier basin, and conversely a geologic-volumetric analysis of the amount of recoverable hydrocarbons in a basin is minimally relevant to a determination of the time rate of production from new discoveries in that basin as a function of economic variables. Consumers want an approach that provides information bearing on the policies or decisions for action they have in mind. Beyond these almost self-evident observations is the consumers' need for an evaluative framework that they can fit to each approach so as to judge how well the model agrees with the real world quantities it represents.

Models and data are inextricably intertwined in this setting, for it is from measurements and manipulation of observed quantities that model assumptions spring. For the most part, measurement is beyond the control of the modeler, and resource data measurement errors are a principal source of difficulty in attempts to validate model assumptions.

It is not unfair to say that *none* of the models presented here matches the predictive ability of really good models in the natural sciences. (For invidious comparisons see D. A. Freeman.[5]) Nor does the quality of currently available data allow the assumptions of any one of them to be tested with the same degree of precision as assumptions that underlie really good models in physics, chemistry, and biology. The data do not yet seem to permit discovery of startlingly simple and accurate mathematical descriptions of how mineral deposits are formed, discovered, and produced. This does not mean that good model building and testing tactics need be abandoned however, but just that we cannot expect even good tactics to yield models that stand up to the best in the physical sciences.

Once a decision is made about how to define the attribute or attributes to be observed and measured, a good tactic is to let the character of the data, along with a dose of innovating thinking, be the principal determinants of the particular mathematical assumptions made. In contrast, many of the models reviewed here are models in search of an application, mathematical forms used more or less successfully as models in quite different contexts, lifted out of these contexts and fitted to oil and gas data. This is particularly true of regression models, a favorite of econometricians.

The types of models discussed unfold on different scales: volume of sediments explored, feet drilled and wells drilled, discoveries in order of occurrence, and time. They vary in level of aggregation, in mathematical form, and are fitted to data in different fashions. Although no clearly discernible intellectual threads neatly tie them *all* together, some of the model types examined can, with some mental stretching, be logically connected.

In this part the principal approaches to projecting amounts of oil and gas remaining to be discovered are described, several of which logically elide with projections of future supply over time. They may be loosely classified as:

- Life cycle
- Rate of effort
- Geologic-volumetric
- Subjective probability
- (Play or province) discovery process
- Econometric

Each projection method is discussed in succeeding chapters. Throughout, the presentation draws heavily on published papers and documents cited, the principal ones being:

Life cycle

Hubbert 1962, 1966, 1974
Moore, C. L. 1966
Ryan, J. M. 1965, 1966
Wiorskowski 1977

Rate of effort

Hubbert 1956, 1972, 1974
Hartigan–Bloomberg 1976

Geologic-volumetric

Zapp 1962
Hendricks 1965
Mallory 1975
Jones 1975

Discovery Process

Arps–Roberts 1958
Ryan, J. T. 1973
Barouch–Kaufman 1976a, b
Drew–Schuenemeyer–Root 1978
Ecko–Jacoby–Smith 1978
Smith–Ward 1981

Subjective probability

USGS, Circular 725 1975
Energy, Mines, and Resources
(Canada) 1977

Econometric

Erickson–Spann 1971
MacAvoy–Pindyck 1973, 1975
Khazzoom 1971
Attanasi 1979

Even though the particular studies reviewed in this part are only a small fraction of the literature describing methods for estimating undiscovered oil and gas and are focused exclusively on North America, they represent reasonably well the principal ideas behind methods currently in use and those under development.

Although for ease of exposition a section is devoted exclusively to each method, in practice two or more are sometimes blended and a hybrid analysis performed. For example, in a recent study of the future supply potential of Nigeria, geologic-volumetric and subjective probability methods were used to appraise the aggregate amount of remaining undiscovered oil in Nigeria and this appraisal in turn combined with a life-cycle model of production to produce projections of future production.[6] Industry practice is to tailor the methods applied to a given region or basin to the quality and detail of information available about it. Analysis requiring aggregate data (geologic-volumetric, for example) may be performed as a check on a highly disaggregated analysis (using a discovery process model, for example).

REFERENCES

1.   R. L. Winkler, *The American Statistician* 32, no. 2 (May 1978): 54.
2.   R. M. Hogarth, "Cognitive Processes and the Assessment of Subjective Probability Distributions," *Journal of the American Statistical Association* 70 (1975): 271–94.

3. L. J. Savage, "Elicitation of Personal Probabilities and Expectations," *Journal of the American Statistical Association*, 66 (1971): 783–801.

4. P. G. Moore, "The Managers' Struggles with Uncertainty," *Journal of the Royal Statistical Society*, Series A, 140 (1977): 129–65.

5. D. A. Freeman, "Statistics and the Scientific Method," in *Proceedings of the Social Science Research Council Conference on Analyzing Longitudinal Data for Age, Period and Cohort Effects* (forthcoming).

6. William D. Dietzman et al., "Nigeria—An Assessment of Crude Oil Potential," Analysis Memorandum prepared for Energy Information Administration, June 7, 1979.

10 LIFE-CYCLE MODELS OF PETROLEUM PRODUCTION

One of the most widely publicized exercises in resource estimation is that of M. King Hubbert.[1] In a March 1956 address before a meeting of petroleum engineers at San Antonio, Texas, under the auspices of the American Petroleum Institute, he forecast that the amount of oil that would be ultimately produced from the coterminous forty-eight states onshore and offshore would be about 150 billion barrels (BB). This estimate was not far from those of several other studies conducted at roughly the same time. Hubbert quotes Pratt[2] as reporting the highest estimate in an opinion survey of twenty-three experts to be 200 BB of crude oil plus natural gas liquids; Pratt's own estimate was 170 BB of total liquid hydrocarbons. Pogue and Hill estimated ultimate production at 165 BB of crude oil production.[3] What remains to be discovered and produced is directly derivable from an estimate of ultimate production. Hubbert summarized his view of a prevailing complacency as follows:

At that time [1955–56], the U.S. petroleum industry had been in operation for just under a century during which the cumulative production amounted to only 52 billion barrels, and annual production had just reached 2.5 billion barrels per year. The intuitive reaction to these facts was one of complacency: If a century has been required to produce the first 52 billion barrels and if future production would amount to two or three times past production, then obviously there was little to worry about at present. This attitude was shared among the informed members of the petroleum industry – petroleum geologists, production engineers, research scientists, and corporate officials alike. Common expressions of opinion among such individuals at that time were: "A shortage of oil won't occur during my lifetime," or "My grandchildren may have to worry about oil shortages; I won't have to." At the same time, among the favorite clichés of the public relations or propaganda arm of the U.S. petroleum industry were variations on the theme: "The United States will be able to produce all the oil that it will need in the foreseeable future," this notwithstanding the fact that the United States had already become dependent upon foreign sources for more than 10 percent of its petroleum requirements.[4]

What was distinctive about Hubbert's approach to resource estimation was his use of the concept of *life cycle* of a fossil fuel to generate not only an estimate of what ultimately will be discovered and produced from a producing region, but the time path of rates of discovery and production:

> it is clear that the production curve of a fossil fuel for a given area has no unique shape, but all such curves have an important element in common: They all rise gradually from zero in their initial phases, and they all decline gradually to zero in their terminal phases. The reason for this is that in the initial phases there are problems of exploitation and development before production can occur, and these necessarily require time. In the terminal phase there is a gradual decline in exploratory success plus an increasing difficulty in extracting the remaining fuel from a greatly diminished supply. Hence in any region comparable in size to a U.S. state, the exploitation of its coal or oil in such a manner that the production curve either ascends or descends vertically is a virtual impossibility.
>
> Again, the larger the region for which production statistics are combined, the more do minor irregularities which may occur in small areas tend to cancel one another. Hence, for large areas, such as the whole United States or the whole world, the combined production statistics are far more likely to give a single cycle having but one principal maximum than a multiple-cycle curve such as that of Illinois.[5]

Following a procedure described shortly, Hubbert used the life-cycle concept to forecast that U.S. domestic petroleum production would soon peak and then decline, the peak being likely to occur in the time interval 1966 to 1971. At the time he made it, this prediction provoked considerable controversy. It proved to be close to the mark! Figure 10-1 displays Hubbert's forecasts based on data through 1956.[6]

Life-cycle models, like most statistical time series models, divorce themselves from the physics and engineering of discovery and from geological description and do not incorporate economic effects. The class of models is based on the assumption that there is a relatively simple functional relation between time and amounts of oil and gas discovered and produced per unit of time. The amounts of oil and gas in place and the proportions of them that are recoverable are parameters to be inferred from observation of what has been discovered and produced per unit time to date.

Figure 10–1. Comparison of Complete Cycles of U.S. Crude Oil Production Based upon Estimates of 150, 200, and 590 Billion Barrels for Q_∞.

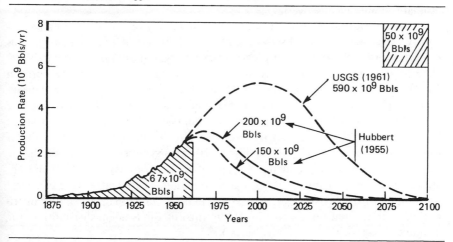

Source: M. King Hubbert, "U.S. Energy Resources: A Review as of 1972," in *A National Fuel and Energy Policy Study, Part I*, Committee on Interior and Insular Affairs, U.S. Senate, Serial no. 93–40, 1974, Figure 25.

Hubbert chooses a logistic function, a logistic growth curve, as a life-cycle model. Defining D_∞ as ultimate cumulative discoveries and D_t as cumulative discovery at time t, he proposes

$$D_t = D_\infty [1 + \alpha e^{-\beta(t - t_0)}]^{-1} \qquad (10\text{-}1)$$

as a "law" of growth of cumulative production. Since the rate of change $dD_t/dt \equiv D_t'$ of D_t with respect to t as a proportion of D_t equals $\beta - (\beta/D_\infty)D_t$, the proportional rate of growth of D_t is a linear function of D_t, and this may be viewed as the primitive assumption or postulate underlying a logistic model of growth. More crudely, if time is discretized and the change in D_t from t to $t + h$ is measured, then the assumption is that the change $D_{t+h} - D_t$ as a proportion $(D_{t+h} - D_t)/hD_t$ of hD_t is a linear function of D_t as $h \to 0$.

Defining $\xi = \log_e \alpha$, it can be shown that the time rate of discovery D_t', the derivative of D_t, is symmetric about the point $t - t_0 = \xi/\beta$ and possesses a single maximum. Cumulative discoveries approach zero as $t \to +\infty$, and as $t \to +\infty$ they approach an asymptote D_∞. In between D_t rises monotonically with increasing t at an increasing rate until $t - t_0 = \xi/\beta$, at which point the rate of increases decreases monotonically. The graph of D_t is, as a consequence, S-shaped. Choice of a logistic function as a model for D_t implies that the rate of discovery will decline over the interval $t \geq t_0 + (\xi/\beta)$ in precisely the same fashion that it rose over the interval $t \leq t_0 + (\xi/\beta)$. This symmetry of the rate of change of D_t about a single maximum means that D at $t = t_0 + (\xi/\beta)$ equals $(1/2)D_\infty$; that is, at time t cumulative discoveries equal precisely one-half of ultimate cumulative discoveries.

Cumulative production Q_t at time t is functionally related to cumulative discovery, since cumulative discoveries D_t at t must equal the sum of Q_t plus proved reserves R_t at t. Since proved reserves begin at zero and ultimately decline to zero, and cumulative discoveries begin at zero and ultimately approach D_t, Hubbert argues that the asymptote for D_t as $t \to +\infty$ must be the same as that for Q_t. If a logistic function is specified as a model for D_t, with α and β identical to that for D_t, shifted by a time interval T to the right as appears in Hubbert,[7] then the maximum of discovery rate D_t' will occur $(1/2)T$ units of time earlier than that of Q_t' (see Figure 10-2). (Symmetry of rate of change about a single maximum and the S-shape of D_t between asymptotes 0 and D_∞ are distinguishing attributes of the logistic function. These features are of course not unique to the logistic function: any symmetric unimodal probability distribution concentrated on $-\infty$ to $+\infty$—the normal or the Student distribution, for example—possesses the same attributes.)

Hubbert fits the logistic model to cumulative *discovery* data and then shifts the fitted discovery model by 10.5 years to obtain a "fitted" model for cumulative production. A comparable display of actual versus fitted *rates* of proved discoveries and of production based on annual data through and including 1961[8] shows considerable variability from year to year in the series. Cumulative data series, are, by virtue of cumulation, smoother.

Hubbert's conclusions from analysis of data through 1961 were that

1. The peak in the rate of proved discoveries,[a] when smoothed to eliminate annual oscillation, had already occurred during the second half of the 1950s.
2. The peak of proved reserves would probably occur about 1962.
3. The peak in the rate of production would probably occur about 1967-1969.
4. The value of Q_∞ would probably be about 170 billion barrels.

These conclusions are visually summarized in Figure 10-4.

A reanalysis done ten years later using data for 1900-1971 to fit the cumulative discovery function using 1930-1971 data, produced negligible change in Hubbert's estimation of D_∞ and Q_∞, estimated at 172 BB again, and of the turning point in U.S. production, estimated to peak in 1967-68. Annual crude oil production peaked in 1970, reaching a maximum of 3.3 BB per year and has since declined.

---

a. The data series for "discoveries" used by Hubbert does *not* represent discoveries made in each year; cf. Part II discussion of "discoveries."

Figure 10-2. Fitting of the Curves of Cumulative Production, Cumulative Proved Discoveries, and Proved Reserves of U.S. Crude Oil as of 1962 by Means of Logistic Equations.

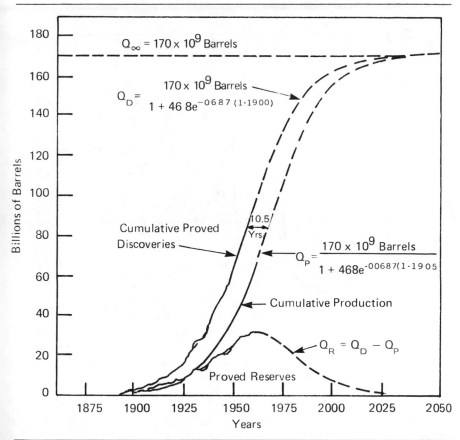

Source: M. King Hubbert, "Energy Resources: A Report to the committee on Natural Resources: National Academy of Sciences," National Research Council Publication 1000–D, 1962, Figure 27. Reprinted with permission.

Although he announced no explicit measure of uncertainty or variability for these forecasts, in 1962 Hubbert stated: "if a contingency allowance were to be made of how much the actual figure of Q_∞ might exceed the present estimate of 175 billion barrels, a figure higher than an additional 50 billion barrels would be hard to justify."[9] A comparison[10] of Hubbert's 1974 point forecasts for ultimate cumulative production of crude oil and of natural gas with the means of probabilistic forecasts of amounts of undiscovered recoverable crude oil and natural gas done in 1975 by the U.S. Geological Survey's oil and gas branch[11] shows Hubbert to be in close agreement (see Table 10-1). (The methods em-

Figure 10-3. Comparison of Annual Rates of Production and of Proved Discovery of U.S. Crude Oil as of 1962 with Theoretical Rates (Dashed Curves) Derived from Logistic Equations.

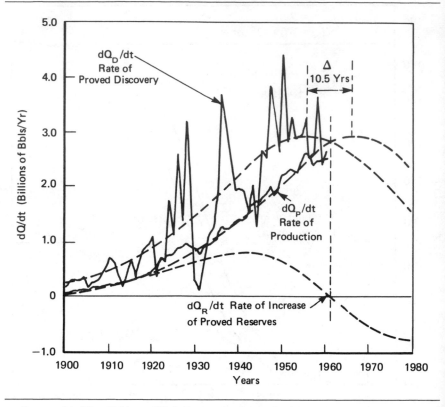

Source: M. King Hubbert, "U.S. Energy Resources," 1974, p. 84. Reprinted with permission.

ployed by the oil and gas branch are dramatically different in concept and are discussed later; see Chapter 13.)

Hubbert's analyses of oil and gas resources rudely shook those who visualized very, very large future supplies of *cheap* petroleum from as yet undiscovered oil and gas in U.S. provinces, provoked considerable controversy, and contributed in an important way to a critical reexamination of currently available U.S. resource estimates. A glance at Table 10-1 shows that a number of estimates of undiscovered oil and gas made between 1956 and 1974 differ by an order of magnitude or more from Hubbert's, and the discrepancies naturally provoked a close study of his use of life-cycle models.

Figure 10-4.   Estimate of the Complete Cycle for U.S. Crude Oil
Production as of 1962.

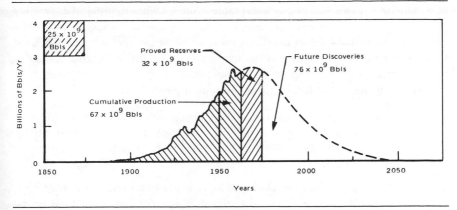

Source: M. King Hubbert, "Energy Resources," 1962, Figure 39. Reprinted with
permission.

Table 10-1.   United States, Onshore and Offshore.

	Lower 48 States	*Total United States*
Crude oil, BB		
Hubbert (1974)	172	212
Circular 725	176	218
Natural gas, TCF		
Hubbert (1974)	1,050	1,184
Circular 725	1,144	1,242

Criticisms were of two kinds. The first suggested that a logistic function is an
inaccurate specification of the structure of production and discovery over time,
because the observed time series of production and of discovery do not behave
past the turning point as the logistic function says it should. The second claimed
that the methods used by Hubbert for estimating parameters of the logistic
model were not robust. In extensive published interchanges between Hubbert
and his critics, questions about model validation and adequacy of methods for
parameter estimation that continually surface are not unique to resource estimation but rather are typical of a scientific inquiry into the structure of observed
data and its causes. Of all possible functions that exhibit initial growth and ultimate decay as a function of time, is a logistic function as employed by Hubbert
an acceptable choice for a model of the time rate of production and of discov-

ery? Can any alternatives be shown to be better? Can any simple function that excludes geology, engineering, and economics adequately represent the evolution of these time series? What plausible tests of a logistic model serve to validate or invalidate it in a scientifically acceptable manner? What method of estimating parameters is most appropriate? Is the method precise in its specification of a parameter estimate? Does it provide a measure of variability or error associated with each parameter estimate? Is the method resistant, or will a small change in the data induce a large change in parameter estimates? Is the method robust?

J.M. Ryan challenged Hubbert's predictions, arguing first:

> There is, in general, no particular *a priori* reason for choosing one asymptotic growth curve over another, although different curves may yield substantially different results and the answers one obtains are a function of the particular curve which is chosen.[12]

and, after Hotelling,[13] that the choice of a particular functional form in such instances is difficult to base on an analysis of causes defined more than thirty years earlier as

> a study of the tendencies manifested repeatedly in the past upon the repeated occurrence of conditions which we term essential, and in spite of the variation of other conditions which we consider non-essential. Such an analysis is not provided by the mere determination of a curve of some assumed type which conforms to the general course of past events.

Ryan continued his challenge:

> There is no fundamental law of physics insuring that cumulative discoveries or cumulative production will follow a logistic pattern in the future. The logistic equation is not the logical consequence of physical concepts such as the laws of conservation of mass or energy. Nor is it a reasonable projection of the consequence of the future economic, political or technological developments which will largely determine future discovery rates. This curve was chosen instead on the purely empirical ground that it provided a good fit to past data. The fact that a particular analytical function may appear to the eye to provide a good fit to past data provides no assurance that it will continue to do so in the future. As will be noted later other functions which also provide a good fit to historical data and which can be used with equal justification lead to radically different conclusions when projected into the future.[14]

Ryan questions the appropriateness of choosing a rate function with a single peak:

> There have been several apparent discovery peaks in the past and there may well be others in the future. Today we have no reliable means of determining whether the 1955–56 peak was a real inflection point. It would be unwise to

assume that an all-time peak in discoveries occurred in 1955–56 when similar indications of past peaks would have led to mistaken results." [15]

He claims that Hubbert's method for fitting a logistic function to the data does not discriminate well among widely varying values of D_∞. Hubbert used trial-and-error graphical method based on the relation

$$\log\left(\left[D_\infty/D_t\right] - 1\right) = -\xi - \beta\left(t - t_0\right) \qquad (10\text{-}2)$$

and the observation that for a given D_t the right-hand side is linear in t and so plots as a straight line on semi-log paper. By this method, values of D_∞ between 170 and 250 BB are acceptable, the higher figure being 40 percent larger than Hubbert's choice of 170 BB.

Hubbert's counter to Ryan's criticisms emphasizes the empirical accuracy of his predictions thus far:

1. By a refinement of Hubbert's graphical method, D_∞ is constrained to lie between 150 and 180 BB.
2. The peak of proved discoveries not only appears to have occurred about 1955–56, but there is additional evidence that proved reserves peaked about 1960–61 and declined thereafter; this occurs halfway between the empirical time lag of ten to eleven years between cumulative discovery and cumulative production as a logistic model for both would predict.
3. Given that cumulative oil production is logistic in character, by virtue of symmetry Q_∞ must be twice cumulative production (or cumulative discoveries) up to the peak or turning point in rate of production (or discoveries). As of 1955–56, the predicted turning point for discoveries, cumulative discoveries were 82.3 and 85.3 respectively, suggesting a Q_∞ of about 165–171.

Hubbert, in his 1974 review of his own earlier analysis, using data through 1971, does not modify his conclusions in any major way.

In none of these exchanges is a comparison made, based on accepted statistical principles, of the relative qualities of fit to the data or of the predictive accuracies of possible alternative models. A more thorough study of comparative predictive accuracies and qualities of fit of a variety of structurally distinct growth models is that of Wiorkowski,[16] to be discussed shortly.

C. L. Moore forecasts future production, gross additions to reserves, and discoveries of crude oil and natural gas to 1980 based on an analysis done in the same spirit as Hubbert's but differing in the particulars.[17] He chooses to fit a Gompertz function to the historical time series and estimates its parameters by the method of least squares.

The Gompertz function is named after Benjamin Gompertz (1779–1865), an English mathematician who constructed it as a description of a "law" of human mortality.[18] Properly scaled, the Gompertz function is known to modern statisticians as an *extreme value* distribution; it has the following form:

$$Q_t = Q_\infty e^{-\beta e^{-\alpha t}}, \qquad \alpha, \beta > 0 . \tag{10-3}$$

The function Q_t satisfies the (defining) differential equation:

$$d\left[\log Q_t\right]/dt = Q_t'/Q_t = \alpha \log Q_\infty - \alpha \log Q_t, \tag{10-4}$$

so that the proportional rate of change of Q_t is a linear (decreasing) function of (increasing) $\log Q_t$. This contrasts sharply with the logistic function, whose proportional rate of change is a linear function of Q_t in place of $\log Q_t$. Whereas the logistic function has its inflection point at t^0 such that $Q_{t^0} = (1/2)Q_\infty$, that for the Gompertz function is at t^0 such that $Q_{t^0} = Q_\infty/e$, e being the Naperian logarithmic base 2.718. Consequently, if estimates of Q_∞ are computed by exploiting these relations for the logistic and Gompertz model between the value of Q_t at its maximum rate of change and Q_∞, the Gompertz model will give an estimate of Q_∞ that is $e/2 = 1.36$ times larger than a logistic function.

The rate of change of the Gompertz function with respect to time t is, in contrast to the logistic function, not symmetric, rising to a single maximum more quickly than it descends. In this respect it at least qualitatively overcomes an objection to the use of a logistic function often preferred by economists, who argue that a logistic function with a given set of parameter values must, a priori, be an inappropriate model for *both* production and discovery. Production from known fields is economically less risky than exploration, and, as a region approaches exploration maturity, a shift in investment from exploration to production is likely to occur more and more rapidly as returns to production appear increasingly more desirable than that to exploration, causing the rate of exploration past its peak to accelerate its decline.

In 1970, P. F. Moore produced the forecasts for crude oil shown in Table 10-2. The forecast of 620 BB for ultimate crude oil production is clearly inconsistent with a forecast of 587 BB *in place* that will ultimately be discovered; the amount ultimately produced must be less than that discovered and, if recovery factors remain in the range of those dictated by current technology, it must be significantly less. Wiorkowski observed that, as shown in Figure 10-5 (and Figure 10-6 for gas, both from Moore), the last eight years of data for cumulative discoveries of oil in place deviates significantly and in an orderly way from Moore's projection and concluded that "This pattern is a clear signal that we are imposing an incorrect model." [19]

Wiorkowski was the first to address the problem of choice of a growth function by carefully comparing performance of a large set of such functions on the same data using statistical methods. He defines a very general family of growth functions by generalizing a function appearing in the biological literature (the Richards function[20]). The logistic and Gompertz functions are special cases.

Table 10-2. Crude Oil Forecasts by P.F. Moore in 1970, Billion Barrels.

Time	Cumulative Oil in Place	Cumulative Gross Additions to Reserves	Cumulative Crude Oil Production	Recovery Factor
1970	431	129	90	129/431 = .30
2000	529	224	201	224/529 = .42
Ultimate	587	353	620[a]	353/587 = .60
1920–1970	342	a	a	
1970–2000	98	95	111	

a. This number is computed using a Gompertz function as Moore fit it into the series.

Source: Peter F. Moore, Appendix F in *Future Petroleum Provinces in the United States*, ed. Ira Cram; National Petroleum Council, Washington, D.C., 1970.

Whether a logistic, a Gompertz, or some other functional form within this family is a suitable choice of model may be determined by using the data to estimate parameters of the extended family of functions and then identifying the corresponding functional form, *provided that the data enable a clear distinction to be made.*

The defining equation Wiorkowski uses for the extended class of Richards function is

$$Q'_t / Q_t = -\alpha \left[\frac{Q_t^\lambda - 1}{\lambda} \right] - \beta \quad , \tag{10-5}$$

with $-\infty < \lambda < +\infty$. Since

$$\lim_{\lambda \to 0} [Q_t^\lambda - 1]/\lambda = \log_e Q_t, \tag{10-6}$$

$\lambda = 0$ corresponds to choice of a Gompertz function, while if $\lambda = -1, Q_t$ is an exponential function. The logistic function arises when $\lambda = +1$. Certain restrictions on the parameters α, λ, and β must obtain if Q_t is to describe a growth function. Solving the defining equation for Q_t as an explicit function of t gives

$$Q_t = \left[\frac{\alpha}{\alpha - \lambda \beta} + \gamma \lambda e^{(\beta\lambda - \alpha)t} \right]^{-1/\lambda}, \tag{10-7}$$

with γ a fourth parameter. Wiorkowski shows that $\gamma(\beta\lambda - \alpha) \leq 0$ and $\gamma \geq 0$ are necessary for Q_t to be a growth function and that

$$Q_\infty = \left[\frac{\alpha}{\alpha - \lambda \beta} \right]^{-1/\lambda} \tag{10-8}$$

is the function's asymptotic value as $t \to \infty$. In terms of Q_∞ the appropriate function to use is (10-9).

$$Q_t = Q_\infty \left[1 + \gamma \lambda Q_\infty^\lambda e^{-\alpha t Q_\infty^\lambda} \right]^{-1/\lambda} , \tag{10-9}$$

with $\alpha > 0$ and $-\infty < \lambda < +\infty$. This family of functions is very flexible; it is S-shaped, can be right or left skewed or symmetric, and has an inflection point at t^α such that $Q_t = Q_\infty [1 + \lambda]^{-1/\lambda}$ provided that $\lambda > -1$. For $\lambda < -1$ there is no inflection point.

Again, the advantage of using a function form that encompasses a variety of special cases with strikingly different properties is that it *may* be possible to discriminate among candidates for a growth function lying within the class: exponential, logistic, Gompertz, or some other. It is possible to enlarge the family of functions further by considering alternatives with defining equations different than those discussed thus far. Wiorkowski uses one additional three-parameter function, the Weibull distribution, whose defining equation[b] is of the form:

$$\frac{Q_t'}{Q_\infty - Q_t} = \frac{\alpha}{\beta} \left[\frac{t - \alpha}{\beta} \right]^{\gamma - 1} , \tag{10-10}$$

as a device for testing choice of the "best" member of the extended Richards family of functions against a structurally different alternative. He argues:

> if widely discrepant estimates of the ultimate recovery are obtained from fitting different families of curves, then there is strong evidence that the family of functions is dominating the data so that the resulting estimates are unusable. Conversely, if the estimates of the ultimate recovery are approximately the same, then one feels more confident that the data are dominating the models and that usable estimates of ultimate recovery are being provided.[21]

Using a nonlinear least-squares method, he fit the Weibull distribution and the extended Richards family to five data series: crude oil production, (derived) crude oil discovery, ultimate crude oil recovery adjusted for revisions, and associated and nonassociated natural gas. The most striking result of Wiorkowski's analysis is that both the extended Richards function and the Weibull distribution fit the data series for cumulative oil production equally well, *individual* residuals from the two choices being identical to two decimal places. Yet the former gives an estimate of ultimate production Q_∞, nearly twice as large as the latter: 446 BB versus 235 BB. He concludes that:

> both models fit the data equally well. The most reasonable conclusion is that the data are inadequate to be used for predicting ultimate production. This result seriously damages Hubbert's methods since he depends heavily on pro-

b. The solution of which is $Q_t = Q_\infty \left[1 - \exp\left\{ -\left(\frac{t - \alpha}{\beta} \right)^\gamma \right\} \right]$. $\tag{10-11}$

Figure 10-5. Analysis and Projection of Historic Patterns of U.S. Crude Oil Discovery and Recovery. (Note that recent discoveries are underestimated because of lack of data.)

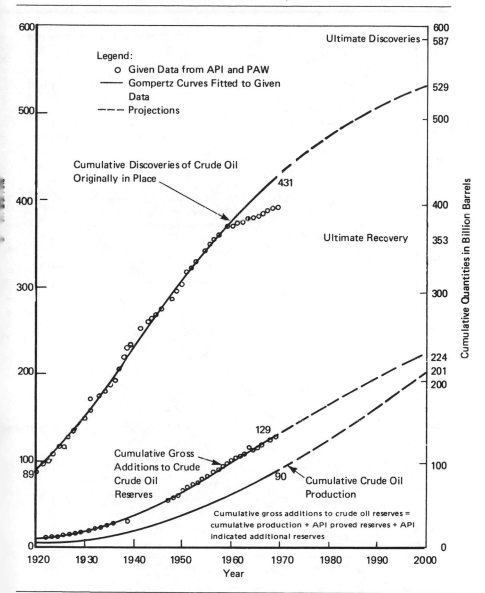

Figure 10-6.   Analysis and Projection of Historic Patterns of Discovery and Recovery of U.S. Net Natural Gas. (Note that recent discoveries are underestimated because of lack of data.)

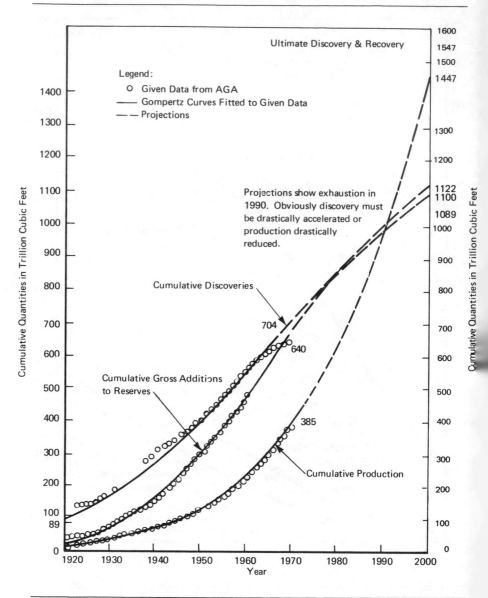

Source: Moore, Appendix F, p. 136, Figure F-2, titled "Analysis and Projection of Historical Patterns of Discovery and Recovery of U.S. Net Natural Gas." Reprinted with permission.

duction data.[c] Further, an approximate 95 percent confidence interval on the shape parameter (λ) doesn't include $\lambda = 1$, which corresponds to Hubbert's assumed logistic curve.[22]

In contrast to the production series, the discovery series and ultimate recovery series yield estimates for Q_∞ that are reasonably close for both the Weibull distribution and the extended Richards family (see Table 10-3). The figures in parentheses in the table are approximate 95 percent confidence intervals for Q_∞, computed under the assumption that the model for uncertain Q_t is

$$\widetilde{Q}_t = H(t, \theta) + \widetilde{\epsilon}_t, \qquad t = \tau, \qquad \tau + 1, \ldots, \tag{10-12}$$

and that the $\widetilde{\epsilon}_t$ are mutually independent and identically distributed normal random variables with mean zero and variance σ^2. In this form, the growth function chosen is H with parameters θ. The value of H at t is the expectation of $\widetilde{Q}_t$.

Wiorkowski pools his estimates for oil discovery and ultimate oil recovery and forecasts 155 BB for the former and 161 BB for the latter. Although these two estimates are close, it is peculiar that the forecast for ultimate oil discovery should be less than that for ultimate oil recovery or production. This may be due in part to differences in the lengths of the series used to estimate discoveries and production (see the discussion of time series) and adjustments for revisions.

Taking into account his forecast of extensions and revisions, he forecasts 7 BB of new oil to be found in the coterminous forty-eight states onshore:

Ultimate production	161 BB
	(157–165)
Cumulative production plus proved reserves as of 1977	–128
Additional future production	33
	(20–37)
Extensions and revisions	–26
New oil in new fields	7 BB

In yet another study of the same time series of production and discovery data to 1976, Mayer et al.[23] using up-to-date statistical methods, conclude that

- Hubbert's results are not greatly affected by consideration of alternative models of cumulative discovery or issues of statistical methodology.

- Hubbert's approach is weakened by the fact that direct analysis of cumulative production cannot support his estimates of ultimate production.

c. Hubbert in fact does not "depend heavily" on production data; but he does depend heavily on discovery data.

Table 10-3.   Weibull and Richards Function Parameter Estimates by Wiorkowski in 1979, Q_∞ , Billion Barrels.

	Crude Oil Discovery	Ultimate Oil Recovery	Crude Oil Production
Weibull	145.5	158.4	235
	(132–158)	(156.6–159.4)	(211.4–258.6)
Richards	155.6	164.6	445.6
	(152.2–159.8)	(161.7–166.3)	(369.6–522.4)

Source: Adopted from J.J. Wiorkowski, "Estimating Volumes of Remaining Fossil Fuel Resources: A Critical Review," in Journal of the American Statistical Association 76, no. 375 (September 1981), p. 546, Table 5.

REFERENCES

1. M. K. Hubbert, "Nuclear Energy, and the Fossil Fuels," in Drilling and Production Practice (Washington, D.C.: American Petroleum Institute, 1956), pp. 7–25, and "U.S. Energy Resources: A Review as of 1972," in A National Fuel and Energy Policy Study, Part I, Committee on Interior and Insular Affairs, U.S. Senate, Serial no. 93–40 1974.

2. W. E. Pratt, "The Impact of Peaceful Uses of Atomic Energy on the Petroleum Industry, in Peaceful Uses of Atomic Energy, Vol. 2, Joint Committee on Atomic Energy, U.S. Congress, 1956, pp. 89–105.

3. S. E. Pogue and K. E. Hill, Future Growth and Financial Requirements of the World Petroleum Industry, Petroleum Department, The Chase Manhattan Bank, New York, 1956.

4. M. King Hubbert, "U.S. Energy Resources, A Review as of 1972: A Background Paper," Serial no. 93–40 (92–75) Part 1 prepared at the request of H. M. Jackson, Chairman, Committee on Interior and Insular Affairs, U.S. Senate, (Washington, D.C.: U.S. Government Printing Office, 1974), p. 66.

5. Ibid., p. 53.

6. Ibid., p. 75.

7. M. King Hubbert, "Energy Resources, A Report to the Committee on Natural Resources," National Academy of Sciences, National Research Council Publication 1000–D, 1962, 141 pp.

8. Ibid.

9. Ibid.

10. Personal communication with M. K. Hubbert, November 1975.

11. U.S. Geological Survey, Oil and Gas Branch, Geologic Estimates of Undiscovered Oil and Natural Gas in the United States, 1975.

12. Quoted from p. 1713 of J. M. Ryan, "National Academy of Sciences Report on Energy Resources: Discussion of Limitations of Logistic Projections," AAPG Bulletin 49 (1965). See also J. M. Ryan, "Limitations of Statistical Methods for Predicting Petroleum and Natural Gas Reserves and Availability," Journal of Petroleum Technology, (1966): 281–86.

13.   H. Hotelling, "The Economics of Exhaustible Resources," *Journal of Political Economy* 39, no. 2 (1931): 137–75.

14.   Ryan, "National Academy Report on Energy Resources," p. 1713.

15.   Ryan, "Limitations of Statistical Methods."

16.   John J. Wiorkowski, "Estimating Volumes of Remaining Fossil Fuel Resources: A Critical Review," *Journal of the American Statistical Association* 76, no. 375 (September 1981).

17.   C. L. Moore, "Projections of U.S. Petroleum Supply to 1980," U.S. Department of the Interior, Office of Oil and Gas, Washington, D.C., 1966.

18.   B. Gompertz, "On the Nature of the Function Expressive of the Law of Human Mortality and on a New Mode of Determining the Value of Life Contingencies," *Philosophical Transactions of the Royal Society of London* 155: (1925): 513–85.

19.   Wiorkowski, "Estimating Volumes of Remaining Fossil Fuel Resources," p. 540.

20.   F. J. Richards, "A Flexible Growth Curve for Empirical Use," *Journal of Experimental Botany* 10 (1959): 290–300.

21.   Wiorkowski, "Estimating Volumes of Remaining Fossil Fuel Resources," p. 544.

22.   Ibid., p. 546.

23.   L. S. Mayer, R. Silverman, S. L. Zeger, A. G. Bruce, "Modeling the Rules of Domestic Crude Oil Discovery and Production," Resource Estimation and Validation Project, Department of Statistics and Geology, Princeton University, February 1979.

11 GEOLOGIC-VOLUMETRIC METHODS

Geologic-volumetric analysis is commonly used to appraise the mineral potential of large regions and until recently was a principal paradigm for arriving at point forecasts of undiscovered oil and gas in petroleum provinces. Among widely cited point forecasts of amounts of oil and gas remaining to be discovered in U.S. territories generated by this method are an early one, Day (1909), and the more recent ones by Duncan and McKelvey, Hendricks and Schweinfurth, Pratt, Weeks and Stebinger, Pratt, and the U.S. Geological Survey (USGS)[1]; see Table III-1. Most of these forecasts are of ultimately recoverable oil; some are accompanied by a forecast of oil in place.

A geologic-volumetric appraisal of a petroleum basin begins with an analysis of geological, geochemical, and geophysical data whose aim is to determine, first, the yield in barrels per unit area or volume of unexplored productive sediment in that basin and, second, the volume of productive sediment remaining to be explored. Key input data include the yield per unit area or volume from that portion of the basin already explored and the area or volume of sediments already explored. The geologic and volumetric components of this method are described by Rose as follows:

Geologic — the consistent application of experienced, professional exploration judgment to geological, geochemical, and geophysical data, so as to appraise those geological factors that control the occurrence and volume of oil and gas;

Volumetric — the determination of sedimentary volumes of basins or basin areas multiplied by some area-form or volume-form of petroleum-yield figures.[2]

113

In essence, this approach to forecasting undiscovered oil and gas is an "extra-polation of data on the abundance of mineral deposits from explored to unexplored ground as the basis of either the area or the volume of broadly favorable rocks." [3] Once a yield per unit area or volume for unexplored sediments and the volume of unexplored sediments are estimated, a point forecast of the volume of undiscovered mineral remaining is computable by multiplying these two numbers.

This is a deceptively simple description of a method that is in fact complex and sophisticated, with as many potential variations as there are geologists who employ it. On the surface it appears simple because the results of an elaborate chain of geological reasoning are packed into a simple algebraic model relating yield and volume.

Descriptive analogy plays a key role in reasoning about geologic variables that determine yield, and this heuristic is the hinge on which swings the appraisal of frontier provinces—provinces that are totally unexplored or very sparsely explored. Modern geological thinking is guided by two fundamental principles, both of which play an important part in the process of constructing geological analogies between mature and frontier petroleum and metallogenic provinces. They are the principles of *uniformitarianism* and of *uniqueness*. Before Hutton,

> deep river valleys were believed to be clefts in the rocks formed by earth-quakes; and mountain ranges were supposed to have arisen with tumultuous violence. To the uprooting of such fantastic beliefs came the Scottish geologist Hutton, whose *Theory of the Earth*, presented in 1785, marked a turning point in geological thought on this subject. Hutton argued that the *present is the key to the past* and that, given sufficient *time*, the processes now at work could have produced all the geologic features of the globe. [4]

Although subject to widely varying interpretations when applied to particular environments, this is an unchallenged tenet of modern geological reasoning.

A complementary postulate universally adopted in one form or another by modern geologists is that of *environmental uniqueness*. According to Nairn, this principle

> states that no environment (interpreted in the very widest sense) in geological history is ever exactly duplicated. . . . The advantage of a uniqueness principle is that it increases awareness of changes which may have occurred in the earth's history which being neglected can lead to uniformitarianism being questioned. [5]

Because of the uniqueness of spatially distinct geological environments, the construction of geological analogies for prediction of volumes of petroleum in place in frontier provinces is a delicate intellectual art far removed from an exercise in deductive logic. Pushed to its extreme, the principle of uniqueness would declare the use of analogy useless. In a discussion of the use of geological theory in prospecting for stratigraphic traps, Moore comments cogently on the exploitation of analogies:

> The geologist is forced to reason from his knowledge of analogous configurations and in fact a knowledge of process plays only a minor role in his predictions. If this use of analogs were not so familiar to us it might seem surprising that it worked. One might suppose that the variety of configurations produced in the course of the earth's evolution would be so vast that it would be hopeless to select the right analog. The principle of uniqueness enunciated by Nairn would seem to reinforce this supposition—especially if it be remembered that for economic success not only must the stratigraphic configuration be predicted correctly but also the geochemical and structural conditions.
>
> However, in practice it seems that sufficient regularities exist in nature to make prediction by analogy effective.[6]

Frontier provinces are by definition provinces for which the kind of data needed for accurate classification of its detailed geological characteristics is sparse or nonexistent. Drawing an analogy between a frontier province and well-explored prototypes is further complicated by the imperfect nature of the devices at the geologist's disposal for measuring physical properties of the geological environment. Yet surprisingly, the concept of geological analogy has worked amazingly well:

> The interaction of numerous processes simultaneously or in succession is what makes the geological phenomenon so difficult to interpret. The information we seek about one process has its signal drowned by the noise generated by countless other processes. Nearly all data from the real world are noisy.
>
> ... by super-imposing analogs from numerous prototypes having a common broad environment we may filter out what is adventitious and local and disclose what is fundamental and of general application. The product of such a second level of abstraction may be called a *template*, "an instrument used as a guide in bringing any piece of work to the desired shape"—the work being the construction of interpretative models ...
>
> When the noise due to local effects is removed, the remaining message appears at first disappointingly simple. But experience seems to show that these concepts have a power out of all proportion to their degree of elaboration.[7]

All petroleum geologists agree that in order for petroleum to be present in a prospective sedimentary unit within a basin:

- The basin must contain an adequate volume of reservoir rock with favorable reservoir properties.
- The reservoir rock must be situated so as to allow trapping of petroleum.
- The basin must also possess a favorable source for generation of petroleum.
- Migration paths from the source rock to the traps must be present.

- The timing of migration must be favorable for subsequent preservation in traps.

Despite agreement on these characteristics there is no unique taxonomy of descriptive geologic attributes useful for creating analogies with well-explored sedimentary units, in particular when the analogy is to be used for prediction. Jones asserts that assigning

> geologic parameters to a productivity prediction scheme is very difficult primarily because productivities of basins and basin segments can be distressingly different despite many geologic similarities, or they can be identical despite gross geologic differences.[8]

He illustrates by comparing properties of two highly productive basin segments, the Asmari Fold Belt in Southwest Iran and an East Texas field. Both possess large traps, a large volume of reservoir rock, favorable source, and a "very effective" migration system (see Table 11-1). No geological expertise is necessary to see the striking difference in character of geological variables for these two highly productive areas. The aim is to "look past large differences in the specifics of the geology and to determine what prerequisites of large oil accumulations are common to East Texas and Iran."[9]

The control sample for a volumetric analysis of an unexplored basin generally consists of one or more well-explored basins. Each variable that influences the

Table 11-1. Some Geologic Differences between Asmari Fold Belt of Southwest Iran and Area of East Texas Field.

Southwest Iran	East Texas Field
1. Located in an orogenic foredeep	1. Located on flank of broad regional uplift
2. Traps are huge compressional anticlines	2. Trap formed by slight bowing of gentle regional unconformity
3. Reservoir locally dips > 50°	3. Reservoir dips uniformly < 50 feet per mile
4. Large reservoir thickness ($\approx$ 1,000 feet or 300 meters)	4. Thin reservoir (100 feet, or 30 meters)
5. Oil column several thousand feet	5. Oil column $\leq$ 200 feet (60 meters)
6. Very low initial porosity	6. Very high initial porosity
7. Very high fracture permeability	7. No fracture permeability
8. No unconformities	8. Trap at intersection of two unconformities
9. Fifteen to twenty similar traps	9. Only one trap
10. Perfect evaporite seal	10. Chalk seal
11. Source rocks obvious	11. Source rocks not conspicuous
12. Short-distance vertical migration	12. Moderate to long-distance lateral migration

Source: R.W. Jones, "A Quantitative Geologic Approach to Prediction of Petroleum Resources," in *Methods of Estimating the Volumes of Undiscovered Oil and Gas Resources*, ed. J.D. Haun (Tulsa, Okla.: American Association of Petroleum Geologists, 1975).

amount of hydrocarbons in place or recoverable per cubic mile or per unit area is measured for basins in the control sample and, after modification in light of differences between basins in the control sample and the unexplored basin, these measurements are used to compute a forecast of the volume of hydrocarbons in place or recoverable in the latter. The particulars are best illustrated by example, so let us examine two approaches to volumetric analysis that are rather different in character.

A volumetric model is a function f whose argument is composed of variables that through f determine the total amount of hydrocarbons in place or recoverable in the basin or sedimentary unit being studied. Although the choice of variables and the particular form of this function varies from application to application, certain variables appear in all such models, and f is generally formulated as a product of these variables. For example, Jones[10] models reserves per cubic mile Y, as follows:

$$Y = f(R, T, S, M; \alpha) = \alpha \times R \times T \times S \times M \ , \qquad (11-1)$$

where α is an experience-based parameter (26 BB/cubic mile), R is the fraction of the basin that can contain producible petroleum, T is the fraction of R that is in trap position, and $S \times M$ is the fraction of trap capacity ($R \times T$) that contains petroleum; that is,

$$S \times M = \frac{\text{producible petroleum in traps}}{\text{trap capacity}} \ . \qquad (11-2)$$

Jones interprets $R \times T$ as "the fraction of rock volume which would hold producible petroleum if all traps were full, and . . . $S \times M$ defines how efficiently source and migration systems in given rock volume exploit available trap capacity."[11] The parameter α is a scale constant chosen so rescaled values of R and T for most basins lie in the interval $0-10$, rescaled values of $S \times M$ lie in the interval $0-100$, and reserves per cubic mile are measured in units of thousand barrels.

Jones highlights the wide variation of $R, T, S,$ and M across basins with a display of values of these variables for ten well-explored basins (see Table $11-2$). Just how different in gross composition basins can be is strikingly illustrated by comparing values of $R, T,$ and $S \times M$ for the Los Angeles Basin with those for the Dezful Embayment: Reserves per cubic mile are almost the same for both, but R and T differ between these basins by one order of magnitude.

Combining geologic analogy, a model for volumetric analysis, and using values of $R, T, S,$ and M for well-explored basins as a control sample, values of these variables for a frontier basin can be forecast, either as point forecasts or as subjectively assessed uncertain quantities, uncertainty being encapsuled in the form of expert judgment expressed in terms of probabilities (see Chapter 13).

A different approach to volumetric analysis is taken by Mallory.[12] Whereas Jones's approach is designed to be applied to frontier provinces, Mallory's aim is to study partially explored basins as well. This raises an additional difficulty.

Table 11–2. Use of 0–10 (Apx) Rating Scale in Analysis of Variations in Reserves per Cubic Mile.

Selected Basins	(1) R, Fraction of Basin	(2) T, Fraction of R	(3) R, Rating Scale	(4) T, Rating Scale	(5) S × M (% Full)	RTSM (Col. (3) × (4) × (5), Reserves/Cubic Mile/ Million Barrels)
Los Angeles	0.0073	0.036	7.3	9.3	40	2,700
Carrizo–Cuyama	0.0068	0.024	6.8	6.3	10	430
Salinas	0.0062	0.046	6.2	12.0	5	380
Big Horn	0.0042	0.009	4.2	2.3	10	100
Powder River	0.0039	0.003	3.9	0.8	10	30
Portion Powder River						
Cretaceous	0.0016	0.006	1.6	1.6	40	100
Michigan	0.0009	0.002	0.9	0.6	30	16
Paradox	0.0009	0.005	0.9	1.3	15	18
Williston (U.S.)	0.0008	0.0003	0.8	0.1	95	8
Iran (Dezful embayment)	0.0008	0.15	0.8	39.0	90	2,800

a. Constant a (26×10^6) in the factor model was proportioned among various factor values as follows: R was multiplied by 10^3, T was multiplied by 2.6×10^2, and $S \times M$ was multiplied by 10^2.

Generally the most favorable prospective sediments in a basin are drilled first, so the yield, measured in volume of petroleum per unit rock volume of *unex-*plored rock volume, is likely to decrease as the proportion of total prospective rock volume explored increases. How one models yield as a function of this proportion is both critical and subject to considerable controversy.[13]

Mallory's method is labeled ANOGRE, for accelerated national oil and gas resource evaluation. It is designed to provide an appraisal of undiscovered oil and gas in the lower 48 states.[14] U.S. petroleum regions are classified areally by geologic provinces, and stratigraphic units within each province are classified vertically (according to the classification built by the American Association of Petroleum Geologists' (AAPG) Committee on Statistics of Drilling.)

Reasoning by geological analogy, it is assumed that the amount of hydrocarbons found in the volume of rock already drilled within a stratigraphic unit is functionally related to the amount of hydrocarbons in the volume of rock within that unit not already drilled. (A key assumption is that the volume of rock condemned by a dry hole is a specific fixed number for a given stratigraphic unit.) Defining

$V_{drilled}$ = the volume of rock tested by development wells in known pools, and the volume of rock drilled and found barren;

$V_{potential}$ = the volume of rock that seems capable of producing but has not been drilled;

HC_{known} = volume of hydrocarbons discovered;

$HC_{unknown}$ = computed volume of hydrocarbons yet to be found;

the basic functional relation between the amount of $HC_{unknown}$ remaining to be discovered in a stratigraphic unit is of the form

$$HC_{unknown} = g(HC_{known}, V_{drilled}, V_{potential}), \qquad (11\text{-}3)$$

and it is assumed that the ratio $V_{drilled}/HC_{known}$ of volume of rock drilled and condemned to the amount of hydrocarbons discovered in this volume of rock equals the ratio of undrilled potential rock to the amount of undiscovered hydrocarbons $V_{potential}/HC_{unknown}$—adjusted by multiplication by a numerical factor of $f > 0$:

$$\frac{V_{drilled}}{HC_{known}} = \frac{V_{potential}}{HC_{unknown}} \times f , \qquad (11\text{-}4)$$

The number f is chosen to account for the empirical observation that exploratory drilling early in the exploration cycle tends to exhaust the richest sediments, so the yield per unit volume of unexplored rock will, on the average, decline as the unexplored volume of rock capable of producing decreases. Mallory proposes to choose f subjectively, suggesting that f usually lies between 0.5 and 1.0 and stating, "Qualitatively, 1.0 seems optimistic but not unreasonable; 0.5 seems conservative; less than 0.5 seems pessimistic."[15]

In March 1974 the USGS proffered revised estimates of crude oil, natural gas liquids, and natural gas. Before the Interior and Insular Affairs Committee of the U.S. Senate, V. McKelvey, the USGS Director, announced that the new estimates:

> are lower than those previously issued by the Survey, particularly for petroleum liquids which were based on studies made in the late 1960s. Since then, new geophysical data have been acquired, and while the estimates of the amount of oil that might be recovered in terrain as yet unexplored by the drill are highly speculative, the prospects have been judged somewhat less optimistic than those of previous estimates. The targets, however, for substantial discoveries are still large.[16]

In commenting about the latest downward adjustments in estimates, then Department of the Interior Secretary Rogers C.B. Morton said that "the prospects for discovery of petroleum on the Outer Continental Shelf are still bright, and fully warrant continued investigation and exploration of this great frontier."[17]

McKelvey emphasized that:

> In developing estimates of undiscovered resources, it should be kept in mind that we are trying to appraise the unknown, and this is particularly true of the Atlantic Shelf, where not a single hole has been drilled. The acquisition of additional information may well warrant a substantial modification of our estimates. Moreover, we have been unable as yet to assess the effect of the recent increases in price on reserves and resources. If prices remain high, it is almost certain that secondary and tertiary recovery of oil will increase, and along with production from previously uneconomic sources, will add to our recoverable reserves.[18]

The estimates, which include and extend those of the American Petroleum Institute and of the American Gas Association, show the following:

- Cumulative U.S. crude oil and natural gas liquids production through 1972 amounted to 115.27 billion barrels (BB) of oil and 437.72 trillion cubic feet (TCF) of gas.

- Measured reserves of crude oil and natural gas liquids are 48.3 BB, and measured reserves of gas are 266.1 TCF.

- Indicated and inferred reserves are estimated to be in the range of 25–40 of crude oil and natural gas liquids, and 130–250 TCF of gas.

- Undiscovered recoverable resources are estimated to be in a range of 200 to 400 BB of crude oil and natural gas liquids, and 1,000 to 2,000 TCF of gas.[19]

In spite of a downward revision in these USGS estimates from those made in the late 1960s, the new USGS estimate of the amount of undiscovered oil and natural gas liquids remaining in U.S. provinces (200–400 BB) was nearly an order of magnitude larger than that being cited by major oil companies. It was derived by application of the ANOGRE geologic-volumetric method with the factor f chosen to range from 0.5 to 1.0.

The large difference between government and industry forecasts of undiscovered recoverable petroleum in U.S. territories prompted John Moody, then a vice president of Mobil Oil Corporation, to question the government forecast. M. King Hubbert gives a cogent account of a conference convened to identify the source of the discrepancy:

On June 5, 1974, a conference was held at the offices of the National Research Council under the auspices of the National Research Council Committee on Mineral Resources and the Environment for the purpose of trying to resolve the very large differences between recent estimates of the magnitudes of oil and gas resources of the United States.

Specifically, in his letter of April 8, 1974, addressed to V. E. McKelvey, Director of the U.S. Geological Survey, John Moody of Mobil Oil Corporation pointed out that the Geological Survey estimates given in the News Release of March 26, 1974, were very much larger—in some cases by an order of magnitude—than the corresponding estimates by Mobil. The purpose of this conference was principally to try to discover the sources of such large discrepancies in petroleum estimates.

In his reply of May 20, 1974, to Moody's letter, McKelvey briefly summarized the premises of the Geological Survey method of estimation. In this letter the following statement identifies what appears to be a major source of the discrepancies under consideration:

In essence, the [Survey] estimates were made by an analysis of each potential province and stratigraphic unit, utilizing the basic assumption that, volume for volume, unexplored favorable rock contains half as much to as much petroleum as explored rock, with the result modified by other factors described in the enclosed.

The enclosure referred to was a "Synopsis of Procedure" by W. W. Mallory of the USGS in Denver, outlining the procedure used by the Survey in its estimates. In the main text of this outline the ratio referred to in McKelvey's letter is not explicitly discussed, but in a footnote on page 1 the following statement occurs:

Few would contest the assumption as stated, but the proper fractional relation between the two sides of the equation is open to conjecture. The fraction can range from one (or greater) to zero. Precedents exist for both 1.0 and 0.5. Qualitatively, 1.0 seems optimistic but not unreasonable; 0.5 seems conservative; less than 0.5 seems pessimistic.

In the discussion of Mallory's presentation during the meeting, the validity of this range of between 0.5 and 1.0 was seriously questioned by several conferees. One oil company conferee suggested that a better figure might be between 0.05 and 0.1.

It was admitted by Mallory that the factor had been arrived at by subjective judgment. Its use was justified by the precedent referred to, namely that 1.0 had been used in a prior Survey analysis by A. D. Zapp (1962) and later 0.5 by T. A. Hendricks (1965).

In the discussion, I pointed out that there was no necessity for leaving the choice of a factor of such fundamental importance to subjective judgment when a much more precise figure can be obtained from the existing data on drilling and discovery. It had been my intention to show how this could be done during my own presentation later in the day. However, so little time was left when I was invited to review my methods that this important question was inadvertently overlooked.[20]

Hubbert challenges the assumption that, given the history of U.S. exploration up to 1965, the appropriate value of f for the United States considered as one enormous petroleum-producing region should lie in the range 0.5 to 1.0, asserting that f should be approximately 0.1. The corresponding estimate of undiscovered petroleum decreases four to eight times and hence is compatible with industry forecasts. In place of subjective judgment Hubbert uses a rate-of-effort model to forecast the ultimate amount of oil discoverable by unlimited drilling (see the discussion of rate-of-effort methods of forecasting) as 167 BB. As of 1965, 136 BB had been discovered ($HC_{known} = 36 \times 10^9$) by 1.5 billion feet of drilling, leaving an estimates 32 BB remaining to be discovered ($HC_{unknown} = 32 \times 10^9$). Earlier studies of the depth (cumulative feet) of exploratory drilling necessary to explore all possible petroleum producing areas of the coterminous United States with a density of one well for each 2 square miles[21] had established that 5 billion feet of drilling would be required. As of 1965, an increment of 3.5 billion feet would be required. If average well density is one well in each 2 square miles, a well, 2,640 feet deep will on the average test 1 cubic mile of rock. That is, the volume of rock, measured in cubic miles, tested by h feet of drilling is $V = h/2,640$. Reasoning in this fashion, Hubbert forecasts that the 3.5 billion feet of added drilling necessary to find 32 BB of remaining undiscovered oil will test a volume $V_{potential}$ of 1.33 million cubic miles of rock. The 1.5 billion feet drilled up to 1965 tested a volume $V_{explored}$ of 0.57 billion cubic miles.

Returning to Mallory's formulation of the relation between HC_{known}, $HC_{unknown}$, $V_{explored}$, $V_{potential}$, and f:

$$f = \frac{HC_{unknown}/V_{potential}}{HC_{known}/V_{explored}}$$

$$= \frac{32 \times 10^9/1.33 \times 10^6}{136 \times 10^9/0.57 \times 10^6} = 0.10 \qquad (11-5)$$

This is how Hubbert concludes that an f of 0.1 is more reasonable than an f between 0.5 and 1.0. He recognizes that this is an f "averaged" in some fashion across widely disparate petroleum provinces and points out:

> It should be kept in mind that this is a ratio of *average* future-to-past discoveries per cubic mile and that it applies only to the future and past discoveries as of 1965. For an earlier or a later date a somewhat different ratio would be expectable.[22]

Hubbert calculates an f of 0.08 based on cumulative feet drilled and amounts discovered as of 1945. He observes finally:

> From this analysis it is evident that the magnitude of the future-to-past discovery ratio R can be determined within narrow limits from the drilling and discovery record in any region that has reached a mature state of development. For the conterminous United States and adjacent continental shelves, as of 1965, the magnitude of this ratio was very close to 0.10. Were this measured value of the ratio R to be used in the Geological Survey estimates instead of the subjectively estimated values of between 0.5 and 1.0 cited in McKelvey's letter of May 20, 1974, and also given by Mallory in his outline, the principal discrepancies between the Survey estimates and those obtained by Mobil and by Jodry would largely be eliminated.[23]

A change in well density or in the amount remaining to be discovered HC_{known} will lead as well to a different f. The assumption that the ultimate amount Q_∞ of oil discoverable by unlimited drilling is 167 BB is critical to Hubbert's numerical analysis since he works backward from it to an estimate of f. Although Hubbert uses a forecast of Q_∞ as a device to compute a narrow range of values for f, the aim of Mallory's method is to choose f judiciously so as to produce a forecast of Q_∞. Short of accepting a particular value for Q_∞ directly, Mallory's method requires a value for f. The lively controversy about choice of a value for f given particular values of the variables $V_{drilled}$ and HC_{known} thus revolves about properties of f regarded as a function of these two variables.

In terms of quantities considered directly observable as exploration progresses, $x = V_{drilled} / V_{potential}$ and $s = HC_{known}$, and the unknown parameter, $S = HC_{known} + HC_{unknown}$, the function f can be written as:

$$x \left(\frac{S}{s} - 1 \right) = f(s, x; S) \ , \tag{11-6}$$

a function with domain $x \geq 0$, $0 \leq s \leq S$ and range $0 \leq f \leq \infty$. This equation *defines* f so that as x and s change, assignments of values to S must conform to it and cannot be made arbitrarily. Since x and s are observables, assignment of a value to S determines f, and conversely. Written in this fashion, the defining equation for Mallory's model shows clearly that the model is incomplete in a formal sense: A forecast of $S - s = HC_{unknown}$ cannot be made from observation of s and x alone; a value for f must be assigned using an exogenous source of information.

REFERENCES

1. T.A. Bryant, "Policy Implications of Oil and Natural Gas Reserve Estimates," M.S. thesis, Massachusetts Institute of Technology, Cambridge, Mass. 1976 provides reference sources for forecasts appearing in Tables III-1 and III-2.

2. P. Rose, "Procedures for Assessing U.S. Petroleum Resources and Utilization of Results," in *First IIASA Conference on Energy Resources*, ed., M. Grenon (Luxenburg, Austria: IIASA 1975), pp. 229–48.

3. V. E. McKelvey, "Mineral Resource Estimates and Public Policy," *American Scientist* 60 (1972): 37.

4. C. O. Dunbar, *Historical Geology* (Wiley: New York, 1948).

5. A. E. M. Nairn, "Uniformitarianism and Environment," *Paleogeography, Paleoclimatology, Paleoecology* 1 (1965): 10–11.

6. Peter F. Moore, "The Use of Geological Models in Prospecting for Stratigraphic Traps," unpublished manuscript, 1974, p. 4.

7. Ibid.

8. Quoted from p. 189 of R. W. Jones, "A Quantitative Geologic Approach to Prediction of Petroleum Resources," in *Methods of Estimating the Volume of Undiscovered Oil and Gas Resources*, ed. J. D. Haun (Tulsa, Okla.: American Association of Petroleum Geologists, 1975), pp. 186–95.

9. Ibid., p. 189.

10. Ibid.

11. Ibid., p. 190.

12. P. F. Mallory, "Accelerated National Gas Resource Appraisal (ANOGRE)," in *Methods of Estimating the Volume of Undiscovered Oil and Gas Resources*, ed. J. D. Haun (Tulsa, Okla.: American Association of Petroleum Geologists, 1975), pp. 23–30.

13. See M. King Hubbert's discussion in "Ratio between Recoverable Oil per Unit Volume of Sediments for Future Exploratory Drilling to That of the Past in the Coterminous United States," Appendix to Section II of Report of Panel on Estimation of Mineral Reserves and Resources, *Mineral Resources and the Environment* (Washington, D.C.: National Research Council, February 1975) and discussion on pp. 1 and 2 in *Methods of Estimating the Volume of Undiscovered Oil and Gas Resources*, ed. J. Haun (Tulsa: American Association of Petroleum Geologist, 1975).

14. "Accelerated National Oil and Gas Resource Appraisal" (ANOGRE), pp. 23–30 in *Methods of Estimating the Volume of Undiscovered Oil and Gas*, ed. J. Haun (Tulsa, Okla.: American Association of Petroleum Geologists, 1975).

15. Ibid.
16. U.S. Geological Survey News Release, March 24, 1974. Cf. Appendix, pp. 263–264 of *U.S. Energy Resources, A Review as of 1972*, by M. King Hubbert.
17. Ibid., p. 263.
18. Ibid., p. 264.
19. Ibid., p. 264.
20. M. King Hubbert, "Recoverable Oil per Unit Volume of Sediments," unpublished memo.
21. A. D. Zapp, "Future Petroleum Producing Capacity of the United States," U.S. Geological Survey Bulletin 1142–H (Washington, D.C.: U.S. Government Printing Office, 1962). T.A. Hendricks, "Resources of Oil, Gas, and Natural Gas Liquids in the United States and in the World," U.S. Geological Survey Circular 522 (Washington, D.C.: U.S. Government Printing Office, 1965).
22. Hubbert, "Recoverable Oil Per Unit Volume of Sediments."
23. Ibid.

12 RATE-OF-EFFORT MODELS

Rate-of-effort models are only a short step away intellectually from life-cycle models: Increments of additions to total amount of hydrocarbons discovered, to production, or to reserves are regarded as a function of cumulative exploratory effort rather than of time. Exploratory effort is generally measured in number of wildcat or exploratory wells drilled.

Arps and Roberts' study[1] of Cretaceous oil production of the east flank of the Denver–Julesburg basin is one of the earliest attempts to use a rate-of-effort-like model to project future discoveries in a petroleum province. Consider a province that contains $N(A)$ fields of size (areal extent) A, and suppose that $N(A, W)$ fields have been found by the first W wildcat wells. They postulate that the number of fields $\Delta N(A, W)$ found by the next increment ΔW of wells drilled is proportional to the total area $A[N(A) - N(A, W)]$ of fields of size A remaining to be discovered:

$$\frac{\Delta N(A, W)}{\Delta W} \propto A[N(A) - N(A, W)] \quad . \qquad (12\text{-}1)$$

This postulate, together with appropriate initial conditions, leads to:

$$N(A, W) = N(A)[1 - e^{-cAW}] \quad ;$$

c is a constant, so the number of discoveries $N(A, W)$ of size A from W wells is an exponential function of W. Strictly speaking, this model is more than a synthetic rate-of-effort model, as it is derived from a specific assumption about the interaction between field size A, number of wells drilled, and number of fields of size A. That is, it characterizes the discovery process in terms of numbers and sizes of fields, and in this respect must be distinguished from rate-of-effort models that do not incorporate such features. A further distinction is that this model and its successors are designed to portray the evolution of discovery in a single province, whereas other rate-of-effort models are designed to treat aggregates of provinces, such as the coterminous United States onshore. Consequently the Arps-Roberts model can be viewed as a *discovery process model* as well (see Chapter 15).

The statement of Arps and Roberts' basic postulate serves as a foil for comparison with assumptions that underlie more synthetic models designed to study discoveries and additions to reserves for the United States as a whole. These "macromodels" are synthetic in that their structure is not a logical derivative of assumptions about physical features of the exploration process but rather are redolent of more traditional economic time series models. Perhaps the simplest is that of Zapp.

In 1962 Zapp[2] asserted that a well density of one well for each 2 square miles drilled to 20,000 feet or to basement rock, whichever is shallower, is required to test all potentially productive onshore and offshore U.S. regions. Exploration drilling at this density requires 5 billion feet of drilling according to his calculation. As reported by Hubbert,[3] Zapp estimated in 1959 that by 1958 approximately 0.98 billion feet of drilling had taken place and that this footage would increase to 1.1 billion feet by the end of 1960. Cumulative production through 1960 was 65 billion barrels (BB), proved reserves were 32 BB, and, according to Zapp, there were 33 BB additional in discovered fields, totaling 130 BB. Hubbert[4] cites Zapp as assuming 170 BB discoverable under then current economic and technological conditions by the next 2 billion feet of exploratory drilling and then adding 290 BB he classified as "submarginal resources." The grand total for all 5 billion feet of exploratory drilling is 590 BB, a figure that at the time represented an "official" estimate by the U.S. Geological Survey.

As Hubbert[5] points out, Zapp's projections implied that the average rate of discovery per foot drilled by the next 4 billion feet drilled subsequent to 1960 would be nearly the same as the rate of discovery per foot for the first billion feet drilled. In Hubbert's words:

> it becomes obvious that each category of Zapp's estimate implies an average rate of discovery per foot. The first block of 130 billion barrels discovered by 1.1 billion feet of drilling would be at an average discovery rate of 118 bbl/ft. The second block would be at a rate of about 90 bbl/ft., and the third, at the highest rate of all, of 145 bbl/ft. The average rate for the total discoveries of 590 billion barrels with 5 billion feet of drilling would be 118 bbl/ft., the same as for the 130 billion barrels already discovered.
>
> This shows clearly the nature of the hypothesis upon which this unprecedentedly large estimate was based. Zapp assumed that the average discoveries per foot of exploratory drilling during the future 4 billion feet of exploratory

drilling would be approximately the same as that for the first 1 billion feet. For simplicity, this hereafter will be referred to as the Zapp hypothesis.[6]

The hypothesis underlying Hubbert's analysis of discovery rates is that the average rate of discovery per foot of drilling declines monotonically with increasing cumulative footage drilled, and he gives an exhaustive account of why, in his opinion, empirical evidence dictates rejection of Zapp's hypothesis.[7]

As an alternative to his life-cycle model of discovery and production, Hubbert estimated Q_∞ for the coterminous forty-eight states using an exponential model for cumulative amount discovered Q_h on a scale of cumulative feet h of exploratory drilling, asserting that such a model has two advantages over one built on a scale of time:

> In the latter case, there is no readily assignable upper limit to the time t during which exploration and discovery may be continued. Also, the rate of discovery with respect to time, dQ/dt, is subject to wide fluctuations in response to extraneous conditions such as economic or political influences. In fact, the rate of discovery with respect to time may be increased to a maximum or shut down completely in response to managerial or political fiat, or to the changes in the economic climate.
>
> Such is not the case for the curve of dQ/dh as a function of cumulative depth of exploratory drilling. In the first place, a practical limit can be set to the density of exploratory wells in any given region, and hence to the cumulative depth of exploratory drilling. Second, the amount of oil discovered per unit of depth of exploratory drilling is almost exclusively a technological variable and is highly insensitive to economic or political influences. For example, the officials of a large oil company may authorize its staff to double the amount of exploratory drilling in any given year and consequently to increase the discoveries per year; no oil company management, however, can successfully order its staff to double the quantity of oil to be found per foot of exploratory drilling.[8]

The data to which he fit parameters α and β of

$$Q_h = \alpha e^{-\beta h} \ , \qquad \alpha, \beta > 0 \ , \tag{12-2}$$

are displayed in Figure 12-1 adapted from Hubbert. The lower shaded area of each column represents the sum of cumulative discoveries plus proved reserves; the upper shaded area is an adjustment for the growth of reserves as a function of elapsed time from date of discovery (see the discussion of revisions of reported reserves in Part II).

The rate of discovery per foot drilled fluctuates rather widely over the first 0.6 billion feet drilled. As h increases past 0.4 billion feet a precipitous decline in the rate takes place. Hubbert's calculations show that two-thirds of the 143 BB found by 1971 were discovered by the first increment of 0.4 billion feet drilled (an average of 235 barrels/foot) while one-third was discovered by the

Figure 12-1.   Average Discoveries of Crude Oil per Foot for Each
10^8 Feet of Exploratory Drilling in the Conterminous United States
from 1860 to 1971.

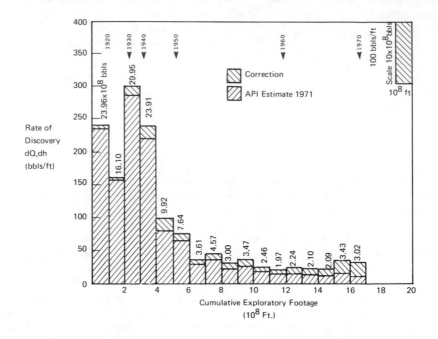

Source: M. King Hubbert, "U.S. Energy Resources: A Review as of 1972," in *A National Fuel and Energy Policy Study; Part I*, Serial no. 93–40, Committee on Internal and Insular Affairs, U.S. Senate, 1974, p. 120, Figure 49.

next increment of 1.3 billion feet (an average of 38 barrels/foot). To Hubbert, "It is obvious that the curve of discoveries per foot is roughly a negative exponential decline curve." However, the apparent increase in rate over the first 0.4 billion feet of drilling suggests that there may be alternative models that will fit the historical data considerably better, the topic following discussion here of fitting the model[9] to the data.

If $dQ_h/dh = Q'_h = \alpha \exp\{-\beta h\}$ is the rate of discovery per foot of drilling, then the amount Q_{h_0} discovered by h_0 cumulative feet drilled is $Q_{h_0} = (\alpha/\beta)[1 - \exp\{-\beta h_0\}]$ and the ratio $Q'_0/Q'_{h_0} = \exp\{-\beta h_0\}$. Hubbert estimates α and β by choosing values of these parameters to match the rate of discovery Q'_h of 30.2 barrels/foot and the actual amount discovered $Q_h = 143$ BB at $h = 1.7$ billion feet. The result is $\alpha = 182$ barrels/foot and $\beta = .106 \times 10^{-8}$; that is, the rate declines by 10.6 percent for each 10^8 feet drilled. For these values of α and β, $Q_\infty = \alpha/\beta = 172$ billion barrels, leaving $172 - 143 = 29$ BB remaining to be found as of 1971, projections that are in close agreement with those derived by Hubbert using a life-cycle model. Figure 12-1 from Hubbert displays the function $\alpha \exp \{-\beta h\}$ with α and β as estimated by Hubbert against the data.

In an introduction to their study of discovery rates, reserve additions, and production using a somewhat different model, Bromberg and Hartigan comment on Hubbert's analysis:

> The discovery rates are far from exponential in decline and the fitting procedure (area and most recent value fitted exactly) is quite arbitrary; a least-squares fit gives a substantially smaller estimate of recoverable oil. It should be remembered that the most recent rates are poorly known due to the large correction factors. It seems quite questionable to associate oil reserves allocated to a given year of discovery with exploratory drilling in that year; the deep-pool and wildcat drilling of a given year is plausibly associated with oil found in that year; but the extension drilling of a given year will extend the fields discovered in previous years, and extension drilling has been much more important in increasing oil reserves. Finally, in recent years, most important increments in reserves have been due to changes in recovery factors, not due to drilling.
>
> Despite our criticism of some of the details of Hubbert's [and Pelto's approach], we believe that approaches of this type for carefully selected data series will be fruitful for projecting reserves. We have used similar techniques on different series, and we also have had to ignore rather large departures from simple models.[10]

Wiorkowski comments in a similar vein:

> The problem with his approach is immediately evident. $Q'_h = \alpha \exp \{-\beta h\}$ obviously does not fit the data. In the cumulative footage range $0 \leq j \leq .5$, [it] grossly underestimates the data and misses totally a discovery peak at $h \simeq .4$. In the range $.5 \leq h \leq 1.5$, it systematically overestimates the observed data. Similar problems exist for the gas data.[11]

Data series for discoveries, for additions to reserves from extension well drilling, from revisions of earlier reserve projections for discovered fields, and for production exhibit considerable fluctuation. It is surprising that only recently have these series been treated as statistical series, that is, explained by models that explicitly characterize the nature of fluctuations about a trend by postulating a probability law for them.

BROMBERG AND HARTIGAN MODEL

In a 1975 report to the Federal Energy Administration, Bromberg and Hartigan project future additions to reserves from new discoveries, extensions, and revisions using such a model.[12] An innovative feature of their study is the assignment of a subjective probability distribution to parameters of the model that are not known with certainty. Expert opinion can be introduced into the analysis in a logically coherent fashion in the form of this "prior to the data" distribution, although they use one that makes ex post parameter estimation correspond to an ordinary least-squares procedure.

Like Hubbert's model, theirs projects an exponential decline in additions to reserves from extensions, from revisions, and from discoveries per unit of effort:

$$\frac{dR_t}{dE_t} = \beta e^{-\alpha E_t} , \tag{12-3}$$

where E_t is cumulative effort up to time t; R_t is cumulative reserves found by time t; and α and β are fixed parameters. If this model were to hold exactly, reserves remaining to be found at time t are

$$R_\infty - R_t = (\beta/\alpha)\left[e^{-\alpha E_t} - e^{-\alpha E_\infty}\right] . \tag{12-4}$$

Recognizing that (12-3) does not hold exactly when applied to these data series and that α, β, and E_∞ are not known with certainty, Bromberg and Hartigan propose a model for discrete increments $\triangle R_i = R_i - R_{i-1}, i = 1, 2, \ldots, n$ of additions to reserves as a function of corresponding discrete increments $\triangle E_i = E_i - E_{i-1}$ of effort: $\triangle \tilde{R}_i$, $i = 1, 2, \ldots$, are uncertain quantities related to $\triangle E_i$, $i = 1, 2, \ldots, n$ by

$$\log\left(\triangle \tilde{R}_i / \triangle E_i\right) = \log \beta - \alpha E_i + \sigma \tilde{Z}_i , \tag{12-5}$$

where $\tilde{Z}_i = 1, 2 \ldots, n$ are mutually independent normal random variables, each with mean zero and variance one.

In other words, $\triangle \tilde{R}_i / \triangle E_i$ is a random proportion of $\beta \exp\left\{-\alpha E_i\right\}$ with expectation $\beta \exp\left\{-\alpha E_i + (1/2)\sigma^2\right\} = M(E_i)$ and variance $M^2(E_i)\left[\exp\left\{\sigma^2\right\} - 1\right]$. Since ultimate reserves

$$\tilde{R}_\infty = \tilde{R}_n + \sum_{i=n+1}^{\infty} \beta e^{-\alpha E_i + \sigma \tilde{Z}_i} \triangle E_i , \tag{12-6}$$

given $\alpha, \beta, \tilde{R}_n = R_n$ and $E_{n+1}, \ldots, E_\infty, \tilde{R}_\infty$ has expectation

$$E(\tilde{R}_\infty | \alpha, \beta; R_n) = R_n + \sum_{i=n+1}^{\infty} M(E_i)\triangle E_i = R_n + \beta e^{\frac{1}{2}\sigma^2} \sum_{i=n+1}^{\infty} e^{-\alpha E_i} \triangle E_i \tag{12-7}$$

and variance

$$\text{Var}\,(\widetilde{R}_\infty | \alpha, \beta; R_n) = \sum_{i=n+1}^{\infty} \text{Var}\,(\Delta \widetilde{R}_i)(\Delta E_i)$$

$$= \beta^2 e^{\sigma^2} [e^{\sigma^2} - 1] \sum_{i=n+1}^{\infty} e^{-2\alpha E_i}(\Delta E_i)^2 \quad . \tag{12-8}$$

Bromberg and Hartigan assert that for large n and α positive, $\widetilde{R}_\infty$ is approximately normal.[a]

In order to project $\widetilde{R}, E_{n+1}, \ldots, E_\infty$ are assumed known and for $i > n$, they set

$$\Delta E_i = \frac{1}{n} \sum_{\varrho=1}^{n} \Delta E_\varrho = \delta \quad . \tag{12-9}$$

If δ is small, then given $\widetilde{R}_n = R_n$,

$$E(\widetilde{R}_\infty) \cong R_n + (\beta/\alpha)\,[e^{-\alpha E_n} - e^{-\alpha E_\infty}] \quad , \tag{12-10}$$

and

$$\text{Var}\,(\widetilde{R}_\infty) \cong \delta\,(\beta^2/2\alpha)e^{\sigma^2}\,[e^{\sigma^2} - 1]\,[e^{-2\alpha E_n} - e^{-2\alpha E_\infty}] \quad , \tag{12-11}$$

so that given an amount R_n discovered by effort E_n, both the mean and variance of the amount $\widetilde{R}_\infty - R_n$ remaining to be discovered declines approximately exponentially with total effort E_n expended by time period n for fixed values of α, β, and $E_{n+1}, \ldots, E_\infty$.

As mentioned earlier, Bromberg and Hartigan's study differs from earlier applications of exponential rate-of-effort models in two ways. It has an explicit characterization of random fluctuations about a trend and it introduces a probability distribution for α, β, and σ^2 encapsulating a priori judgments about these parameters. With this model, the relevant probability distribution for projecting the uncertain quantity $\widetilde{R}_\infty$, given that at time period n only cumulative effort E_n and cumulative reserves R_n are known with certainty, is the *predictive* distribution for $\widetilde{R}_\infty$: the distribution of $\widetilde{R}_\infty$ given $R_n, E_1, \ldots, E_n$ but unconditional on the uncertain quantities $\widetilde{\alpha}, \widetilde{\beta}$, and $\widetilde{\sigma}^2$. Although in principle this distribution can be computed directly, it is a messy computation and they "Monte Carlo" it. The first step in their procedure is the following:

1. Assign a uniform prior density to $\widetilde{\alpha}$, $\log \widetilde{\beta}$, and $\log \widetilde{\sigma}$: the density is proportional to $d\alpha\,(d\beta/\beta)\,(d\sigma/\sigma)$ with $\sigma > 0$ but α and β unrestricted in sign.

---

a. In fact, the distribution of $\widetilde{R}_\infty$ for large n depends on the relative magnitudes of σ, αE_i and ΔE_i for $i = n+1, \ldots, \infty$. The (asymptotic) distribution of $\widetilde{R}_\infty$ may be far from normal and under certain conditions may possess a fat right tail. This behavior is apparently exhibited in the simulation of $\widetilde{R}_\infty$ conducted by the Bramberg and Hartigan, although the cause is obscured by the "mixing" effect of the posterior distribution for parameters α, β and σ^2. Unnormalized sums of lognormal random variables have been studied by Barouch and Kaufman.[13]

The justification for this diffuse prior density, one that carries little a priori information about $\tilde{\alpha}$, $\tilde{\beta}$, and $\tilde{\sigma}$—is that subsequent to observing $(\Delta R_i, \Delta E_i)$, $i = 1, 2, \ldots, n$, the distribution of $\tilde{\alpha}$, $\tilde{\beta}$, and $\tilde{\sigma}$ is essentially determined by the data, and the parameters of this distribution can be computed by ordinary least-squares procedures. The posterior distribution that results is bivariate normal for $\tilde{\alpha}$ and $\log \tilde{\beta}$ given $\tilde{\sigma} = \sigma$, and $1/\tilde{\sigma}^2$ is gamma. The remaining steps of Bromberg and Hartigan are

2. Project a sequence of future values of effort $E_{n+1}, \ldots, E_\infty$.

3. Generate a random sample of values of $\tilde{\alpha}$, $\tilde{\beta}$, and $\tilde{\sigma}$ from their joint (posterior) distribution.

4. Given $\tilde{\alpha} = \alpha$, $\tilde{\beta} = \beta$, and $\tilde{\sigma} = \sigma$ from step (3), generate a random sample of values of $Z_{n+1}, \ldots Z_N$, N chosen so that the contribution to $\tilde{R}_\infty$ from terms $N + 1, N + 2, \ldots$ is negligible.

5. Compute a value $R_\infty = R_n + \sum_{i=n+1}^{N} \beta \exp\left\{-\alpha E_i + \sigma Z_i\right\} \Delta E_i.$

$$(12\text{-}12)$$

6. Repeat steps (3)–(5) until the relative frequency distribution of R_∞ values so generated is judged to be an accurate representation of the probability distribution for $\tilde{R}_\infty$.

The predictive distribution of $\tilde{R}_\infty$ so generated is Bromberg and Hartigan's projection or forecast of R_∞.

The model is applied to historical data series for the coterminous United States onshore and offshore, including the Gulf of Mexico but excluding the Atlantic Shelf, Pacific offshore, and Alaska. The exploration and production history of these regions is short by comparison and $\tilde{R}_\infty$ for each of them is better appraised by use of expert opinion (see Subjective Probability Methods in Chapter 13).

Bromberg and Hartigan's original intention was to explain additions to reserves using cumulative drilling activity as a measure of effort. They abandoned this approach after observing that:

> A plot of yearly reserves added per well drilled versus cumulative wells drilled demonstrates a surprising upturn at about the mid 1950's . . . simultaneously, drilling activity has steadily declined since that time. . . . Thus, it is apparent that in recent times drilling effort alone cannot explain the trend in additions to reserves.[14]

Noting that as of 1973 revisions accounted for approximately 72 percent of additions to reserves of 3.15 BB and extensions for about 18 percent and that additions from revisions have been increasing in the last twenty-five years while additions from new oil exhibit a decline, they chose to model additions from revisions, extensions, new pools, and new fields separately. Data series for recovery factors are particularly shaky and compound the difficulties in separating additions due to changes in recovery factors from revisions due to a change in original oil-in-place estimates as field information from development and extension drilling accrues. They say,

> In the absence of a sufficiently long series of recovery factors, it was decided not to break the revisions into the two components for predictive purposes. The projection of future recovery factors would in any case be a risky business. Nevertheless, the evidence is compelling that the principal component in reserve additions is due to increases of recovery factors.[15]

Their investigation of reserve additions from revisions, extensions, and discoveries led them to conclude that measuring effort in cumulative feet drilled or in cumulative number of wells drilled led to essentially similar relations between effort and reserve additions. Although cumulative additions to reserves exhibit proportionately mild fluctuations, within each type of reserve addition, rates of change of additions as a function of drilling effort display significant fluctuations and behave quite differently. Figures 12-2 through 12-4, from their study, document Bromberg and Hartigan's conclusion that

> The exponential fit to all three curves is quite poor and could not be used for extrapolation. If these rates were to be extrapolated, the most plausible rule is to use the last 10 years data, and to project a constant rate, in million barrels per well,

Extensions:	.8	95 percent interval	(.6, 1)
New pool:	.13		(.09, .18)
Wildcat:	.04		(.025, .06)

> In view of the wide range of these discovery rates, it would be necessary to consider the different types of discovery separately; to predict future discoveries it would be necessary also to predict the total future amounts of drilling in the various classes. This type of prediction was used by the National Petroleum Council (1973), but their prediction period was only the next fourteen years. It is risky indeed to extrapolate the "constant" discovery rate much beyond this period of time; for example, a great burst of drilling activity like the 1950's peak should be expected to depress the discovery rates again.
>
> The more drilling in a given year, the more oil discovered, and also the more wells drilled cumulatively, the less oil found per well. But the fluctuations of the various discovery rate series prevent exponential projection, and make necessary instead projection of a shaky constant discovery rate and very speculative estimates of the amounts of drilling in various categories. In view of the weak relationship between discovery rate and cumulative drilling, we decided not to project on a cumulative drilling base, but to project extensions, new pool and new field discoveries on a time base, Figures 27-30 [Figures 12-5 through 12-8 here]. These curves are much more regular than the

discovery rates as a function of drilling. New pools and new fields were similar enough to be combined. It must be expected that part of the decline in discoveries per year is due to the decline of drilling activity, so that the extrapolation will give an underestimate of the amount of oil remaining to be found, should the decline in drilling be arrested as can be expected.[16]

Using time as a surrogate for effort they model the logarithm of the change $\Delta D_t = D_t - D_{t-1}$ in new discoveries (new pools plus new fields), measured in billion barrels from year $t - 1$ to year t, as a linear function of time plus a random error term fit $\log \Delta \widetilde{D}_t$ with

$$\log \Delta \widetilde{D}_t = -.60 - .026\,(t - 1945) + .29\widetilde{Z}_t \ , \tag{12-13}$$

the $\widetilde{Z}_t$ being independent normal random variables with mean zero and variance one. The parameter values -0.60, -0.026, and $.29$ are the expectations of $\widetilde{\beta}, \widetilde{\alpha}$, and $\widetilde{\sigma}^2$, respectively, after observation of the data, and again, because the prior distribution was chosen in a very particular way, these values correspond to ordinary least-squares estimators for β, α, and σ^2. The fit of this model to the historical data series (Figure 12-7) is quite good.

Letting e_t denote cumulative extensions at the end of year t, the model fit to the change $\Delta \widetilde{e}_t = \widetilde{e}_t - e_{t-1}$ in extensions from t to $t - 1$ in billions of barrels is

$$\log \Delta \widetilde{e}_t = .77 - .046\,(t - 1945) + .26\widetilde{Z}_t \ . \tag{12-14}$$

The data series for $\log \Delta e_t$ exhibits systematic curvature at its beginning (1946) and at its end (1973), so that a strictly concave function of time (such as a quadratic) would appear to fit better and would lead to a smaller projection for cumulative extensions e_∞. Bromberg and Hartigan did not fit a quadratic function of time, "in view of the recent declines in drilling."

The data series for changes Δr_t in cumulative revisions suggests that $\log \Delta r_t$ is approximately a linear function of $\log \Delta P_t$, ΔP_t the change in cumulative production, and the least-squares fit,

$$\log \Delta \widetilde{r}_t - \log \Delta P_t = -.84 - .0003\,(P_t - P_{1945}) + .4\widetilde{Z}_t, \tag{12-15}$$

shows the coefficient of $P_t - P_{1945}$ to be so near zero as to be effectively negligible. That is, $\Delta \widetilde{r}_t$ is approximately a random proportion $\exp \{0.4\widetilde{Z}_t\}$ of ΔP_t. This is not unreasonable as the process of producing a field or pool is a key source of information for computing revised estimates of oil in place and of recoverable oil; the larger the rate of production, the larger the expected rate of revisions.

The predictive distribution for ultimate cumulative revisions $\widetilde{e}_\infty$ is generated by the procedure described earlier, assuming that ΔP_t for 1974 and ensuing years is the average yearly production of 2.25 BB over the time span 1945 to 1973. Since production cannot continue once reserves are exhausted, a stopping rule for the simulation of $\tilde{e}_\infty$ is imposed: Production continues so long as reserves are positive. Even though it is assumed that future increments of production per year are known with certainty and are exactly equal to the average production over 1945 to 1973, the stopping rule introduces uncertainty about total production, since production stops at a "random" time.

Table 12-1 is a summary of properties of marginal predictive distributions for discoveries, revisions, extensions, and production. The mean of the distributions for discoveries, extensions, and revisions are not reported; however, the mean of future oil reserves is near 70 BB. Bromberg and Hartigan's projections say that there is a 0.95 probability that (in billion barrels):

- Future discoveries lie in the interval (5.7, 23.7).
- Future extensions lie in the interval (8.5, 22.6).
- Future revisions lie in the interval (16.5, 151.7).
- Future production from the coterminous United States plus the Gulf of Mexico lies in the interval (62.9, 205.9).

They also project that the median of future revisions is four times as large as the median of future discoveries and nearly four times as large as the median of future extensions. The obvious conclusion is that revisions will play the major role in determining future reserves and total future production. They interpret this as suggesting that "large improvements in reserves might be possible through technological advances in recovery techniques."[17] Unfortunately, no technological breakthrough of major importance in recovery technology has occurred as yet and the effect of large price increases on reserves generated by secondary and tertiary recovery in well-explored regions does not appear to be substantial.[18]

Table 12-1. Bromberg and Hartigan Forecasts, Billion Barrels.

1973 Cumulative production		103.1	
1973 Reserves		25.7	
		Fractiles[a]	
	.025	.50	.975
Future discoveries	5.7	10.2	23.7
Future extensions	8.5	12.9	22.6
Future revisions	16.5	40.7	151.7
Total production (excluding Alaska and offshore)	166	194	309

a. The 0.025 fractile of future discoveries is the value such that the probability that future discoveries is less than or equal to this value is 0.025 and the probability that it is greater than this value is 0.975. Interpret other fractile values in a similar way.

Source: Bromberg and Hartigan, "U.S. Reserves of Oil and Gas."

The particular choice made by Bromberg and Hartigan for future rates of effort is

$$\Delta E_j = \frac{1}{n} \sum_{i=1}^{n} \Delta E_i \quad \text{for } j = n + 1, \ldots, \infty, \tag{12-16}$$

The average of rates of effort observed at $1, 2, \ldots, n$. Choosing future ΔE_is to be equal allows a natural translation from effort E_i to time t, since future E_is are then linearly increasing with time. The effect of a change in the constant value of future ΔE_is then appears as a change in the coefficient α. A variable pattern of future ΔE_is would require the models adopted for discoveries to be slightly recast to account for this variability. In particular, the distribution of $\tilde{R}_\infty$ given $(\Delta R_i, \Delta E_i)$, $i = 1, 2, \ldots, n$ varies with variations in the pattern of $E_{n+1}, \ldots, E_\infty$ since it depends on both ΔE_j and E_j, $j = n + 1, \ldots, \infty$. Increasing returns to effort from substantial future price increases may increase future rates of effort until resource exhaustion dictates a reduction in rate. This scenario is, in the judgment of many, more likely than a price-cost scenario leading to a constant rate of effort, and it would lead to a different distribution of $\tilde{R}_\infty$ than the latter scenario. Just how sensitive properties of $\tilde{R}_\infty$ are to shifts in $\Delta E_{n+1}, \ldots, \Delta E_\infty$ away from Bromberg and Hartigan's assumption is an empirical issue that is best resolved by simulating the effects.

Choosing ΔP_t as the explanatory variable for $\Delta \tilde{R}_t$ and assuming that the ΔP_ts are known with certainty disguises the essential fact that *both* $\Delta \tilde{R}_t$ and $\Delta \tilde{P}_t$ are uncertain, a priori. If the ultimate objective is to predict the flow of supply, then ΔP_t ought to be treated as an uncertain quantity along with $\Delta \tilde{R}_t$, $\Delta \tilde{e}_t$, and $\Delta \tilde{D}_t$.

Figure 12-2. Extension to Oil Reserves per Outpost Well.

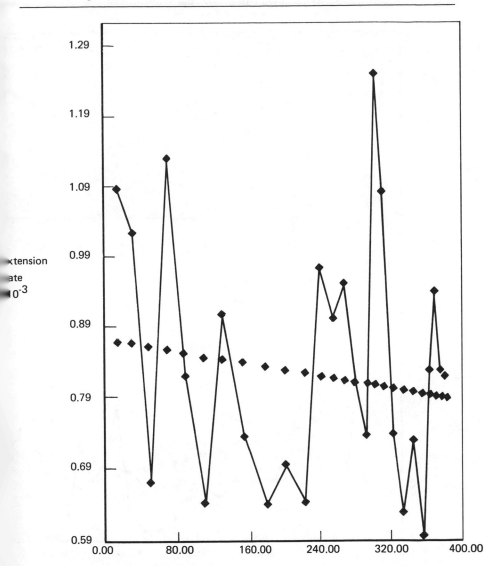

Source: L. Bromberg and J. A. Hartigan, "U.S. Reserves of Oil and Gas," Department of Statistics, Yale University, New Haven, Conn., 1975, Figure 20.

Figure 12-3.   Wildcat Oil Discoveries per Wildcat Well.

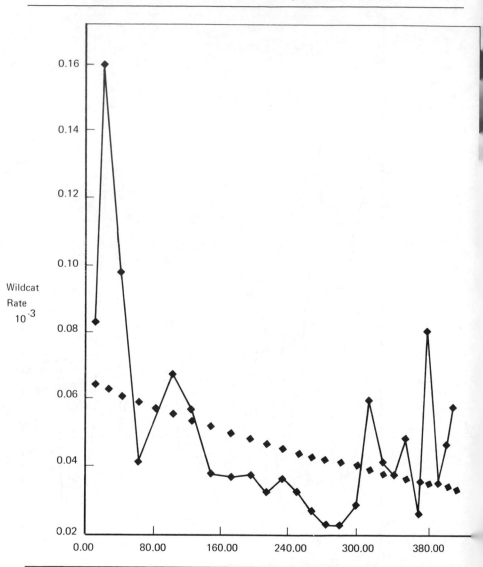

Source: Bromberg and Hartigan, "U.S. Reserves of Oil and Gas," Figure 21.

Figure 12-4. New Pool Discoveries per New Pool Well.

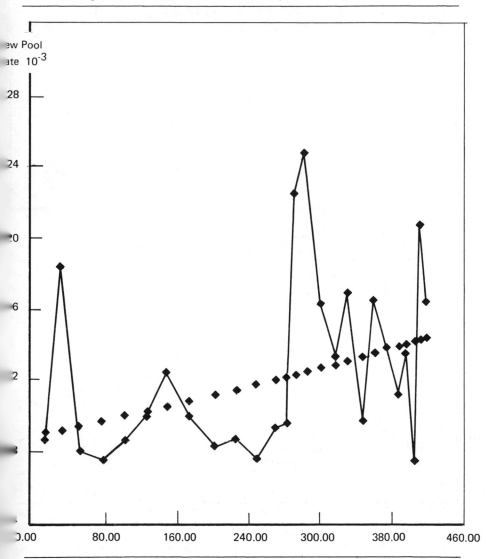

Source: Bromberg and Hartigan, "U.S. Reserves of Oil and Gas," Figure 22.

Figure 12-5.   Oil Extensions against Time, Exponential Fit.

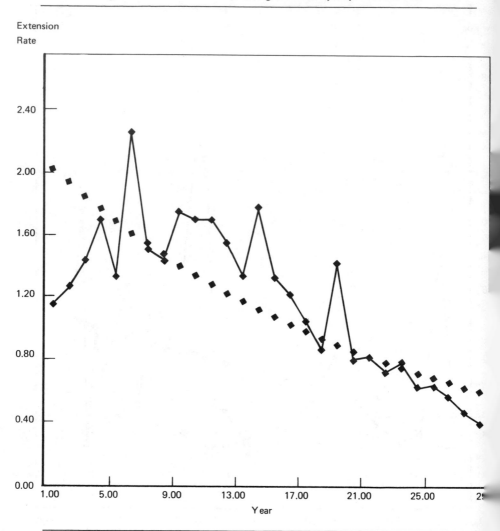

Source: Bromberg and Hartigan, "U.S. Reserves of Oil and Gas," Figure 27.

Figure 12-6. Log Oil Extensions against Time, Linear Fit.

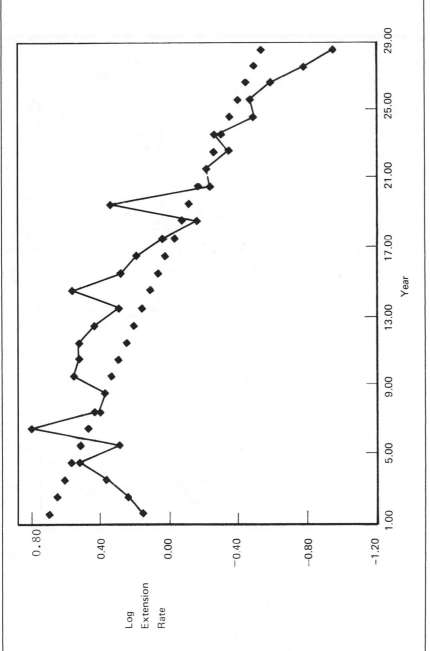

Source: Bromberg and Hartigan, "U.S. Reserves of Oil and Gas," Figure 28.

Figure 12–7. Oil Discoveries against Time, Exponential Fit.

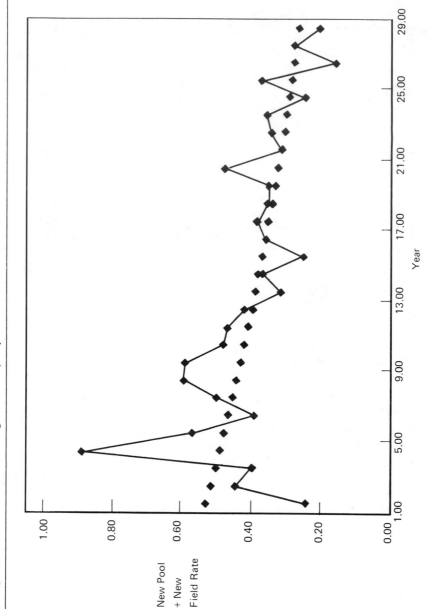

Figure 12-8. Log Oil Discoveries against Time, Linear Fit.

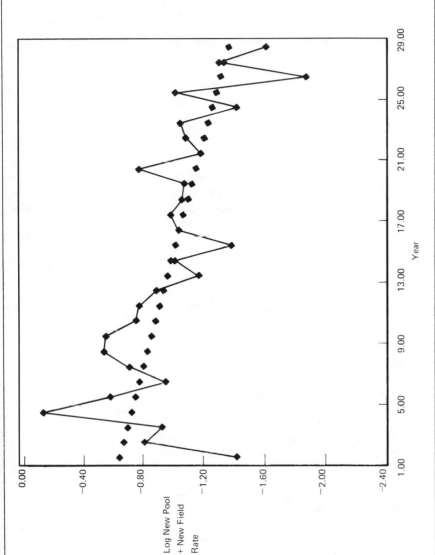

Source: Bromberg and Hartigan, "U.S. Reserves of Oil and Gas," Figure 30.

REFERENCES

1. J. J. Arps and T. G. Roberts, "Economics of Drilling for Cretaceous Oil Production on the East Flank of the Denver–Julesburg Basin," *AAPG Bulletin*, 42, no. 11 (1958): 2549–66.
2. A. D. Zapp, "Future Petroleum Producing Capacity of the United States," U.S. Geological Survey Bulletin 1142–H (Washington, D.C.: U.S. Government Printing Office, 1962).
3. M. K. Hubbert, "U.S. Energy Resources: A Review as of 1972," in *A National Fuel and Energy Policy Study, Part I*, Serial no. 93–40, Committee on Interior and Insular Affairs, U.S. Senate, 1974, pp. 101–103.
4. M. K. Hubbert, unpublished USGS paper presented to National Resources Subcommittee of the Federal Council on Science and Technology, February 1961.
5. Hubbert, "U.S. Energy Resources," p. 103.
6. Ibid., p. 103.
7. Ibid.
8. Ibid., p. 101.
9. M. K. Hubbert, "Degree of Advancement of Petroleum Exploration in the United States," *AAPG Bulletin* 51, no. 11 (1967): 220–27.
10. L. Bromberg and J. A. Hartigan, "Report to the Federal Energy Administration: United States Reserves of Oil and Gas," Department of Statistics, Yale University, New Haven, Conn., 1967.
11. John J. Wiorkowski, "Estimating Volumes of Remaining Fossil Fuel Resources: A Critical Review," *Journal of the American Statistical Association* 76, 375 (1981).
12. L. Bromberg and J. A. Hartigan, "U.S. Reserves of Oil and Gas," report to the Federal Energy Administration, Department of Statistics, Yale University, New Haven, Conn., May 1975.
13. E. Barouch and G. M. Kaufman, "On Sums of Lognormal Random Variables," Massachusetts Institute of Technology Sloan School of Management Working Paper 831-76, 1976.
14. Bromberg and Hartigan, "U.S. Reserves of Oil and Gas," 1975, p. 14.
15. Ibid., p. 15.
16. Ibid., p. 15.
17. Ibid., p. 16.
18. A recent study of the economics of secondary and tertiary recovery in the Permian basin by the Interagency Oil and Gas Project projects large price rises as adding to reserves by lengthening the producing life of installed wells, but generating little economic stimulus for installation of additional secondary and tertiary capacity. See "Future Supply of Oil and Gas from the Permian Basin of West Texas," Geological Survey Circular 828, U.S. Department of Interior, 1980, 57pp.

13 SUBJECTIVE PROBABILITY METHODS

As early as the eighteenth century, mathematicians and philosophers[1] studied the implications of interpreting the meaning of probability for an uncertain event as a quantitative measure of an individual's personal degree of belief in the likelihood that that event will occur. The development of this point of view rests on assumptions or postulates describing behavior in the face of uncertainty, a principal one being the following: *There is only one form of uncertainty, and all uncertainties can be compared.*[2] In more concrete terms this fundamental assumption means that if E_1 and E_2 are (for you) *any* two uncertain events, then in your judgment either E_1 is more likely than E_2, or E_2 is more likely than E_1, or E_1 and E_2 are equally likely. In addition, if E_3 is a third uncertain event, E_1 is more likely than E_2, and E_2 is more likely than E_3, then E_1 is more likely than E_3. Thus, uncertain events are comparable and comparisons must be "coherent" with respect to the relation "is more likely than."

Coherent comparability (Lindley's[3] terminology for these two ideas) can be used to define the probability for any uncertain event, and leads to a representation of it in the following form: Let E be an uncertain event and $\bar{E}$ the event "not E." Consider two lotteries, one of which gives you a valuable prize if $\bar{E}$ occurs and nothing if E occurs; the other yields the *same valuable* prize or nothing in a different way. A needle pointer is perfectly centered in a black circle 1 foot in circumference drawn on a board that is perfectly smooth and exactly level. The needle pointer is balanced in such a way that when spun it is equally likely that it will stop pointing at any given place on the circumference of the black circle as at any other. You mark off a segment of the circumference in red. In this lottery you receive the valuable prize if, when spun once, the pointer comes to rest at a point marked red; otherwise you receive nothing. By definition your personal or subjective probability for the event E is the proportion of the circumference that you must mark red in order to render you indifferent between this lottery and the lottery giving you the same valuable prize if the event E actually obtains.

This is only one among many constructs that can be used to define how judgments can be scaled. It is an intellectually operational definition and if coherent comparability is accepted, it is possible to show that there exists a *unique* proportion of the circumference of the circle rendering you indifferent between the "real" lottery and this artificially constructed one. Of course, there is no need to refer to the scaling process that constitutes the definition when actually assessing uncertain events. Further development of this point of view shows that probability so defined possesses the essential properties of "mathematical" probability and consequently the rules of probability apply to judgments scaled in this fashion.

In 1960 C.J. Grayson published the first orderly presentation of how personal probabilities can be used in oil and gas exploration.[4] He focused on a specific decision problem faced by oil and gas operators: whether or not to drill a well in a specific location. In practical terms the drilling decision problem is one of the most important faced faced by explorationists. Wildcat well drilling in particular is characteristically high risk with large attendant uncertainties, and, because of the frequency with which such decisions are made in the face of widely varying information quality and type, a method for systematically quantifying expert geologic judgment as an aid to logical choice is a natural replacement for hunches and intuition.

Although each individual drilling decision is only a very small component of the total exploratory effort generating supply of oil and gas from a petroleum basin, the methods Grayson proposed for quantifying geologic opinion are applicable to aggregates of uncertain quantities as well, and today improved versions of his methods form part of the kit of analytical tools used by many explorationists.

The first official government projections of undiscovered hydrocarbons expressed as personal probabilities appeared only in the late 1970s, the two most notable examples being *Oil and Gas Resources of Canada 1976* and the U.S. Department of the Interior Circular 725 prepared by the U.S. Geological Survey (USGS).[5] The former is methodologically close in spirit to discovery process modeling, whereas the latter is principally geologic-volumetric in character.

CIRCULAR 725

With the exception of the forecasts of Hubbert, projections of undiscovered oil and gas published by USGS personnel from the end of the Second World War until 1975 conformed to one another in several ways. Geologic-volumetric methods were applied to very large geographic regions, and the resulting projections, derived by a small number of geologists over a short time, were expressed as point estimates (single numbers). No formal accounting for uncertainty appeared. Relative to annual consumption during this period, the amounts of recoverable hydrocarbons projected as remaining to be discovered were comfortably large—several hundreds of billions of barrels—and the forecasts tended to increase over time. Although the USGS projections for the United States were at variance with the petroleum industry's view of what remained to be discovered, order-of-magnitude differences between industry and government forecasts generated little publicity outside industry circles.

The 1973 oil embargo changed this and focused public, executive branch, and congressional attention on the USGS projections. A resource analysis group was formed. One of its principal responsibilities was to update past projections of amounts of crude oil, natural gas, and natural gas liquids recoverable from new discoveries in petroleum provinces in the coterminous United States and in Alaska, onshore and offshore, to 200 meters deep. The USGS prepared Circular 725 entitled "Geological Estimates of Undiscovered Recoverable Oil and Gas Resources in the United States," which summarizes the results: *recoverable* undiscovered oil lies in the interval 50 to 127 billion barrels (BB) with probability 0.95 and has an expectation (mean) of 82 BB.

This projection is strikingly different from earlier projections published by the USGS in the amount of geological study devoted to it and in the methods employed. More than seventy government geologists participated in a work-up of approximately 100 petroleum provinces of the United States. Geological opinion was elicited, then expressed in terms of subjective or personal probabilities for uncertain events; for example, "The amount of undiscovered *recoverable* oil remaining in the province is greater than one billion barrels." Probability distributions for amounts of undiscovered recoverable oil, natural gas, and natural gas liquids in regions such as onshore Alaska were computed by combining the probability distributions for each province within a region.

Circular 725 is the first U.S. government mineral resource appraisal couched explicitly in subjective probability terms, a radical methodological departure from past practice. The principal conclusions are summarized in Figures 13-1 and 13-2 and in Table 13-1, from the circular. A mean of 82 BB of undiscovered recoverable oil and a mean of 484 trillion cubic (TCF) of undiscovered recoverable natural gas are amazingly close to current industry point forecasts of these quantities. Figure 37 (Figure 13-3) from the circular effectively illustrates the decline over time of USGS projections for undiscovered hydrocarbons to amounts close to industry projections. Public perception followed quickly upon publication of Circular 725, a summary of its conclusions appearing on the front pages of the *New York Times*, the *Boston Globe*, and other metropolitan newspapers.

Figure 13–1.   Estimated Range of Undiscovered Recoverable
Resources Crude Oil and Natural Gas in the United States.

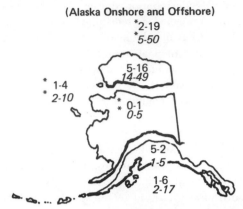

(Alaska Onshore and Offshore)
*2-19
*5-50

Undiscovered Recoverable Oil 12-49 Billion Barrels
Undiscovered Recoverable Gas 29-132 Trillion Cu Ft.

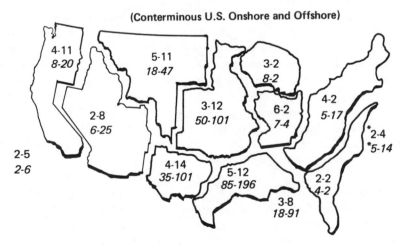

(Conterminous U.S. Onshore and Offshore)

Undiscovered Recoverable Oil 36-81 Billion Barrels
Undiscovered Recoverable Gas 286-529 Trillion Cu Ft.

* Marginal Probability Applied
** For Regional Distribution of Inferred Reserves, See Tables 4 and 5

Source: U.S. Geological Survey, Oil and Gas Branch, "Geological Estimates of Undiscovered Recoverable Oil and Gas Resources in the United States," U.S. Department of the Interior Circular 725, 1975.

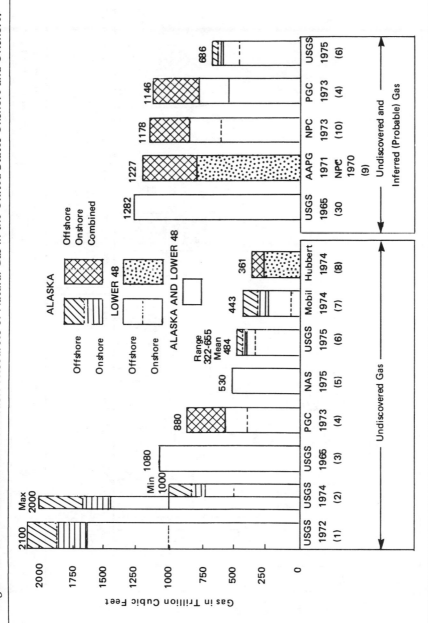

Figure 13-2.   Undiscovered Recoverable Resources of Natural Gas in the United States Onshore and Offshore.

Source: USGS, Circular 725.

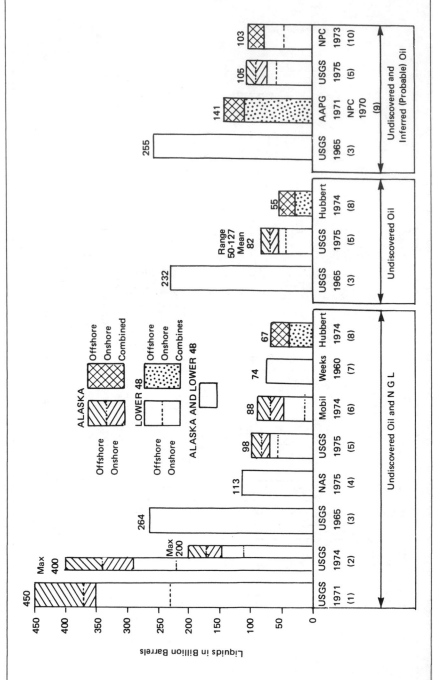

Figure 13-3.   Undiscovered Recoverable Resources of Liquid Hydrocarbons in the United States Onshore and Offshore.

Table 13-1.   Production, Reserves, and Undiscovered Recoverable Resources of Crude Oil, Natural Gas, and Natural Gas Liquids for the United States, December 31, 1974.[a]

| Area | Cumulative Production | Reserves of C | | | Undiscovered Recoverable Resources |
| | | Demonstrated | | | |
		Measured	Indicated	Inferred	Range 95%-5%
Crude Oil, Billion Barrels					
Lower 48 onshore	99.892	21.086	4.315	14.3	20-64
Alaska onshore	0.154	9.944	0.013	6.1	6-19
Total onshore	100.046	31.030	4.328	20.4	37-81
Lower 48 offshore	5.634	3.070	0.308	2.6	5-18
Alaska offshore	0.456	0.150	Negligible	0.1	3-31
Total offshore	6.090	3.220	0.308	2.7	10-49
Total onshore and offshore	106.136	34.250	4.636	23.1	50-127
Natural Gas, Trillion Cubic Feet					
Lower 48 onshore	446.366	169.454		119.4	246-453
Alaska onshore	0.482	31.722		14.7	16- 57
Total onshore	446.848	201.176		134.1	262-506
Lower 48 offshore	33.553	35.811		67.4	26-111
Alaska offshore	0.423	0.145	n.a.	0.1	8- 80
Total offshore	33.976	35.956		67.5	42-181
Total onshore and offshore	480.824	237.132		201.6	322-655
Natural Gas Liquids, Billion Barrels					
Total onshore and offshore	15.730	6.350	n.a.	6	11-22

a. Onshore and offshore to water depth of 200 meters.
Source: USGS, Circular 725.

Explicit assumptions about prices, costs, and technology (as they bear on the recovery factor in particular) underlie the projections of the circular. A wise consumer of these forecast numbers will know what they are and why they were made.

Even though all energy mineral data and data on individual mineral deposits in particular are tainted by economic effects, methods for projecting future amounts of undiscovered recoverable hydrocarbons discussed thus far—life-cycle, rate-of-effort, and geologic-volumetric—leave unsaid explicit assumptions about the influence of economic variables on observables and projections alike.

The principal surrogate is the recovery factor. It reflects the joint effects of prices, costs, exploration, development, and production technology on amounts of undiscovered recoverable hydrocarbons, so projections of these amounts can be expected to be very sensitive to the values chosen for them. Not only does a change in recovery factor influence what will be produced from known deposits, but it partially determines the minimum size of undiscovered deposit that is economically vaible. A shift upward in the recovery factor translates a fraction of subeconomic deposits into economically producible deposits and provides added incentive for exploration by decreasing the economic breakeven size for undiscovered deposits.

The economic and technological backdrop for Circular 725 forecasts is . . .

that undiscovered recoverable resources will be found in the future under conditions represented by a continuation of price/cost relationships and technological trends generally prevailing in the recent years prior to 1974. Price/cost relationships since 1974 were *not* taken into account because of the yet undetermined effect these may have on resource estimates. If fundamental changes in cost/price relationships are imposed or if radical improvements in technology occur, estimates of recoverable resources will be affected accordingly.

These assumed conditions permit the appraisal of recoverable oil and gas resources to be made on the basis of: (1) relevant past history and experience concerning recovery factors; (2) the geology favorable to the occurrence of producible hydrocarbons; and (3) the size and type of reservoirs which have been found, developed, and produced.

Recoverable resource potential as reported here for the frontier basins of Alaska, and to some extent the offshore areas of the lower 48 states, is especially uncertain. Many of the frontier basins will have very severe economic constraints under which oil and gas may be recovered. A certain amount of the recoverable oil and gas in basins used for analogs, but which lie in areas of favorable economics (such as the lower 48 states), may not be economically recoverable in the Arctic or offshore basins; this fact was taken into consideration in the estimating process . . .

The economic recovery factor used was based on a current national average of approximately 32 percent for oil and 80 percent for natural gas (McCulloch, 1973). The sub-economic portion of the remaining resources for oil is estimated to be an additional 28 percent recoverable, for a total of 60 percent recovery (Geffen, 1975). Sub-economic identified resources of crude oil were calculated on the following assumptions: (1) that [,] on the average, 32 percent of original oil-in-place is recoverable if there are no substantial changes in present economic relationships and known production technology, and (2) that ultimately the recovery factor could be as large as 60 percent. By definition, the sum of cumulative product to date, plus the current estimate of demonstrated reserves, will account for 32 percent of the original oil-in-place in *known* fields; an increase to 60 percent will allow another 28 percent to be recovered. As indicated in Figure 13, that 28 percent which is currently considered sub-economic amounts to about 120 billion barrels. The inferred reserves are made up partly of revisions of current estimates partly of "undis-covered" oil from future extensions and new pools in known fields. Assuming that all the inferred is "undiscovered," the 23.1 billion barrels economically recoverable at the 32 percent recovery factor would have a sub-economic component of about 20 billion barrels. Thus, the sub-economic identified category [sic] was estimated to be 120 to 140 billion barrels. Similarly, the sub-economic component of undiscovered resources was estimated to include 44 to 111 billion barrels.

It is extremely optimistic to assume that 60 percent of the oil-in-place will eventually be recovered. If [this] becomes a reality, it is likely to occur only through gradual development over an extended time period. The remaining 40 percent of oil-in-place is not included as it is considered to be nonrecov-erable, [just] as coal which is too thin to mine is excluded from recoverable resources.[7]

Thus, an average 32 percent recovery factor is assumed for recovery of original oil in place from future discoveries regarded as producible under price-cost con-ditions prevailing just prior to 1974-prior to the fourfold increase in world oil prices that followed the Organization of Petroleum Exporting Countries (OPEC) embargo. Altering the assumption made about prices to reflect the structural shock caused by OPEC's actions raises a host of very difficult questions about methods, data, and effects. Extrapolation of past history becomes less tenable, and of necessity expert opinion, even as it bears on partially explored regions, takes on a more central role. With no empirical exploration, development, and production experience in the range of prices that followed the embargo, con-struction of projections of supply has become doubly difficult. Interaction be-tween economists, engineers, and geologists is a necessity. Hence, the decision to base the USGS's first effort at employing subjective probability to forecast recoverable hydrocarbon resources on preembargo economic conditions is under-standable. This decision, at least as it applies to mature and partially explored provinces, is supported by the belief of many geologists that the largest fields in these provinces were discovered as of 1974.

Individual petroleum provinces, 102 of them, constitute the basic units ana-lyzed. The study provides a review of the geologic framework within which the provinces sit, defines region and province boundaries, and, based on planimeter-ing of tectonic maps, presents point estimates of sedimentary rock area and vol-ume for onshore and offshore regions (to 200 meters); see Table 13-2. Although not reported in Circular 725, estimates of sedimentary rock area and volume for individual provinces were made.

The descriptive template used to appraise the "quality" of a generic province conforms to currently accepted geologic theory, which is touched on in Circu-lar 725:

A petroliferous province must have: (1) an adequate thickness of sedimentary rocks; (2) source beds containing considerable dispersed organic matter; (3) a suitable environment for the maturation of organic matter; (4) porous and

Table 13-2. Sedimentary Rock Area and Volume by Regions.

Region	Onshore Area in 1,000 mi²	Onshore Volume in 1,000 mi³	Region	Offshore (Water depths 0–200 meters) Area in 1,000 mi²	Offshore Volume in 1,000 mi³	Total Area in 1,000 mi²	Total Volume in 1,000 mi³
1. Alaska	252.2	644.7	1A. Alaska	318.1	501.7	570.3	1,146.4
2. Pacific Coastal States	125.5	192.1	2A. Pacific Coastal States	18.4	32.0	143.9	224.1
3. Western Rocky Mountains	329.9	549.1				329.9	549.1
4. Northern Rocky Mountains	360.6	591.6				360.6	591.6
5. West Texas and Eastern New Mexico	193.4	283.8				193.4	283.8
6. Western Gulf Basin	238.7	774.8	6A. Gulf of Mexico	112.8	570.0	351.5	1,344.8
7. Mid-Continent	446.6	324.2				446.6	324.2
8. Michigan Basin	122.0	108.0				122.0	108.0
9. Eastern Interior	166.2	204.0				166.2	204.0
10. Appalachians	205.4	501.4				205.4	501.4
11. Eastern Gulf and Atlantic Coastal Plain	109.4	127.8	11A. Atlantic Coastal States	102.3	233.0	211.7	360.8
Total lower 48 onshore	2,297.7	3,656.8	Total lower 48 offshore	233.5	835.0	2,531.2	4,491.8
Total onshore United States	2,549.9	4,301.5	Total offshore United States	551.6	1,336.7	3,101.5	5,638.2

Source: USGS, Circular 725, p. 17, Table 3.

permeable reservoir beds; (5) hydrodynamic conditions favorable for both early migration and ultimate entrapment of oil and gas; (6) a favorable thermal history; (7) adequate trapping mechanisms; and (8) suitable timing of petroleum generation and migration in relation to the development of traps. Many other features are favorable but not absolutely necessary. Examples of favorable indications in unexplored basins are: the presence of oil and gas seeps, a varied sequence of rock types, some organically rich marine sediments as source beds for the generation of oil and associated gas, non-marine organically rich sediments for genesis of non-associated gas, structural features that show progressive growth through geologic time, unconformities, and the presence of evaporite deposits. For areas in an early stage of exploration, important indicators are: shows of oil and gas in non-commercial wells, presence of saline or sulfate water in potential reservoirs, commercial production, a favorable ratio between wells drilled and oil and gas discoveries, and traps that are detectable by conventional geological and geophysical methods.[8]

No one particular method was employed to generate the data used by individual teams of geologists in the course of quantifying their judgments in the form of probabilities. Rather,

> Estimates of recoverable oil and gas resources are based upon a series of resource appraisal techniques. These techniques all have the common characteristic of having been selected on the basis of the *available* geologic information for each province or region.
> The techniques used include: (1) an extrapolation of known producibility into untested sediments of similar geology for a well-developed area; (2) volumetric techniques using geologic analogs and setting upper and lower yield limits through comparisons with a number of known areas; (3) volumetric estimates with an arbitrary general yield factor applied when direct analogs were unknown; (4) Hendricks' (1965) potential-area categories; and (5) comprehensive comparisons of all known published estimates for each area to all estimates generated by the above methods.[9]

An outline of the study design for Circular 725 appears in Figure 13-4, which displays the major steps taken to generate fractiles for amounts of undiscovered recoverable oil (or gas) in each province, *conditional on oil (or gas) appearing in commercial quantities.* A distinctive feature of a frontier province is that it may contain no economically viable deposits. The probability of this event, a priori, is greater than zero and so must be assessed. For each frontier province a marginal probability for the event that commercial oil (or gas) is found was assessed and combined with probabilities for amounts conditional on the presence of commercial oil (gas) to form a marginal probability distribution for amounts of undiscovered oil (gas).

Figure 13-5b, from Circular 725, graphs this distribution for central Alaska onshore. The probability of no commercial oil is 0.30; the probability of com-

Figure 13-4. Study Design.

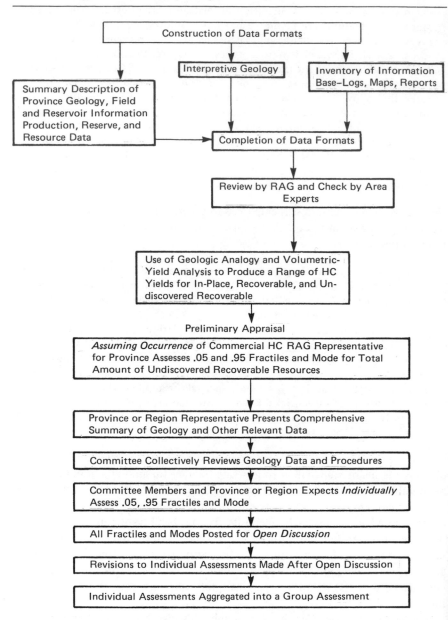

Figure 13.13. Probability Distributions by Monte Carlo Analysis of Undiscovered Recoverable Resources for Alaska; Aggregate Probability Distributions for Three Onshore Subregions and the Total Alaska Onshore.

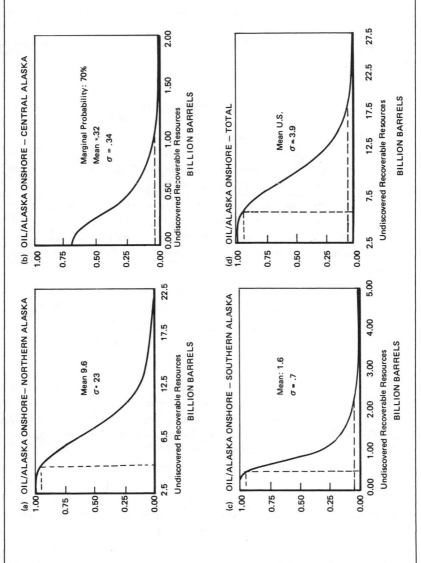

(a) OIL/ALASKA ONSHORE— NORTHERN ALASKA

Mean 9.6
σ - 23

Undiscovered Recoverable Resources
BILLION BARRELS

(b) OIL/ALASKA ONSHORE – CENTRAL ALASKA

Marginal Probability: 70%
Mean -.32
σ = .34

Undiscovered Recoverable Resources
BILLION BARRELS

(c) OIL/ALASKA ONSHORE – SOUTHERN ALASKA

Mean: 1.6
σ = .7

Undiscovered Recoverable Resources
BILLION BARRELS

(d) OIL/ALASKA ONSHORE – TOTAL

Mean U.S.
σ = 3.9

Undiscovered Recoverable Resources
BILLION BARRELS

Source: USGS, Circular 725, p. 24, Figure 8.

mercial oil in some positive amount is 0.70; the probability of more than 0.5 billion barrels is 0.25; the probability of more than 1 billion barrels is 0.05. The graph presents the probability that recoverable reserves are greater than x for all values of $x > 0$ (and the probability that $\tilde{x} = 0$ when this event has nonzero probability). In more formal terms, each of these graphs represents the right tail of a probability function for the amount of recoverable reserves $\tilde{x}$. Similar graphs were constructed for all provinces.

The research design required each of the geologists expressing a formal opinion about a particular province to assess two fractiles and a mode. Several methodological questions arise:

1. The value assumed by $\tilde{x}$ can in principle be any number greater than or equal to zero. In most cases it is reasonable to assume that, conditional on $\tilde{x} > 0$, $\tilde{x}$ possesses a probability density on $(0, \infty)$ which is unimodal, positively skewed, and whose first two moments (mean and variance) exist. What is the appropriate functional form for the density of $\tilde{x}$, and how are numerical values of its parameters to be chosen?

2. Geologists' opinions about any given province differ. These differences are reflected in differences between recorded values of 0.05 fractiles, of 0.95 fractiles, and of modes across geologists appraising that province, even after considerable group discussion. How should these differences be reconciled, if at all?

3. If a group appraisal in the form of a single probability distribution for each province is desirable, how should individual assessments be aggregated?

4. Given that (3) is resolved, how should a probability distribution for undiscovered recoverable resources in a region composed of more than one province be computed?

On the surface, question (4) is a mechanical issue: If uncertain quantities $\tilde{x}_1$, $\tilde{x}_2, \ldots, \tilde{x}_N$ representing undiscovered recoverable resources in provinces labeled $1, 2, \ldots, N$ are assumed to be mutually independent, then there are straightforward methods for computing the probability distribution of the sum $\tilde{R} = \tilde{x}_1 + \tilde{x}_2 + \ldots + \tilde{x}_N$. This assumption of independence was made and the density of $\tilde{R}$ approximated by "Monte Carloing" 10,000 values of each $\tilde{x}_1, \tilde{x}_2$, $\ldots, \tilde{x}_N$ and computing the resulting cumulative frequency distribution for $\tilde{R}$. The bottom right-hand graph in Figure 13-5 is an example; it displays the probability that an uncertain quantity $\tilde{R}$ composed of the sum of undiscovered recoverable oil in three Alaskan onshore provinces is greater than R for all $R > 0$.

Is the assumption of mutual independence of the $\tilde{x}_i$s composing $\tilde{R}$ tenable? There is no compelling evidence against it as a reasonable working assumption, and it vastly simplifies both the assessment procedure and the ensuing computations. However, such assumptions must be carefully examined, a priori, and the study design should provide for evidence in favor of it or against it. Although physical evidence may support independence, dependencies can creep in through the process by which judgments are formed, shared, and quantified if one or more individuals participate in the assessment of more than one of a collection of provinces composing a regional aggregate.

Individual $\tilde{x}$s were assumed to be lognormal. That is, each $\tilde{x}$ possesses a lognormal density indexed by two parameters: the mean $E(\log_e \tilde{x}) = \mu$ of $\log_e \tilde{x}$, and the variance $\text{Var}(\log_e \tilde{x}) = \sigma^2$ of $\log_e \tilde{x}$. This particular choice of functional form for the distribution of $\tilde{x}$ seemed reasonable, because (1) it is unimodal and positively skewed with a relatively "fat" right tail; (2) it is sufficiently flexible in shape to allow a satisfactory match between it and a broad range of judgments; and (3) empirical evidence suggests that the relative frequency distribution of sizes of mineral deposits of a particular type in a particular place is approximately lognormal in shape.[a] Since the lognormal distribution is indexed by two parameters, and since two fractiles and a mode were assessed by each geologist, choice of any pair from these three assessed points corresponds to a choice of numerical values for μ and σ^2. An alternative would be to do a balancing act, choosing μ and σ^2 so the resulting distribution provided 0.05 and 0.95 fractiles and a mode not "too far" from the assessed values of these quantities.[b] Yet another alternative is to fit the three assessments exactly by choice of a three-parameter distribution for $\tilde{x}$; for example, with a three-parameter lognormal distribution.

With the recognition that geologists *may*[c] exhibit the same cognitive biases repeatedly observed in experimental studies of how people appraise uncertainty, the lognormal distribution corresponding to each of the three pairs (0.05 fractile and 0.95 fractile, 0.05 fractile and mode, and 0.95 fractile and mode) was computed, and that distribution possessing the largest 0.05 to 0.95 fractile spread was adopted. Marginally relevant experimental evidence gathered by Alpert and Raiffa and Capan[11] and others, suggests that most assessors are on the average too conservative in the numerical values they assign to extreme fractiles. In an experimental setting, it is not unusual to observe events like "the true value of uncertain quantity $\tilde{x}$ exceeds the assigned 0.95 fractile" occurring with a relative frequency of 15 to 20 percent when a large group is asked to assign fractile val-

---

a. The raw (unnormalized) sum of lognormal $\tilde{x}$s is *not* lognormal, although it can be shown that under certain conditions, in its asympototic region the sum possesses a density that is approximately three-parameter lognormal in shape.[10]

b. An exact match between assessed values and corresponding values from a fitted lognormal distribution occurs very rarely in practice.

c. "May" rather than "do," since there is no published study of subjective probability assessment by geologists in which (1) the task characteristics closely match what must be done "on the job," and (2) the research design allows identification of distinguishable sources of miscalibration or bias. There is, however, extensive evidence that, in less realistic settings, geologists and petroleum engineers assess like everyone else; that is, they announce probabilities for uncertain events that on the average are miscalibrated.

ues to each of a set of distinct uncertain quantities. Similarly, such experiments show that the event "the true value of the uncertain quantity $\tilde{x}$ lies in the 0.25 to 0.75 fractile interval" often occurs with a relative frequency substantially less than 50 percent. There is a learning-curve effect: Discussion, feedback, and repetition of assessment exercises all reduce miscalibration. One lesson that emerges is that doing what comes naturally when assessing uncertainties is not necessarily enough. Probability is a skill like golf, tennis, or bowling that is best learned by carefully controlled practice.

For many of the provinces studied for Circular 725, fractile assessments varied considerably from geologist to geologist, even though the substantive geologic information available was shared by all. A single group opinion was desired, so a simple method for aggregating individual judgments was employed: average the 0.05 fractiles and average the 0.95 fractiles.

Ideally (see Steiner[12]) the choice of method for aggregating opinions should take into account the nature of the task, the resources at the group's disposition, and the group's structure. Hogarth constructs a framework for evaluating the desirability of particular methods, the underlying theme being that "when experts interact, the aggregation of opinion is a cognitive and social process rather than a mechanical algorithm."[13] Many considerations bear on the choice of method, among which are these:

- Is acceptance by the group of a final group opinion necessary?
- Can opinions be easily verified by an "objective" external reference?
- Is the group's mix of substantive types of expertise well balanced and appropriate?
- How many assessors should be in the group?

Among frequently proposed methods Hogarth discusses are

- Representation of group assessments as opinion pools by weighted averaging of individual fractiles or distributions
- Delphi-like procedures
- Open, unstructured discussion between assessors

Participants in the Circular 725 study engaged in open discussion both of substantive geology and of individual fractile (and modal) assessments, their judgments subsequently represented as an opinion pool. At issue is whether or not this interaction "biased" the resulting appraisals through the effect of social pressures on group members whose judgments were not in conformity with judgments of the majority or with those of a clearly perceived peer group leader. The importance of social pressures in influencing opinions has been extensively studied in artificial experimental settings, and these experiments support the intuitively acceptable conclusion that in circumstances where expert opinions can be evaluated by comparison with actual outcomes soon after the opinions are stated, the pressure to conform is less than when there is a substantial hiatus between expression of opinions and observation of the events about which the opinions are expressed. Hogarth argues that using such evidence to downgrade the value of interaction among group members may be unwarranted since there is considerable evidence that, in some settings, an interactive group predicts better relative to certain preset norms than various combinations of individual assessments when no interaction is allowed. He concludes that "there are many reasons which cause interaction processes to be dysfunctional but there is no reason why they *must* be so." [14]

Unfortunately, the process of eliciting opinion was mismanaged. Several of the groups of geologists participating in this assessment were subject to severe "dominant personality" effects, open discussion of assessments allowed intrusion of peer group pressure to conform, and in-group differences in geologists' assessments, although recorded, were suppressed in the final report in favor of a single "consensus" distribution. The failure to report differences in opinion between geologists is particularly distressing, as lack of consensus in judgment about recoverable hydrocarbons in a petroleum province often indicates competing descriptive hypotheses about the geology of the province, one or possibly none of which may be correct. The consumer of the projections reported, given only the consensus distributions, has been denied valuable information.

A further weakness in the results stems from the group leader's unwillingness to consider asking each geologist to assess enough fractiles for each uncertainty quantity so that the shape of the assessed distribution would be evident. The procedure adopted for fitting a distribution to two fractiles and a model value is an avoidable ad hocery.

Schemes for minimizing adverse effects of social pressures have been proposed, a notable example being that of Press. It is "a new type of controlled feedback data collection protocol, which we call qualitative feedback, in contradistinction to the quantitative feedback of Delphi [-like protocols]." [15] The basic idea is as follows:

> With strictly qualitative feedback there is no inherent reason, save perhaps logic, that will force the group towards consensus. Furthermore, if the qualitative feedback protocol is applied only in contexts involving relative values or preferences, individual judgments can never be "wrong," but can only be faultily reasoned, in that they ignore certain important arguments relating to the issues involved. [16]

Press's protocols can be used in eliciting opinions from geologists and look promising as a device for inferring group judgment.

CANADIAN RESOURCE APPRAISAL

Probabilistic projections of the sum of remaining reserves, discovered resources, and undiscovered oil and gas potential for Canadian petroleum provinces or regions are shown in Table 13-3 of the Energy, Mines, and Resource Bureau appraisal. As always, *reserves, resources,* and *potential* each means what it is defined to mean in the particular study at hand. Here, *resource* includes "all conventional oil and gas accumulations known or inferred to exist. *Reserves* comprise that portion of the resource that has been discovered. (Discovered resources in frontier provinces are a special case and are not included in the

Table 13-3. Summary of Oil and Natural Gas Resources of Canada, 1975.[a]

	Likelihood of Existence		
	"High" 90% Probability	*50/50 Chance* 50% Probability	*"Low"* 10% Probability
Region	Oil Resources, Billion Barrels		
Atlantic Shelf South	1.2	1.9	3.0
Labrador-East Newfoundland Shelf	1.7	2.6	4.5
Northern Stable Platform Basins	0.01	0.6	3.2
St. Lawrence Lowlands	0.04	0.09	0.2
Western Canada	10.9	11.7	13.5
Mainland Territories	0.3	0.5	1.0
Mackenzie Delta-Beaufort Sea	4.3	6.9	12
Sverdrup Basin	1.1	2.0	4.0
Arctic Fold Belts	0.5	1.8	4.3
Total Canada (Accessible Regions)[b]	25	30	43
Region	Gas Resources, Trillion Cubic Feet		
Atlantic Shelf South	8.6	13.2	20
Labrador-East Newfoundland Shelf	18	26.7	45
Northern Stable Platform Basins	0.4	2.3	12
St. Lawrence Lowlands	0.7	1.4	3.2
Western Canada	89	97	107
Mainland Territories	6.0	9.7	20
Mackenzie Delta-Beaufort Sea	39	60	99
Sverdrup Basin	21	40	80
Arctic Fold Belts	2.9	11	26
Total Canada (Accessible Regions)[b]	229	277	378

a. Remaining reserves, discovered resources, and undiscovered potential.

b. These columns do not total arithmetically to the Canada totals because individual curves must be summed using a statistical technique described elsewhere in the same report.

Source: Energy, Mines, and Resources Bureau (Canada), *Oil and Gas Resources of Canada, 1976*, EMS Report EP77-1, 1977, p. 2, Table 1.

reserves category but are added to potential.) The word *potential* describes the fraction of resources inferred to exist but not yet discovered. The terms *potential* and *undiscovered resources* are thus synonymous and can be used interchangeably.[17]

Figure 13–6 from the Canadian report, displays graphs of the probability that this sum, measured in billion barrels of oil (trillion cubic feet of gas), exceeds a given value for all values greater than or equal to zero. Aside from the inclusion of remaining reserves and discovered resources, these graphs are similar to Circular 725 graphs displaying expert geologic opinion in the form of probabilities.

Although principal conclusions of the Canadian report and Circular 725 are presented in the same form, right-tail probabilities for amount of resource remaining to be discovered, there are significant differences between the two studies. Differences are foreshadowed in the questions the Canadian study proposed to answer and in the tenets and assumptions about resource estimation explicitly stated in a review of methods employed (presented as an appendix to the report):

1. How large is the resource?
2. Where is the resource located?
3. In what size deposits does it exist?
4. What is its composition and quality?
5. How certain are we of any of these opinions?

In addition, the estimates must be generated in a form amenable to economic analysis with supply rates being the ultimate objective, which in turn can be related to demand forecasts.[18]

Assumptions about resource projections that shape the procedures employed are:

1. Resource estimates can be made on any level of data.
2. Degree of uncertainty related to an estimate must be identified (and incorporated into any subsequent use of the estimate).
3. The probability of existence and the probable size of deposits should be considered separately.
4. The sizes of resource deposits are approximately lognormally distributed in nature.
5. The eventual discovery of a resource is not a requirement in the estimation of the ultimate recoverable resource.
6. No a priori economic considerations should be included in estimations of ultimate recoverable resources.
7. Current and foreseeable technology of recovery is assumed.[19]

Taken literally, the sixth assumption is self-contradictory; the fraction of a deposit "ultimately recoverable" is a function of both technology *and* economics. Even if this fraction is defined as the amount ultimately recoverable with

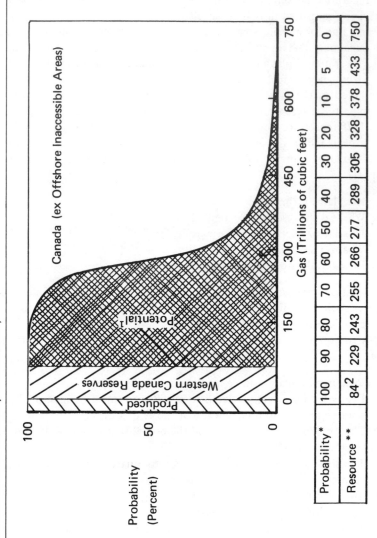

Figure 13-6. Estimates of Oil and Natural Gas Resources Excluding Inaccessible Offshore Areas. (Cumulative Percent Probability Distribution.)

Probability*	100	90	80	70	60	50	40	30	20	10	5	0
Resource**	84[2]	229	243	255	266	277	289	305	328	378	433	750

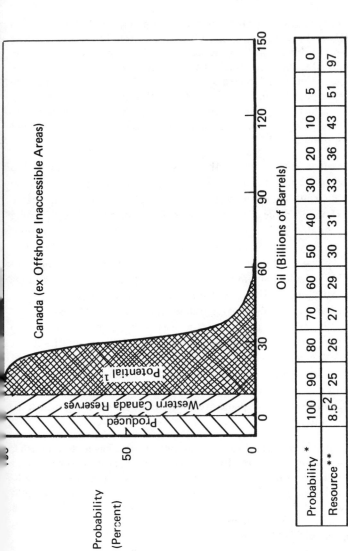

Canada (ex Offshore Inaccessible Areas)

Oil (Billions of Barrels)

Probability *	100	90	80	70	60	50	40	30	20	10	5	0
Resource**	8.5²	25	26	27	29	30	31	33	36	43	51	97

1 Includes discovered resrouces in frontier regions

2 Includes Western Canada reserves plus discovered resources in frontier regions. Past production is not included

*Probability in percent.

**Resource in billion barrels of oil and trillion cubic feet of gas.

Source: Energy, Mines, and Resources Bureau (Canada), *Oil and Gas Resources of Canada*, p. 3, Figure 1.

current and foreseeable future technology for unbounded cost, what can be so recovered is an extrapolation from experience in which costs play a leading role and limit the intensity of the recovery effort. The interpretation of this assertion employed in the study is something less:

> Potential hydrocarbon resources that may be non-economic today or in the near term must not be excluded from ultimate recoverable resources. The economics of supply are subject to radical change and are not entirely predictable. Conceptually, the technologically recoverable resource must include all the prognosticated potential deposits, recognizing that some portion of the resource may never become an economic reserve. Nevertheless, it is necessary to establish some arbitrary minimum cutoff, well below any foreseeable economic limit, in order to exclude hydrocarbons in some accumulations that approach "background" level.[20]

The concept of *petroleum play* underlies the assessment procedures employed in the Canadian study. Inferences are made about the number of deposits in a play and about the size distribution of these deposits. Probabilities for the aggregate volume of hydrocarbons in place are *derived* rather than directly assessed, using a probabilistic play model. This model dictates how probabilities for the number of deposits in the play and for their size distribution are to be combined to produce a probability law for the aggregate volume of hydrocarbons in place in all deposits in it. The logical paradigm underpinning the Canadian study requires the geologist to identify, a priori, the "natural" microunit (the play) considered by oil and gas operators responsible for planning exploration strategy and doing it. In highly speculative frontier provinces (the West Coast Shelf, for example) information is so sparse that identification of individual play possibilities is not feasible. Larger rock aggregates, either sedimentary unit or basin, are appraised using geologic-volumetric methods; yield and volume of sediment are regarded as uncertain quantities, and probabilities are assessed for them. Probabilities for the total volume of hydrocarbons in place are computed by combining probabilities for yield and for volume:

> The volumetric approach makes use of analogues to the basin, area or rock unit under consideration, but is usually done at the basin level. Ultimate yield of recoverable hydrocarbon per unit volume of rock is determined for sedimentary basin geologically similar to the basin under consideration. The yield value is multiplied by the volume of the sediments in the basin. The approach is usually inadequate because analogues are not available for many basins and, where they are, reliable field figures (barrels per cubic mile) are not available. The method is, however, a useful check on the "exploration play" method and, if little data are available, the "volumetric" approach may be the best alternative. It is desirable to describe estimates of the yield and perhaps of the volume by frequency distributions and to incorporate the possibility that the area may be barren of pooled hydrocarbons.

In the "exploration play" approach, estimates are made for individual, demonstrated or conceptual exploration plays in an area. This requires more data than the "volumetric" approach, but answers the question—What are the sizes of the accumulations present?—and is more specific as to where the hydrocarbons may occur. Both of these considerations are necessary for economic analysis.[21]

The exploration play model used in the Canadian study is composed of oil and gas occurrence attributes, which describe geologic conditions that must obtain for hydrocarbons to be present in an anomaly; potential equation variables that jointly determine the quantity of hydrocarbons in a prospect conditional on some positive amount being present; and an equation describing how the amount of hydrocarbon(s) in a prospect depends on potential equation variables. The data form in Table 13-4, from the report, shows oil occurrence attributes and potential equation variables considered in appraisal of the hydrocarbon potential of a typical play.

Oil occurrence attributes or factors are treated as if they were dichotomous; that is, each attribute is "favorable" or "unfavorable" or is "present" or "absent." This is a considerable simplification of reality, adopted as a compromise between descriptive accuracy and simplicity in modeling and assessment. Behind each occurrence attribute is a rich and variegated data set, including core analyses, geochemical arrays, maps and their interpretations, and more, upon which assessment of the likelihood of the presence or adequacy of an attribute is based. A story can be told about each attribute. Squeezing the story into a descriptive dichotomy is bound to distort the plot. The alternative, increasing the number of descriptive categories that each attribute can attain, enormously complicates assessment and would require modification of descriptive categories from play to play to conform to changes in geologic character. Dichotomized descriptions seemingly avoid complexity and allow use of a uniform format for assessment from one play to the next. But an uncomfortable-looking lump has been swept under the rug.

The probability that hydrocarbons are present in a generic prospect is computed by multiplying the probabilities assigned to "present" or "adequate" for each attribute. This computation rests on the assumption that attributes are mutually independent for a given prospect. Current geological reasoning supports this independence assumption so long as it is applied to a single prospect. Then, if p_i is the probability that attribute i is present or adequate for a generic prospect and there are m attributes,

$$\prod_{i=1}^{m} p_i = \text{the probability that the prospect is a deposit.} \qquad (13\text{-}1)$$

The probability that a play contains at least one deposit is equal to one minus the probability that it contains none. Independence of attributes *across* pros-

Table 13-4. Data Form Used in Estimating Play Parameters.

Play/Prospect							Date		Remarks
Potential Equation Variables							"Oil" Occurrence Factors		
Conditional Probability Percent GT.							Presence or adequacy	Marginal Probability	
100	95	75	50	25	5	0			
Area of closure 1	10	18	23	26	33	50	Geometric closure	.8	
Reservoir thickness 10			40			90	Lithofacies	1	
Porosity .08			.12			.16	Porosity	.8	
Trap fill .05			.4			.7	Seal	1	
Recovery			.35				Timing	.5	
Water saturation			.25				Source	.5	
Shrinkage			.7				Preservation	1	
Gas fraction			.5				Recovery	1	
Number of prospects 3	8	15	20	25	40	60			8,000 feet Average depth
							Product	.16	

Source: Energy, Mines, and Resources Bureau (Canada), Oil and Gas Resources of Canada, 1976, p. 67, Figure A2.

pects, an assumption adopted in the Canadian study, precludes learning from experience: The probability that the ith attribute is "present" or "adequate" at the $(n + 1)$st prospect drilled is p_i, irrespective of the outcome of drilling prospects $1, 2, \ldots, n$. Assuming mutual independence between attributes for each prospect as well as *across* prospects for a given attribute, and defining q as

$$q = \prod_{i=1}^{m} p_i \, , \tag{13-2}$$

if there are N prospects, the probability that the play contains at least one deposit is $1 - q^N$.

Functional dependence of an attribute across prospects is logically opposite to mutual independence. An example is to assume that if drilling shows an attribute to be present or adequate at any one prospect, then it is known with certainty that the attribute is present or adequate at all other prospects. Similarly, if the attribute is absent or not adequate at one prospect, then it is known with certainty to be absent or not adequate at all others. In this case, learning from experience is instantaneous upon drilling only one prospect: The first outcome determines with certainty all subsequent outcomes. If p_i is the probability that the ith attribute is present or adequate for any one prospect prior to drilling the first prospect, and attributes are mutually independent, then the probability that all prospects are deposits is

$$q = \prod_{i=1}^{m} p_i$$

irrespective of the number of prospects.

Neither independence nor functional dependence adequately describe how judgments about occurrence attributes typically respond to drilling information. In between these extreme assumptions are many possibilities.

The amount of hydrocarbons in a prospect is a function of potential equation variables, the particular function chosen for the Canadian study being that shown in Table 13-5. Fractiles (of the marginal distribution) for each potential equation variable were assessed and recorded as shown in Table 13-5. Underlying the assessment procedure are three assumptions: First, potential equation variables are assessed conditional on all oil occurrence attributes being present or adequate. Potential equation variables conditioned on this event are assumed to be mutually independent across prospects. How reasonable are these independence assumptions? According to many geologists, asserting independence across prospects for area of closure and for trap fill is plausible. Values assumed by these two attributes are determined in part by the structural geometry of individual prospects and are defined only at individual prospect locations; that is, they can take on nonzero values only at specific locations.

On the other hand, porosity is an attribute that is possibly nonzero over the total (within play) area encompassing the sedimentary unit that is the target for drilling. The definitional "texture" is consequently very different from attributes related to structure. Knowledge of porosity at coordinate location (x, y) can reasonably be considered to influence judgments about porosity at $(x + h, y + k)$, and the smaller h and k are, the stronger the influence is likely to be. Hence, a priori, it seems more reasonable to assume that porosity values $\tilde{\phi}(x, y)$ at (x, y) coordinate points within the play area A are probabilistically dependent than to assume they are independent. Then the probability law assessed for $\tilde{\phi}(x, y)$, $(x, y) \epsilon A$, must incorporate spatial dependence and assessment procedures expanded to include appraisal of the nature of these dependencies. How to do this without imposing an intolerable burden on the geologists who have to make these judgments is an open question for research. Although independence is flawed as a structural assumption, it vastly simplifies assessment and computation of relevant probabilities, and if the assessor is lucky the final results may not prove to be very sensitive to the choice.

The mathematical structure of the Canadian model is simple enough to allow calculation by numerical methods of the probability distribution for the total amount of hydrocarbons in place in a play. As before, define p_i to be the probability that the ith attribute is present or adequate at any one of N prospects for $i = 1, 2, \ldots, m$ (m attributes), and

$$q = \prod_{i=1}^{m} p_i .$$

Assuming mutual independence of all occurrence attributes and potential equation variables, the probability of a play containing n deposits is for $n = 0, 1, 2, \ldots$:

$$P\left\{\tilde{N} = n\right\} = \sum_{N=n}^{\infty} P\left\{\tilde{N} = N\right\} P\left\{\tilde{n} = n \mid N\right\} \tag{13-3}$$

$$= \left[\frac{q}{1-q}\right]^{n} \sum_{N=n}^{\infty} P\left\{\tilde{N} = N\right\} \binom{N}{n} (1-q)^{N} .$$

When the cumulative probability function for the amount in place in n deposits is defined as $F^{*n}(R) = P\left\{\tilde{X}_1 + \ldots + \tilde{X}_n \leq R\right\}$, the cumulative probability function for the amount in place in the play as a whole is

$$P\left\{\tilde{R} \leq R\right\} = \sum_{n=1}^{\infty} P\left\{\tilde{N} = n\right\} F^{*n}(R) \quad \text{for } R > 0 , \tag{13-4}$$

and

$$P\left\{\tilde{R} = 0\right\} = P\left\{\tilde{n} = 0\right\} = \sum_{N=0}^{\infty} P\left\{\tilde{N} = N\right\} (1-q)^{N} . \tag{13-5}$$

Table 13-5. Alternative Equations Illustrating Two Approaches to the Estimation of Potential Resources.

Volumetric Equation

Area potential = Volume of rock · Hydrocarbon yield/Unit volume

Exploration Play Equation

Prospect potential = Area of trap · Reservoir thickness · Porosity · (1 – Water saturation) · Trap fill · Oil fraction · Recovery · Constant

≡ Volume of pores in trap · Hydrocarbon fraction · Engineering factors

Source: Energy, Mines, and Resources Bureau (Canada), *Oil and Gas Resources of Canada, 1976*, p. 66, Figure A1.

Computation of $P\{\tilde{R} \leq R\}$ or, equivalently, of $P\{\tilde{R} \geq R\} = 1 - P\{\tilde{R} \leq R\}$ was done by use of a Monte Carlo procedure, a procedure conceptually simpler than but possibly not as efficient computationally as some alternatives, such as a fast Fourier transform method.

REFERENCES

1. Bernoulli, J. *Ars Conjectandi* (Basil, Switzerland: 1713). P. S. LaPlace, *Essai Philosophique sur les Probabilites*, 5th ed. (Paris: Bochelier, 1825).
2. D. Lindley, *Making Decisions* (New York: Wiley-Interscience, 1971), p. 18.
3. Ibid.
4. C. J. Grayson, *Decision under Uncertainty: Drilling Decisions by Oil and Gas Operators* (Boston, Mass.: Harvard Business School, Division of Research, 1960).
5. Energy, Mines, and Resources Bureau (Canada), *Oil and Gas Resources of Canada, 1976*, Report EP77-1, 1977. The methods employed by R. McCrossan and his group to generate the projections in this report were developed independently of and earlier than those used to generate the projections appearing in Circular 725, U.S. Geological Survey, Oil and Gas Branch, "Geologic Estimates of Undiscovered and Recoverable Oil and Gas Resources in the United States," Circular 725, U.S. Department of the Interior, 1975.
6. Ibid.
7. Ibid.
8. Ibid.
9. Ibid.
10. E. Barouch and G. M. Kaufman, "On Sums of Lognormal Random Variables," Massachusetts Institute of Technology Sloan School of Management Working Paper 831-76, 1976.

11. From, among others, M. Alpert and H. Raiffa, "A Progress Report on the Training of Probability Assessors," unpublished manuscript, Harvard University, 1969, and E. C. Capen, "The Difficulty of Assessing Uncertainty," *Journal of Petroleum Technology* (August 1976): 843–50.

12. I. D. Steiner, "Models for Inferring Relationships between Group Sizes and Potential Group Activity," *Behavioral Science* 11 (1966): 273–83.

13. R. M. Hogarth, "Methods for Aggregating Opinions," in *Fifth Research Conference on Subjective Probability, Utility, and Decision Making*, European Institute of Business Administration and Centre European d'Education Permanente, 1976.

14. R. M. Hogarth, "Cognitive Processes and the Assessment of Subjective Probability Distributions," *Journal of the American Statistical Association JASA* 70, no. 350 (June 1975): 271–89.

15. S. J. Press, "Qualitative Controlled Feedback for Forming Group Judgments and Making Decisions," *Journal of the American Statistical Association JASA* 73, no. 363 (September 1978): 526–35.

16. Ibid.

17. Energy, Mines, and Resources Bureau (Canada), *Oil and Gas Resources of Canada, 1976*.

18. Ibid.

19. Ibid.

20. Ibid.

21. Ibid.

14 ECONOMETRIC METHODS

Immediately following the Arab oil embargo of 1973, the shortage of natural gas in the United States came into sharp focus. The shortage was induced by the Federal Power Commission's (FPC) fixing of natural gas prices at a level close to unregulated prices that had obtained more than ten years earlier.[1] A combination of increasing real finding costs, fixed prices, and inflation generated both increases in demand for and a decline in exploratory drilling for natural gas.

An economic and political debate over the benefits and social costs of attempting to stimulate increased production from known fields and discovery of new reserves of natural gas unfolded. Should natural gas prices be deregulated? If so, at what rate over what period of time? Should regulatory control of field prices be extended to cover intra- as well as interstate markets? At what price and when would a particular price deregulation policy clear markets of excess demand?

Projections of the consequences of adopting a generic price deregulation policy clearly demand an analysis much wider in scope than is encompassed by any of the models of oil and gas exploration, discovery, and production discussed in the book thus far. Modeling the time rates of amounts of natural gas discovered and produced is only one piece of the analytical pie that must be baked. Synergisms between price paths, demand over time, and time rates of drilling exploratory and development wells and of amounts discovered and amounts produced all must be addressed explicitly.

This set of problems lies in the domain of econometrics. An econometric model is a statistical model the form of which is principally dictated by economic theory and which can be designed either to provide insight into the *structure* of supply and demand or to *project* their future time paths or both.

It is somewhat misleading to view econometric modeling as sharply different in concept from other approaches to modeling petroleum supply, since models called "econometric" share common features with several of the approaches discussed earlier. However, most econometric models incorporating a representation of petroleum supply (1) are highly aggregated geographically, (2) explicitly display interactions between demand for and supply of oil and natural gas, and (3) are used to project the effects of changes in prices and in the regulatory regime on both demand and supply. Thus they do differ in form if not in concept from highly disaggregated models, both in level of aggregation and in scope.

Prior to the 1973 Arab oil embargo, the FPC had an operational econometric model of gas supply in place, as described by Khazzoom.[2] Erickson and Spann[3] had constructed an econometric natural gas model that differed in structure from Khazzoom's. A first version of the MacAvoy-Pindyck[4] natural gas industry model followed. These and Fisher's[5] econometric model of the oil and natural gas industry are precursors of a train of econometric and optimization models (linear and nonlinear programming) that incorporate representations not only of oil and gas discovery and production but of the processes by which alternative sources of energy are exploited, for example, the macroeconomic models of Hudson and Jorgenson and of Manne.[6]

The econometric model builder typically begins with a qualitative description of how economic attributes—prices, costs, taxes—are perceived by participants in markets for petroleum to interact with physical variables, which determine current and future stocks and flow. Criteria for decisionmaking on the part of market actors are posed and applied to a representation of how decisions implied by these criteria influence the time rate of generation of flows from both currently available stocks and future stocks elicited by additional investments in exploration, development, and production. Producers are generally assumed to behave optimally relative to some normative criterion, for example to maximize "profit" or "expected net present value," or possibly "expected utility." The qualitative reasoning employed to describe the behavior of individual decisionmakers may be sophisticated, incorporate key features of complex interactions between economic variables and the process of discovery and extraction, and account for expectations about future flows as a function of both current and future prices, costs, taxes, and regulatory regimes. Yet much precision is lost in translation of a descriptively rich verbal argument to models of the form employed by most econometric model builders to date: Relations between prices, costs, and physical variables such as number of exploratory wells drilled, amounts discovered, and success ratios are represented as a system of stochastic equations, generally but not always linear or log-linear in form.

The loss of both structural and predictive precision occurs because of a series of tactical compromises in model design and in data analysis forced on the econometric modeler who chooses, a priori, to represent supply and demand for oil and gas from a large geographic region (National Petroleum Council [NPC] district or coterminous forty-eight United States, for example) as a set of stochastic equations simple enough in form to allow application of a standard kit of econometric tools for estimation of unknown model parameters given the observed data.

The structure of exploration and development decision problems faced by the oil and gas industry cannot be captured with the same accuracy by highly aggregated oil and gas supply models as with a collection of highly disaggregated models. Disaggregated models designed to treat decision problems actually faced by oil and gas operators produce "rational expectations" or predictions about future behavior based on an explicitly stated economic theory of decision in the face of uncertainty, and, as Attanasi points out, "Because properties of these models of rational expectations are known, observed economic behavior can be tested for consistency with these predictions."[7] In contrast, most existing oil and gas supply models use proxy variables to represent currently unobservable future expectations about decisionmaking behavior because the resulting model is simpler in form, requires much less and more easily available input data, and because certain model forms of proxy variables produce "optimal" predictions relative to a prespecified cost or loss assigned to forecast or prediction error. Unfortunately, the criterion function for which, say, a model employing distributed lag proxies for future expectations is appropriate, may not be compatible with the criterion function employed by industry's decisionmakers:

> A characteristic of nearly all expectation-generating schemes is that predictions are made without consideration of the decision makers' criterion function, which embodies attitudes toward risk, and the penalty function, which determines the economic consequences of under- and overestimating the value of an uncertain variable. Whereas a particular expectation-generating scheme may in some sense provide optimal predictions, there is no assurance that loss functions for prediction and decision problems will be the same. In this sense, schemes that are presented in much of the literature . . . are ad hoc.[8]

Our concern here is narrow: How well do aggregated econometric models project future supplies of oil and gas from new discoveries? What particular characteristics of econometric models determine the accuracy (or lack of it) with which numbers of future discoveries, amounts discovered, and amounts produced, are predicted?[9]

As devices for projecting oil and gas supply over a future period of more than a few years, econometric models have not done well, in particular for periods during which both costs of finding and producing and prices have risen sharply. A conjunction of problems with data and certain features of the exploration environment combine to degrade their predictive performance:

- Among problems with data are definitional imprecision, large changes in levels and ranges of values of prices and cost in recent years, and relatively small-sized samples for estimation due to aggregation both geographical and over time (year-to-year measurements).

- Improvements in technology increase success ratios.

- Econometric modelers did not anticipate the saw-toothed shape of amounts discovered over time in regions still possessing unexplored sedimentary horizons (the outer Continental Shelf and the Gulf of Mexico in particular) nor did they anticipate large potential additions to reserves from deep gas plays with the arrival of new technology and higher prices, as in the Overthrust Belt and Tuskaloosa Trend plays.

and finally,

- The functional forms employed in most econometric models do not conform closely to the physical character of exploration, discovery, and production.

The functional forms chosen to represent highly aggregated time series of drilling rates, success ratios, and amounts discovered bear little resemblance to their disaggregated counterparts. The hope is that by combining a sufficiently large number of geologically, technologically, and economically distinct fields into a single aggregate, the behavior over time of the aforementioned quantities will be probabilistically smoothed in sufficient degree to permit accurate representation by simple, estimable functions. The degree to which this obtains poses a problem of structural validation, for, if times series of success ratios, of amounts discovered, and of amounts produced are highly aggregated, then it is difficult if not impossible to distinguish real structure from casual correlations.

A line of thought motivating the structure of equations describing exploration and production in some econometric models of petroleum supply and demand is that an economically rational agent who explores faces a "portfolio choice" problem, somewhat similar to an investor in the stock market. At a given time this agent must decide how to allocate drilling effort over the set of prospects and fields perceived by him as candidates for additional drilling. At one extreme are low-risk development or infill wells to be drilled in known deposits; at the other extreme are rank wildcat wells to be drilled in search of new deposits. The former type of drilling, at the *intensive* margin, is characterized by high success ratios and low expectations of incremental amounts of hydrocarbons found relative to the latter, drilling at the *extensive* margin. How the rational explorer allocates effort to each of the two types of drilling is influenced by levels of and changes in levels of wellhead prices, costs, and by his perception of drilling success probabilities and of amounts that might be discovered conditional on a commercial discovery. An explicit solution to this problem of choice is not possible without access to detailed information about the agent's attitude toward risk, judgments about success probabilities, and judgments about amounts that, conditional on success, individual wells of each type might discover.

The level of aggregation at which most econometric models are pitched smears the portfolio choice problem. Aggregated time series data do not clearly reveal the directional response of drilling to changes in economic variables, much less to changes in perception of geologic potential and, in the absence of specification of the microeconomics of individual sets of known fields and of prospects, the effect of an increase in price relative to cost on the time rate of sizes of deposits discovered is difficult to predict. Over the very long run the expected sizes of discoveries may decrease as a function of exploratory effort, but on the way to depletion the evolution of discovery sizes can see-saw violently. If a price increase motivates a search for geologic anomalies not previously believed to be suitable targets for drilling, then drilling at the extensive margin can result in a sharp increase in amounts discovered over a short period of time before a decline again sets in. The recent intensive exploitation of the Overthrust Belt and Tuskaloosa Trend deep gas plays are examples of exploration aimed at a large class of prospects known to exist but unsought until deep gas prices were deregulated. With 1980 economics in force, proved reserves in the latter play range from 6 to 76 trillion cubic feet,[10] and estimates of 14 billion barrels of recoverable oil and 52 billion cubic feet of recoverable gas have been cited for the Overthrust Belt. During periods when geologic imagination does not successfully conjure up new target types, financial returns to wildcat drilling may be modest by comparison, and even exceeded by allocating well drilling effort away from rank wildcatting to extension, outpost, and development wells. In sum, there is no economic argument, a priori, dictating either an increase or a decrease in average discovery size within most large (highly geologically aggregated) regions over a short horizon (say, five to ten years). Behavior of sizes discovered is as dependent on the industry's geologic template for the region as on economics, a behavioral determinant that is not adequately represented in *any* of the models discussed here. And accentuating the difficulties is the fact, again, that many aggregated yearly time series of data labeled "discoveries" do not in fact represent discoveries in the year in which they were made.

Since model parameters are not known with certainty a priori, they must be estimated from past data. Even during periods when time series of prices and costs remained within a reasonably narrow range and no order-of-magnitude shifts in values occurred (1950–1973), it is not clear that estimated parameter values closely mirrored "true" parameter values. Pindyck's[11] comparative analysis of three econometric supply models of natural gas, to which we turn shortly, demonstrates the sensitivity to shifts in the data of estimated parameters for each of these three models. Even if one accepts the structure of these models as a reasonable representation of oil and gas supply, the soundness of the predictions they generated is thrown into question.

Validation of the predictive accuracy of a model requires, at a minimum, that observational time series data be partitioned, a portion of the data be used to estimate model parameters, and the remainder of the data be retained for comparison with projections based on estimates of parameter values imputed by the initial block of data. This provides significant information about the quality of (reduced form) forecasts. Unfortunately, most published econometric studies use up all the data to estimate parameters.[12]

In succeeding sections of this chapter several prominent econometric models are described in just enough detail to allow comparison of their structure with models discussed earlier. This sets the stage for discussion of their predictive performance. The reader is directed to original sources for structural and economic interpretation of numerical values of parameter estimates for the models discussed.

KHAZZOOM'S FPC MODEL OF GAS SUPPLY

The logic behind the specific functional relations chosen to represent the physical processes at work is fairly straightforward: In Khazzoom's model the discovery process is presented as a schematic black box, the input to which is price.[13] This input generates levels of exploratory well drilling for both associated and nonassociated gas, which in turn determines well successes and failures; each success in a given period is associated with a reservoir size, the sum of which constitutes the amount discovered in that time. Khazzoom displays this process schematically as in Figure 14-1. An almost identical schematic for the impact of oil price on directional drilling for oil can be drawn.

The amount $\widetilde{ND}_t$ of recoverable gas in reservoirs newly discovered during the tth time period (year) is an uncertainty quantity whose expectation is assumed to depend on the regulated ceiling prices C_{t-1} and C_{t-2} of gas during the preceding two years, the prices PO_{t-1} and PO_{t-2} during the preceding two years, and the prices PL_{t-1} and PL_{t-2} of natural gas liquids during the preceding two years:

$$\widetilde{ND}_t = f_1(C_{t-1}, C_{t-2}, PO_{t-1}, PO_{t-2}, PL_{t-1}, PL_{t-2}) + \widetilde{\omega}_t , \qquad (14\text{-}1)$$

where $\widetilde{\omega}_t$ is a random error term with mean zero. Extension and revisions $\widetilde{XR}_t$ in period t are represented as depending on C_t, PO_t, ND_{t-1}, and XR_{t-1}:

$$\widetilde{XR}_t = f_2(C_t, PO_t, ND_{t-1}, XR_{t-1}) + \widetilde{\delta}_t, \qquad (14\text{-}2)$$

where $\widetilde{\delta}_t$ is a random error term with mean zero. Khazzoom studies two specific functional forms for $\widetilde{ND}_t$, one linear and one quadratic form. He treats $\widetilde{XR}_t$ similarly: his equations (2.1)–(2.4) are as follows:

$$\widetilde{ND}_t \qquad\qquad\qquad\qquad\qquad\qquad\qquad\qquad\qquad\qquad\qquad (14\text{-}3)$$

$$= \mu_0 + \mu_1 \sum_{i=1}^{2} C_{t-i} + \mu_2 \sum_{i=1}^{2} PO_{t-1} + \mu_3 \sum_{i=1}^{2} PL_{t-i} + \mu_4 \sum_{i=1}^{2} ND_{t-i} + \widetilde{\omega}_t ,$$

Figure 14–1.   The Discovery Process in Khazzoom's Model of Gas Supply.

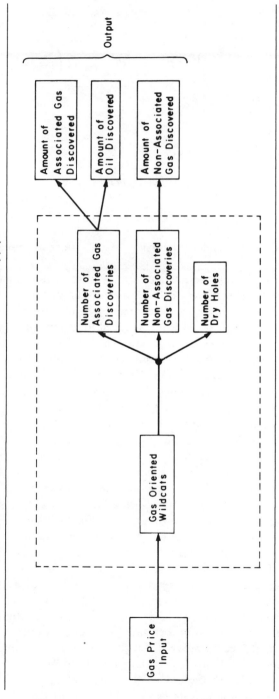

Source: J. David Khazzoom, "The FPC Staff's Econometric Model of Natural Gas Supply in the United States," *Bell Journal of Economics and Management Science* 2, no. 1 (Spring 1971): 51–93. Reprinted with permission.

$$\widetilde{ND}_t = \mu_0 + \mu_1' \sum_{i=1}^{2} C_{t-i} + \mu_2' \left(\sum_{i=1}^{2} C_{t-i} \right)^2 + \mu_3' \sum_{i=1}^{2} PO_{t-1}$$

$$+ \mu_4' \left(\sum_{i=1}^{2} PO_{t-i} \right)^2 + \mu_5' \sum_{i=1}^{2} PL_{t-i} + \mu_6' \left(\sum_{i=1}^{2} PL_{t-i} \right)^2$$

$$+ \mu_7' \sum_{i=1}^{2} ND_{t-i} + \widetilde{\omega}_t' \ . \tag{14-4}$$

$$\widetilde{XR}_t = \pi_0 + \pi_1 C_t + \pi_2 PO_t + \pi_3 PL_t + \pi_4 ND_{t-1} + \pi_5 XR_{t-1} + \widetilde{\xi}_t \tag{14-5}$$

$$\widetilde{XR}_t = \pi_0' + \pi_1' C_t - \pi_2' C_t^2 - \pi_3' PO_t + \pi_4' PO_t^2 t + \pi_5' PL_t$$

$$- \pi_6' PL_t^2 + \pi_7' ND_{t-1} + \pi_8' XR_{t-1} + \widetilde{\xi}_t' \ , \tag{14-6}$$

These equations are "a condensed version of eight finer categories of gas supply."[14]

The salient features of these equations are first, that as functions of the parameters to be estimated from the data they are linear functions and second, that the error terms are additive, a standard and well-understood model form. In econometric parlance, this model is in "reduced form" and of "distributed lag" type, as both new discoveries and extensions and revisions at period t are dependent on past ceiling prices, oil prices, and natural gas prices. The extensions and revisions equation is "autoregressive" as $\widetilde{XR}_t$ is regarded as a function of XR_{t-1}.

The data to which these equations are applied consist of yearly observations from 1961 to 1969 from twenty-one FPC price districts. The FPC price districts do not correspond to a grouping of data dictated by geologic attributes. Khazzoom is sensitive to this distinction, and, in place of incorporating traditional district dummy variables to account for differences in expectations of $\widetilde{XR}_t$ and $\widetilde{ND}_t$ from district to district, he estimated regional effects for groups of districts by grouping together districts that are "geologically similar." The regional effects are incorporated into the intercept terms in Eqs. (14-3)-(14-6).

Parameter estimation is based on yearly data for twenty-one FPC price districts for the time series: 1961-1968 and 1961-1969. Khazzoom points out that these districts accounted for 90 percent of total gas supply in the United States in 1961-1969. He questions the quality of the data on extensions and revisions, hypothesizing that "the revisions data are distorted by factors which we know only vaguely about at present."[15] Subsequent to 1960, the price series constructed for estimation follows 1960 FPC ceiling price guidelines, subsequent amendments, and decisions from area rate proceedings. Ordinary least-squares estimation of parameters is employed "in the absence of an *a priori* reason for expecting random (error) terms to be serially correlated."

Khazzoom addresses specific policy questions with his model: What projections of future supply follow from leaving the ceiling price fixed at 1969 levels?

From planned changes in future ceiling prices? In order to simplify the interpretation of his predictions and sort out the impact of ceiling price changes alone, he abridges the model (14-3) for new gas discoveries to:

$$\widetilde{ND}_t = \beta(C_{t-1} + C_{t-2}) + \delta(ND_{t-1} + ND_{t-2}) + \widetilde{\omega}_t \ . \tag{14-7}$$

Since Eq. (14-3) includes C_{t-1} as an explanatory variable for $\widetilde{ND}_{t-1}$ as well as $\widetilde{ND}_{t-2}$, and $\widetilde{ND}_{t-2}$ appears as an explanatory variable for $\widetilde{ND}_{t-1}$, $\widetilde{ND}_t$ can be expressed as a function of the values of C_{t-1}, C_{t-2}, ... alone by successive elimination of lagged values of new discoveries. For example, as in Khazzoom's Eq. (5.2):

$$E(\widetilde{ND}_t) = \beta C_{t-1} + \beta(1 + \delta)C_{t-2} + \beta\delta C_{t-3} + \delta(1 + \delta)ND_{t-2} + \delta^2 ND_{t-3} \ . \tag{14-8}$$

This equation shows that a rise in the ceiling price in just one year, say year $t-2$, affects the expectation of supply over the entire future time horizon; that is, a 1 cent rise in price at $t-2$ increases $E(\widetilde{ND}_{t-1})$ by βC_{t-2}, increases $E(\widetilde{ND}_t)$ by $\beta(1 + \delta)C_{t-2}$, and in general increases $E(\widetilde{ND}_{t+\tau-1})$ by $\beta f_\tau(\delta)C_{t-2}$, where $f_\tau(\delta)$ is a function of δ. The shape of the response to a price increase at $t - 2$ of $E(\widetilde{ND}_{t+\tau-1})$ for $\tau = 1, 2, \dots$ depends on the particular values assigned to β and δ.

ERICKSON-SPANN MODEL

Erickson and Spann[16] estimate the response of new discoveries of natural gas to changes in price with an econometric model of oil and gas discovery that is different in several respects from Khazzoom's model. In place of representing the amount of new gas discovered in a particular time period as a quantity directly dependent on prices in previous periods, new gas discovered in petroleum administration for defense district i during a time period t is viewed as the product of the number W_{it} of wildcat wells drilled at t, the success ratio F_{it} during period t, and the average size N_{it} of gas discovery in i during period t. In their notation, defining S_{it} as the average size of oil discovery in i during t, P_{it} as the (constant 1947-1949 dollars) deflated wellhead price per barrel of oil in district i at t and similarly G_{it} as the deflated wellhead price per million cubic feet (MCF) of gas, $Z_1, \dots, Z_4$ as district dummy variables, X_{it} as number of Texas shutdown days, $\delta = 1$ if a Texas district and zero otherwise, M_{it} and R_{it} as the percentage of wildcats drilled by all companies and by major companies respectively, in i during t, and D_{it} as average wildcat depth in feet in i during t, the well equation is

$$\widetilde{W}_{it} = f_1(P_{it}, G_{it}, F_{i,t-1}, R_{it}, D_{i,t-1}, \delta X_{it}, Z_1, Z_2, Z_4) \times \widetilde{\epsilon}_{it} \ . \tag{14-9}$$

The success ratio equation is

$$\widetilde{F}_{it} = f_2 (P_{it}, G_{it}, \delta X_{it}, Z_1, Z_2, Z_4) \times \widetilde{\xi}_{it} \quad , \tag{14-10}$$

and the average oil and average gas discovery size equations are, respectively,

$$\widetilde{S}_{it} = f_3 (P_{it}, G_{it}, F_{i, t-1}, \delta X_{it}, Z_1, Z_2, Z_4) \times \widetilde{\theta}_{it} \quad , \tag{14-11}$$

and

$$\widetilde{N}_{it} = f_4 (P_{it}, G_{it}, F_{i, t-1}, \delta X_{it}, Z_1, Z_2, Z_4) \times \widetilde{\eta}_{it} \quad . \tag{14-12}$$

The residual error terms all have expectation one.

The particular functional forms chosen for these equations are multiplicative rather than additive in other than dummy variables Z and X; for example,

$$\log \widetilde{N}_{it} = \beta_0 + \beta_1 \log P_{it} + \beta_2 \log G_{it} + \beta_3 \log F_{i, t-1} + \beta_4 (\delta X_t)$$

$$+ \beta_5 ([1 - \delta] X_t) + \beta_6 Z_1 + \beta_7 Z_2 + \beta_8 Z_4 + \log \widetilde{\eta}_{it} \quad . \tag{14-13}$$

This allows the logarithm of the aggregate amount of gas discovered in district i during period t, $\widetilde{W}_{it} \times \widetilde{F}_{it} \times \widetilde{N}_{it}$, to be represented as a linear function of parameters that must be estimated from the data; that is, $\log \widetilde{W}_{it} \times \widetilde{F}_{it} \times \widetilde{N}_{it}$ is a linear function of $\log P_{it}$, $\log G_{it}$, $\log D_{i, t-1}$, Z_{it}, and X_t. Consequently the unknown parameters can be estimated using standard econometric methods for treating time series cross-sectional data. In this model, certain of the coefficients are interpretable as elasticities; a further advantage is that elasticities are additive. For example, the own-price elasticity of supply for gas discoveries is the sum of the elasticities for wildcat drilling, for the success ratio, and for average gas discovery size. Yearly data for 1946 through 1958–59 are used to estimate model parameters in the 1971 versions of the Erickson–Spann model.

MACAVOY-PINDYCK MODEL

In 1973 MacAvoy and Pindyck[17] presented the first version of an econometric model of natural gas supply designed to deal with many of the same issues examined by Khazzoom and by Erickson and Spann. The model was intended to answer the questions: What is the impact on supply and on regulatory induced shortages in markets of complete deregulation of wellhead prices? What is the impact of increased regulation with prices set relative to "cost of service"? And what of an in-between policy allowing specific patterns of wellhead price increases? Although the functional forms that represent relations between dependent variables such as success ratios, average discovery sizes, and wells drilled and explanatory variables that "drive" the dependent variables are similar to those

Figure 14-2. The MacAvoy–Pindyck Model.

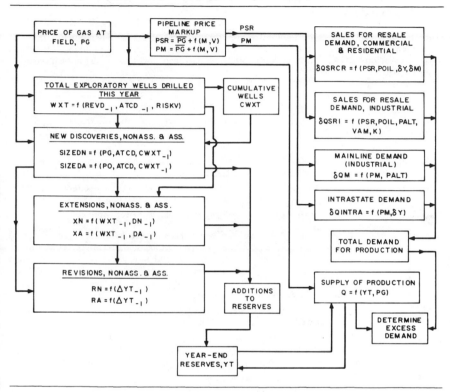

Source: P. MacAvoy and R.S. Pindyck, "Alternative Regulatory Policies for Dealing with the Natural Gas Shortage," *Bell Journal of Economics and Management Science* 4, no. 2 (Spring 1973): 51–98. Reprinted with permission.

employed by Khazzoom and by Erickson and Spann, MacAvoy and Pindyck formulate a more elaborate model, which describes production and demand in both field markets and wholesale delivery markets. Figure 14-2 is a schematic displaying components of their model for a generic region plus the variables on which the dependent variable in each equation of their model explicitly depends. The spatial interconnections between production districts and regional wholesale markets are, for simplicity in display, omitted.

Table 14-1 is a glossary of abbreviations and acronyms for the model. Of particular interest given the theme of this book are field market equations describing exploratory drilling, associated and nonassociated average gas discovery size per exploratory well drilled, and extensions and revisions. Nonassociated gas discovered in field market i during a given year t, DN_{ti}, equals the product of average size of discovery $SIZEDN_{ti}$ and exploratory wells drilled WXT_{ti}. A similar identity for associated gas holds: $DA_{ti} = SIZEDA_{ti} \times WXT_{ti}$. Equations for

Table 14-1.   List of Variables and Data Sources of the MacAvoy–Pindyck Model.

The variables used to estimate the model, including sources of data, are listed below. The list is divided into field market variables and wholesale market variables. Following American Gas Association (AGA) procedures, all offshore data are included in the appropriate on-shore district.

Wells

WXT	Total exploratory wells, successful or otherwise.
CWXT	Cumulative number of exploratory wells (*WXT*), from 1963 to *t*.
REVD	Index of average deflated revenue from gas and oil for each production district.
ATCD	Average drilling cost per well of exploratory and development wells, from *Joint Association Survey* (AGA–API–CPA).
RISKV	Index of the variance in size of discoveries over time by production district. This is a pure cross-section variable that does not change in time.
AFX	Average depth of an exploratory well (averaged over total wells).

Reserves   All data from AGA–API–CPA *Reserves of Crude Oil*. Only available in disaggregated (i.e., associated-dissolved, nonassociated, by production district) form from 1966 except for year-end reserves (*YN, YA, US*) which we have from 1965. (Disaggregated data extrapolated for 1964 and 1965 to provide starting values for lagged and unlagged endogenous variables in the historical simulation.) Data are given in millions of cubic feet.

SIZEDN	Average size of nonassociated discoveries, by production district (averaged over total wells, dry and successful).
SIZEDA	Average size of associated discoveries, by production district (averaged over total wells, dry and successful).
YN	Year-end nonassociated reserves. Reserves as defined by the AGA.
YA	Year-end associated-dissolved reserves. See *YN*.
YT	Year-end total gas reserves. *YT* = *YN* + *YA*.
YO	Year-end oil reserves.
US	Year-end reserves in underground storage other than in their original locations.
XN	Extensions, of nonassociated gas. Includes any newly proved reserves already established in pools and fields.
XA	Extensions of associated-dissolved gas. See *XN*.
RN	Revisions of nonassociated gas. Includes any proved *decreases* in size of proved reserves discovered by drilling of extension or development wells or changes (+ or –) resulting from better engineering estimates of economically recoverable reserves in established pools.
RA	Revisions of associated-dissolved gas. See *RN*.
FN	New-field discoveries of nonassociated (discovered by new-field wildcats).
FA	New-field discoveries of associated-dissolved gas. See *FN*.
PN	New pool discoveries of nonassociated gas.
PA	New pool discoveries of associated-dissolved gas.
DN	= *FN* + *PN*. Total new discoveries, nonassociated.
DA	= *FA* + *PA*. Total new discoveries, associated-dissolved gas.

(Table 14-1 continued on next page)

Production	All data are from AGA–API–CPA. *Reserves of Natural Gas*, and are available in disaggregated form from 1965 for gas (and 1940 for oil). In million cubic feet and thousand barrels.
QN	Production of nonassociated gas, i.e., net production.
QA	Production of associated-dissolved gas, net production.
QO	Production of oil.
QSRI	Sales for resale that will end up as industrial sales are determined by multiplying total sales for resale (from FPC Form 2 Reports) by the fraction of total industrial natural gas consumption (from U.S. Bureau of Mines' *Minerals Yearbook*).
QSRCR	The difference between total sales for resale and industrial sales for resale. (See *QSRI* for sources.)
QINTRA	The difference between total consumption of natural gas by state and total mainline sales plus sales for resale plus field sales in that state. Figures on total consumption by state are taken from Bureau of Mines' *Minerals Yearbook*, as are total field sales. Mainline sales and sales for resale figures are from FPC Form 2 Reports.
QM	Total mainline industrial sales volume by interstate pipeline companies by state and year (1955 to 1970) in million cubic feet, from FPC Form 2 Reports.
Prices	
PG	New contract price of interstate sales of gas at the wellhead, from Table F, FPC, *Sales of Natural Gas* for 1952 to 1971 in cents per million cubic feet [MCF]. (Compiled by Foster Associates, Inc.)
PO	Wellhead price of oil in \$/barrel for 1954 to 1971, from Bureau of Mines' *Minerals Yearbook*.
POIL	Average price in dollars per MCF-energy-equivalent of fuel oil paid by electric power companies by state, from Edison Electric Institute *Statistical Annual of the Electric Utility Industry*.
PALT	Price of alternate fuels in dollars per MCF-energy-equivalent. Weighted average (over kilowatt-hours generated) of prices of fuel oil and coal consumed by electric utility industry in generating electric power, by state and year from 1962 to 1969. (See *POIL* for source.)
PM	Price of mainline industrial sales made on a firm basis, by state and year from 1952 to 1970. Data taken from FPC Form 2 Reports, in \$/MCF.
PSR	Price of sales for resale in \$/MCF from 1956 to 1970, by state and year. Data taken from FPC Form 2 Reports.
M	The distance from the center of a producing region to the population center of the consumption region.
V	The sum of the squares of pipeline diameters entering the state, by state.
VAM	Value added in manufacturing in millions of dollars by state and year, from 1958 to 1969 (interpolated for 1968). Data taken from U.S. Department of Commerce, Bureau of the Census, *Annual Survey of Manufacturers*.
K	New capital expenditures in millions of dollars; see *VAM* for source.
Y	Personal income by state, from 1956 to 1969, in millions of dollars. Data taken from U.S. Bureau of Economic Analysis, *Survey of Current Business*.
N	Population by state and year, from 1955 to 1970, in millions. Data taken from U.S. Department of Commerce, Bureau of the Census, *Current Population Reports*.

Source: MacAvoy and Pindyck, "Alternative Regulatory Policies for Dealing with the Natural Gas Shortage." Reprinted with permission.

WXT, $SIZEDA$, $SIZEDN$, extensions XN and XA of nonassociated gas, respectively, are formulated.

The model geographically spans four regional field markets. Dummy variables are used to account for the field market effect on each of the seven dependent variables just described. Incorporating this effect into the constant term of each equation and dropping market subscripts for simplicity in exposition, the field market equations for exploratory wells drilled, and associated and nonassociated average discovery sizes take the form:

$$WX\widetilde{T}_t = \beta_0 + \beta_1 (REVD_{t-1}) + \beta_2 (ATCD_{t-1}) + \beta_3 (RISKV) + \widetilde{\epsilon}_t^{(1)} \; ;$$

$$(14\text{-}14)$$

$$SIZED\widetilde{N}_t = \gamma_0 + \gamma_1 (PG_{t-1} + PG_{t-2} + PG_{t-3})$$

$$+ \gamma_2 (ATCD_{t-1} + ATCD_{t-2} + ATCD_{t-3})$$

$$+ \gamma_3 (CWXT_{t-1}) + \widetilde{\epsilon}_t^{(2)} \; ; \qquad (14\text{-}15)$$

$$SIZED\widetilde{A}_t = \delta_0 + \delta_1 (PO_{t-1} + PO_{t-2} + PO_{t-3})$$

$$+ \delta_2 (ATCD_{t-1} + ATCD_{t-2} + ATCD_{t-3})$$

$$+ \delta_3 (CWXT_{t-1}) + \widetilde{\epsilon}_t^{(3)} \; . \qquad (14\text{-}16)$$

Thus exploratory wells drilled in year t are assumed to be driven by three variables: by a weighted sum of an index $REVD$ of oil and gas revenue, by average drilling cost $ATCD$ for all wells drilled in a district or field market in the previous year, and by a depletion surrogate. Effects of depletion of the resource base are assumed to be captured by inclusion of cumulative wells drilled up to and including year $t - 1$, $CWXT_{t-1}$, in the equation for discovery sizes.

These equations are "pooled" in that the coefficients of exploratory variables are assumed to be identical across field markets. For example, the coefficients β_1, β_2, and β_3, are constants independent of the field market the equation represents. This implies that given $\beta_2 < 0$, a unit increase in average drilling costs generates the same expected decrease in exploratory wells drilled in year t in all four markets; that is, β_2. Similarly, the coefficients γ_3 and δ_3 of cumulative wells drilled in the average size equations for associated and nonassociated gas equations are the same for all field markets. In each equation heterogenous effects of explanatory variables on the dependent variable across field markets are thrown into the dummy variable coefficients.

The equation for extensions XN of nonassociated gas is of exactly the same form as that of associated gas; the former is

$$X\widetilde{N}_t = \alpha_0 + \alpha_1 WXT_{t-1} + \alpha_2 DN_{t-1} + \widetilde{\omega}_t \ . \tag{14-17}$$

Revisions RN of nonassociated gas are modeled somewhat differently. MacAvoy and Pindyck argue that large short-run reserve increases in a given year increase revisions in the following year, so RN at t should depend on the change $YT_{t-1} - YT_{t-2}$ in reserves from $t - 2$ to $t - 1$:

$$RN_t = \eta_0 + \eta_1 (YT_{t-1} - YT_{t-2}) + \widetilde{\theta}_t \ . \tag{14-18}$$

Associated revisions are modeled similarly.

The description of physical variables that combine to generate supplies of natural gas is completed by specification of production from reserves. In place of representing production from field markets as defined for well- and discovery-size equations, the United States is divided into two regions: Louisiana South and the coterminous forty-eight states excluding Louisiana South, the rationale being the Louisiana South production is mostly offshore and so has rather different cost characteristics from most of (onshore) U.S. production. Production from Louisiana South and from onshore regions are modeled as functions of wellhead price PG and total reserves YT. The equation for Louisiana South production $Q_t^{(L)}$ is of the form:

$$\log \widetilde{Q}_t^{(L)} = a_0 + a_1 \log PG_t^{(L)} + a_2 YT_t^{(L)} + \widetilde{r}_t \ . \tag{14-19}$$

Geographically dividing the remainder of the coterminous forty-eight states into (1) the Permian Basin, (2) Oklahoma, Kansas, and Texas Railroad Commission districts 2, 3, 4, and (3) other producing regions, equations for each are specified. Regional heterogeneity, as in other equations, is captured by use of dummy variables. With the understanding that the regional effect is thrown into the intercept term, production in each of these three regions is represented by equations of the following form:

$$\log \widetilde{Q}_t = b_0 + b_1 \log P_t + b_2 \log YT_t + \widetilde{\mu}_t \ , \tag{14-20}$$

with b_1 and b_2 constant across the three regions defined. In particular a_1 and a_2 for Louisiana South are not constrained to be equal to b_1 and b_2, respectively. Hence the production equations are only partially pooled across the United States.

The Louisiana South equation is estimated using data from 1962 to 1971 and the equations for other regions using 1965-1971. Only after 1965 did production nationwide approach full short-term capacity. Three additional years of data were tacked onto the Louisiana South production series used to estimate model parameters "in order to have sufficient degrees of freedom" in estimation; this is a single time series and so, although pooled data can be brought to bear on parameters of other production equations, only *seven* data points are available for Louisiana South if 1965–1971 data are used!

The authors emphasize the far from ideal quality of the data:

> Data for many variables are either difficult or else impossible to obtain, particularly for years prior to 1966. In addition, many of the data are extremely noisy. As a result, a good deal of compromise was often required in estimating equations between functional forms that were theoretically pleasing and those that lent themselves to the existing data . . . time bounds of the regression are different for different equations, not only because of data limitations, but because of structural changes over time in the industry.[18]

Although not transparent from this presentation, the portion of the MacAvoy-Pindyck model representing reserves interacts very weakly with the block of equations representing production (the system is "block-recursive"). This motivated the authors to estimate the set of equations describing exploration and discovery as if it were functionally independent of the equations describing production. Parameters of each block were estimated by a two-stage least-squares technique.

Wholesale demand for gas, wholesale price markups, and interregional flows of natural gas are modeled in the same spirit.[19] These equations and the supply equations together with an accounting identity for gas reserves at year t as a function of extensions, revisions, reserves at $t - 1$, and changes in underground storate from $t - 1$ to t constitute their model.

FORECASTING PERFORMANCE OF THE MACAVOY–PINDYCK MODEL (VERSION 2)

How well does this model forecast, given parameter values fixed at maximum likelihood estimates based on use of data as described in the preceding sections? This question is addressed by simulating the model over the period 1966–1970, an interval that falls *within the time interval of data used to estimate parameters.* Even though all of the data are "used up" to estimate parameters, a historic simulation of the model provides some interesting insights.

Historical values of exogenous variables are used, expectations of all dependent variables are generated, and root mean square errors for actual versus predicted demand and actual versus predicted supply are calculated. The authors comment on the supply-side simulation results:

> simulated production supply is larger than the actual production levels in 1966, 1967, and 1968, but smaller than actual levels in 1969 and 1970, with a deficit of over two trillion cubic feet in 1970. Most of this error occurs in Louisiana South and offshore Louisiana in the Outer Continental Shelf. There

more is found than forecast, probably because of a stepped-up program of Federal Offshore Leasing. With a continuation of the usual Leasing Program, forecasts in the 1970s should be more accurate. Production demand is underestimated slightly at first, but strongly later, with an error that grows from about 1.9 trillion cubic feet in 1966 to about 3.2 trillion cubic feet in 1969. At least part of the reason for the underestimate of demand is the overestimate of the average wholesale price. The simulated wholesale price is about 4.2 cents too high from 1966 to 1968, and three cents too high in 1969 to 1970. This is because of more gas flowing from small production regions in the simulation than in the real world (the small regions have higher markups). Such an error, while part of the backcast, would not persist over long forecast periods, because of depletion of reserves in small regions.

It would appear that the model's forecasts should be viewed as having a margin of error (probably negative) of at least one or two trillion cubic feet.[20]

A COMPARISON OF MODELS

Pindyck[21] provides a revealing comparison of the structure and predictive performance of the three econometric models just described by reestimating parameters of each over the same data base. Ideally one wishes a model to be robust. Numerical estimates of the parameter of a robust model are not sensitive to outliers—data points that lie far away from measures of central tendency of the data as a whole and that do not conform to overall patterns in the data. If a model is not robust, then shifts in the data may induce large changes in parameter estimates and, at the same time, in the size of errors of projections of dependent variables.

In the present context predictive error can arise either from a misspecification of model structure due to changes in the structure of the oil and gas industry over the last two decades or from lack of robustness of the models relative to the data sets they describe, or from both sources. Current evidence does not allow a clear-cut identification of the relative contributions of misspecification and lack of robustness to predictive error.

In his comparison, Pindyck shows that reestimation of *respecified* Erickson-Spann and Khazzoom models on an updated data set corresponding to that used to estimate the third version of the MacAvoy–Pindyck model leads to large changes in estimated parameter values for the first two models, and in important instances, to changes in signs of coefficient estimates. Sign changes are particularly vexing because the sign of a coefficient in an equation determines the direction of influence of the explanatory variable to which it is attached on the dependent variable.

In order to render the models comparable, Pindyck made some changes in the Khazzoom and Erickson–Spann specification. Natural gas liquids were omitted from Khazzoom's model. This omission turned out to be important. We return to it later. Erickson and Spann's exploratory drilling equation was reformulated to include all exploratory wells drilled in place of new pool and new wildcat wells. The ratio of all exploratory wells drilled in a district in a given year to total exploratory wells drilled in the United States in that year was modeled in place of the two ratios used by Erickson and Spann (the ratio of wildcats drilled by majors to wildcats drilled by all companies and the ratio of wildcats drilled by majors to all wildcats drilled by majors).

Data for 1965 througy 1969 over 18 production districts were used to reestimate parameters of all three models. Substantial changes in both signs of estimates and levels of statistical significance of parameter estimates occur.

- Price variables that were highly statistically significant in Khazzoom's original new discoveries equation become insignificant.

- The quality of fit of Erickson and Spann's success ratio and discovery size equations deteriorates.

- The signs of estimated coefficients of gas price and of average feet drilled reverse in Erickson and Spann's exploratory wells equation. Sign reversals for gas price and oil price occur in the success ratio equation.

- Spann and Erickson's original estimate of the price elasticity of natural gas changes from 0.69 to 2.36.

Two possible explanations for a shift in the perceived dependence of discoveries on price in Khazzoom's model are suggested by Pindyck. First, by estimating over a shorter time period (1964–1969) than that used by Khazzoom (1961–1969), three years in which price and discoveries were most variable are dropped. Second, Khazzoom's discovery equations contain discoveries at year $t - 1$ as an explanatory variable for discoveries at t, and this lagged variable "explains" most of the variance of discoveries in year t. In fact, omission of NGL price from Khazzoom's specification may be the real reason for this shift.

Khazzoom offers a different explanation: The price of natural gas liquids (NGL) is an important determinant of the decision to drill or not to drill for nonassociated natural gas. To exclude NGL prices from the model specification is to leave out a major explanatory variable. The resulting misspecification leads to insignificant price variable coefficients in the drilling equations and the mistaken conclusion that drilling rates do not depend very strongly on price. In initial work on his natural gas model, Khazzoom omitted NGL prices and was led down the same path:

> At the time when I was working on the development of the natural gas model for the U.S., I did not use the price of liquids. The price parameters that I estimated turned out to be insignificant. I inquired from several people at the FPC about the reason why I was not able to capture any significant relationship between gas supply and gas price. I discussed it also with several Commissioners at the time, since insignificance in the relationship between gas

supply and price has far reaching implications for their pricing policies. I learned then that, for several districts, the price of liquids is very significant in the decision on whether or not to drill for non-associated natural gas. Indeed I learned at the time that in setting the ceiling price for natural gas, the FPC deducted, for specific districts known to have substantial amounts of gas liquids—they deducted what they called "credit for liquids." They usually came up with a fairly low price for natural gas in these districts. It is evident that for these districts, the exclusion of the price of liquids in the estimation process would, in effect, be excluding "the" major driving force in the exploration for gas.

I, therefore, included the price of the liquids in those districts (there were about 10 of them). The relation I estimated showed a significant coefficient for gas price.

If my model did specify the relationship correctly, then what Robert Pindyck's standardization did amounts to mis-specifying my model. On that ground it should not come as a surprise that the model did not yield significant parameters for the gas price. Moreover, that specific result does not come as a surprise. As I said, I derived a similar relationship for the period for which I estimated but without the liquids' price, it showed an insignificant relationship between gas supply and gas price. I very much suspect that the fact that Pindyck found the respecified model not to have a significant price relationship is due not so much to the change in the length of the sample period, but rather to the exclusion of the price of liquids from the equation.

The dilemma that Pindyck faced in comparing models is still with us. (The Energy Modeling Forum No. 5 at Stanford has bogged down in its attempt to compare oil and gas models.) The effort has gone a little beyond reporting simulation results which show plots scattered all over the place. They have not been able to track down the reasons for the divergence between the various models.) Pindyck attempted to overcome this problem by standardizing the models. Perhaps standardization is not a bad idea, if the problem of misspecification it creates could somehow be overcome. Moreover, the problem that Pindyck overlooked is the fact that a standardization may not be symmetric. Had he standardized by using Khazzoom's model or Erickson's model, it is not clear that he would have gotten the same results. By the same token, if these alternative standardizations have given the same results as the standardization he used, then we would have had a greater confidence in his results.

In short, what Pindyck ended up with are two models, Khazzoom's and Erickson's, which do not necessarily have a relationship to the original models, since after standardization these models were depicting a reality different from the one that initial models depicted. [Personal communication, June 18, 1981.]

Khazzoom's remarks about how his model and Erickson and Spann's model were modified to conform to MacAvoy and Pindyck's model specification highlight difficulties that arise in attempting to compare econometric models. One-sided standardization of models in this instance severely skewed the results.

Erickson and Spann's original parameter estimates are based on data for the years 1946 to 1959 and Pindyck's reestimation on data for 1964 through 1969. New contract prices in all districts show greater percentage increases from 1946 to 1959 than from 1964 to 1969. Since new discoveries for the later period do not exhibit a large percentage increase relative to the earlier period, Pindyck suggests that a smaller estimate of price elasticity of new discoveries for the 1946 to 1959 period than for the 1964 to 1969 period is to be expected. As the Erickson–Spann equations are logarithmic, price elasticities add across equations and the principal contribution to the price elasticity 2.36 of new discoveries comes from the price elasticity 1.826 of average size of discovery. Both the success ratio equation and the equation for the average size of discovery have low explanatory power when applied to the 1964 to 1969 period (R^2 of 0.274 and 0.268, respectively), which leads Pindyck to question the validity of these two equations and of the estimated value 1.826 of price elasticity of average size of new discovery in particular.

Following this comparison of model structures is a comparison of predictions made by each model. Dependent variables are simulated (Monte Carloed) first over 1965 through 1971, with independent variables taking on values that actually obtained in those years. Then each of the three models is simulated over a longer future period, 1971–1980, for two distinct regulatory policies: a "cost of service" or historical average cost pricing implying annual price increases of 1 cent per thousand cubic feet (MCF) and a policy of partial deregulation in which new contract wellhead prices are allowed to rise by 15 cents in 1974 and 4 cents per year thereafter. (New contract price per thousand cubic feet (MCF) at the wellhead averaged about 18–20 cents in 1969.) Oil prices and other exogenous variables for this time interval are set on a "medium" growth path. State-by-state value added in manufacturing, income, and capital equipment additions as well as population level are exogenous determinants of natural gas demand. The first three are projected to grow at 4.2 percent per year in constant dollars. Prices are assumed to grow at 3.5 percent per year, so in current dollars, value added and capacity grow at 7.7 percent. Population is projected to grow at 1.1 percent. Both crude oil field prices and wholesale prices for alternative fuels are assumed to grow at 5 percent per year in real dollars or at 8.5 percent in current dollars. The net effect is to double oil prices from 1972 to 1983. Average drilling costs grow at 6.8 percent per year.[22]

Hindsight shows some of the exogenous variables to have been set on a growth path far more gentle than that actually experienced. This must be kept in mind when interpreting the results of Pindyck's comparison. A much sharper rise in prices for crude oil and for natural gas relative to drilling costs will magnify differences in projections from the three models compared.

Principal features of this forecasting exercise are as follows:

- The respecified Khazzoom model shows little response of reserve additions to price increases even when partial price deregulation is in force.

This result is not surprising in light of the low estimates of price elasticity that occur when Khazzoom's model is reestimated for 1964–1969. In sharp contrast reestimation leads to large price elasticities for the Erickson–Spann model:

- The respecified Erickson–Spann model exhibits extreme sensitivity to price. When price is partially deregulated new discoveries increase by a factor of 10 from 1971 to 1980, production doubles, and a supply excess of 18 trillion cubic feet by 1980 is projected.

The reestimated MacAvoy–Pindyck model produced projections in between:

- The "cost of service" policy produces a projection of excess demand of 10 trillion cubic feet in 1980, while with partial deregulation excess demand clears by 1979.

To place these conflicting results in a current perspective, the price increases adopted with partial deregulation lead to a price of 75–80 cents per MCF by 1980. Present prices are very different; that is, regulated 1980 prices of $2–2.30 per MCF and deregulated deep gas at $5–7.00 per MCF. Projections of price increases from a 1969 base of about 18–22 cents per MCF to $2–2.30 per MCF and above by 1980 would stretch the models far beyond the upper limit of price intervals used to estimate model parameters.

Pindyck's concluding remarks obliquely imply that the functional form chosen by econometricians to represent exploratory drilling, drilling successes and failures, and field sizes as a function of price incentives may be inadequate:

> The three models that have been examined here are probably representative of the current state of the art of econometric modeling of the natural gas industry, but they provide no consensus on how gas supplies are likely to respond to ceiling price increases. It is clear that a knowledge of the dynamic response of exploration and discovery to changes in the price incentive is crucial to the design of regulatory policy; unfortunately, it represents an area that is still not well understood.[23]

VERSION 4 OF THE MACAVOY–PINDYCK MODEL

In light of large jumps in petroleum prices after 1973 and the evident sensitivity of parameter estimates and projections to the data, MacAvoy and Pindyck in their two 1975 books and Pindyck in a 1978 article produced other versions of the model described earlier, designed to deal with shortcomings of earlier versions.[24] In Pindyck's words:

> Although [the third version] provided a satisfactory overall description of the industry, it did have some shortcomings. First, some of the equations, particularly several of those describing the process of exploration and reserve

accumulation, fit the data poorly. Second, forecasts of the model seemed to be unreasonably optimistic in their implications for the kinds of price increases that would be needed to clear markets and eliminate excess demand for natural gas. In particular, the model predicted increases in onshore new discoveries and in offshore production in response to higher prices that were considerably out of line with engineering and intuitive estimates.

These problems were in large part a result of the limited range of data used in estimation. Most of the equations of the model were estimated using pooled cross-section time-series data, but the considerable amount of cross-sectional variation in prices that existed during the 1960s was not sufficient to capture accurately the effects of price changes in equations describing onshore exploration and new discoveries of gas. Data limitations also proved to be a problem for the equations describing offshore discoveries and production of gas, since these equations were estimated from time-series data for a single region.

There may also have been problems with the specification of some of the model's equations. The representation of price expectations did not capture the effects of the larger price increases that have occurred in the last two years and that are likely to occur in the future. The equations describing extensions and revisions of natural gas (and oil) were not sufficiently rich in detail to describe accurately the response of these components of reserve additions to increase in price.[25]

The fourth version of the MacAvoy–Pindyck model differs from earlier versions in several respects:

- The exploratory wells (WXT) equation is substantially modified.

- Equations describing oil and gas success ratios are introduced, one for the fraction of wells drilled that discover oil and another for the fraction of wells drilled that discover gas.

- An equation for the average size of an oil discovery and an equation for the average size of a gas discovery replace the (two) equations describing average sizes of associated and nonassociated gas in version 2.

- Extensions and revisions are modeled differently and in more detail.

New discoveries of gas are represented as the product of exploratory wells drilled WXT, the fraction FG of WXT that discovers gas, and the average size SZG of new discoveries in place of the product of WXT times the sum $DN + DA$ of the average sizes of nonassociated gas discoveries DN and associated gas discoveries DA.

The ratio of wells that discover oil or gas or both to total exploratory wells drilled takes on values in the interval zero to one by definition. Consequently, representing the log of this ratio as a linear function of explanatory variables plus an error term that can take on values in $(-\infty, \infty)$ misrepresents the success ratio, even though the data may have characteristics that render this misspecification of form unimportant. In order to account for this attribute of success ratios, the ratios of wells that discover gas to total exploratory wells and of wells that discover oil to total exploratory wells, Pindyck states that he used a trinomial logit model for these ratios.

An aside is in order. To define *trinomial logit*, begin with a simpler model: Suppose that $\tilde{x}_1, \ldots, \tilde{x}_n$ are independent dichotomous random variables, each assuming a value of zero or one. The probability that $\tilde{x}_i = 1$ for any $i = 1, 2, \ldots, n$ is conditioned on the values assumed by m independent variables $y_1^{(i)}, \ldots, y_m^{(i)}$ at i. Call this conditional probability $\theta(y_1^{(i)}, \ldots, y_m^{(i)})$. Then c continuous variable linear logistic model for the $\tilde{x}_i$s takes the form:

$$\log_e \left[\theta\left(y_1^{(i)}, \ldots, y_m^{(i)}\right) / 1 - \theta\left(y_i^{(i)}, \ldots, y_m^{(i)}\right) \right]$$

$$= \beta_0 + \beta_1 y_1^{(i)} + \ldots + \beta_m y_m^{(i)} \ . \tag{14-21}$$

Defining

$$y^{(i)} = \left(1, y_1^{(i)}, \ldots, y_m^{(i)}\right) \quad \text{and}$$

$$\beta y^{(i)} = \beta_0 + \beta_1 y_1^{(i)} + \ldots + \beta_m y_m^{(i)} \ , \tag{14-22}$$

the probability

$$\theta\left(y_1^{(i)}, \ldots, y_m^{(i)}\right) = e^{\beta y^{(i)}} / \left[1 + e^{\beta y^{(i)}}\right] \ , \tag{14-23}$$

and the particular value it assumes depends on $y^{(i)}$. With $\theta(\cdot)$ expressed in this form, it is easy to see that given the parameter β and given $y^{(1)}, \ldots, y^{(n)}$, the probability of observing $\tilde{x}_1 = x_1, \ldots, \tilde{x}_n = x_n$ is

$$\prod_{i=1}^{n} \left[\theta\left(y^{(i)}\right)\right]^{x_i} \left[1 - \theta\left(y^{(i)}\right)\right]^{1-x_i}$$

$$= \prod_{i=1}^{n} \left[e^{\beta y^{(i)}} / 1 + e^{\beta y^{(i)}}\right]^{x_i} \left[1/1 + e^{\beta y^{(i)}}\right]^{1-x_i} \tag{14-24}$$

In statistical parlance, the last probability is the likelihood function for β given data $(x_i, y^{(i)})$, $i = 1, 2, \ldots, n$. The unknown parameter β can be estimated, given the data, by exploiting properties of the likelihood function.

If in place of dichotomous values (success or failure, one or zero, and so on) for observed values of the $\tilde{x}_i$, each $\tilde{x}_i$ can assume one of three distinct values corresponding to "well discovers gas," "well discovers oil," "well is dry," then the logit model just described must be slightly extended in form. To this end set $x^{(i)} = (x_1^{(i)}, x_2^{(i)}, x_3^{(i)})$ and let $x^{(i)} = (1, 0, 0)$ if the ith exploratory well discovers gas, $(0, 1, 0)$ if it discovers oil, and $(0, 0, 1)$ if it is a dry hole. Define

$\theta_1(y^{(i)})$ as the probability that $x^{(i)} = (1, 0, 0)$ given $y^{(i)}$ and $\theta_2(y^{(i)})$ as the probability that $x^{(i)} = (0, 1, 0)$. It follows that, given $y^{(i)}$, the probability that $x^{(i)} = (0, 0, 1)$ is

$$\theta_3\left(y^{(i)}\right) = 1 - \theta_1\left(y^{(i)}\right) - \theta_2\left(y^{(i)}\right) . \tag{14-25}$$

The trinomial logit model may now be expressed as

$$\log_e\left[\theta_1\left(y^{(i)}\right)/1 - \theta_1\left(y^{(i)}\right) - \theta_2\left(y^{(i)}\right)\right] = \beta y^{(i)} ,$$

$$\log_e\left[\theta_2\left(y^{(i)}\right)/1 - \theta_1\left(y^{(i)}\right) - \theta^2\left(y^{(i)}\right)\right] = \gamma y^{(i)} , \tag{14-26}$$

where β and γ are vectors of fixed but unknown parameters to be estimated from observed data $(x_i, y^{(i)})$, $i = 1, 2, \ldots, n$. After some algebra,

$$\theta_1\left(y^{(i)}\right) = e^{\beta y^{(i)}}/1 + e^{\beta y^{(i)}} + e^{\gamma y^{(i)}} ,$$

$$\theta_2\left(y^{(i)}\right) = e^{\gamma y^{(i)}}/1 + \theta^{\beta y^{(i)}} + e^{\gamma y^{(i)}} , \tag{14-27}$$

and the probability of observing $x^{(1)}, \ldots, x^{(n)}$ given $y^{(1)}, \ldots, y^{(n)}$ is

$$\prod_{i=1}^{n} \left[\theta_1\left(y^{(i)}\right)\right]^{x_1^{(i)}} \left[\theta_2\left(y^{(i)}\right)\right]^{x_2^{(i)}} \left[1 - \theta_1\left(y^{(i)}\right) - \theta_2\left(y^{(i)}\right)\right]^{x_3^{(i)}} . \tag{14-28}$$

In the equations for success ratios used by Pindyck, the role of y is played by a vector whose components are eighteen dummy variables distinguishing production districts, the arithmetic average $\overline{SRG}_{t-1, t-3}$ of success ratios at years $t - 1$, $t - 2$, and $t - 3$, the negative of the square of this average, and a similar average $\overline{SRO}_{t-1, t-3}$ of oil success ratios at years $t - 1$, $t - 2$, and $t - 3$. With $\overline{SRG}_t$ denoting the proportion of wells during year t that discovered gas and $\overline{SRO}_t$, the proportion that discovered oil for a generic district, Pindyck sets

$$\log_e\left[\widetilde{SRG}_t/1 - \widetilde{SRG}_t - \widetilde{SRO}_t\right]$$

$$= a_0 + \overline{SRG}_{t-1, t-3}\left[1 - \overline{SRG}_{t-1, t-3}\right] - \overline{SRO}_{t-1, t-3} + \widetilde{e}_t , \tag{14-29}$$

where $\widetilde{e}_t$ is an error term with zero expectation. (The district effect is, for simplicity in exposition, assumed to be absorbed into the intercept constant a_0.) This is the gas success ratio equation. The oil success ratio equation is modeled in a similar way. The dependent variables SRG_t (= number of gas discoveries in year t/WXT_t) and SRO_t (= number of oil discoveries in year t/WXT_t) are *not* probabilities but are sample proportions. Consequently the model is, strictly speaking, *not* a trinomial logit model, especially in the presence of an additive error term. The two equations for $\tilde{SRG}_t$ and $\tilde{SRO}_t$ might more appropriately be called "linear regression equations with functionally transformed dependent variables," especially in light of their treatment as regression equations in the course of parameter estimation.[26]

Peculiar features of these equations are that no economic variables such as price or cost appear as explanatory variables and assignment of coefficients of one to the lagged three-year moving average and minus one to its square. According to Pindyck, the coefficients of price variables introduced in version 3 exhibited only slight statistical significance, and, when reestimated on the updated data base for version 4, were all statistically insignificant. He found success ratios to be "better represented by allowing them to adjust to long-run equilibrium values that differ from each of the FPC districts represented in the model."[27] An equilibrium solution to a pair of success ratio equations consists of the pair (g_∞, p_∞) of numbers satisfying the following equations: (Note the asymmetry in these equations: The right-hand side of the gas equation contains a term $g_\infty(1 - g_\infty)$, and the product term on the right-hand side of the oil equation is $p_\infty(1 - g_\infty)$.)

$$\log(g_\infty/1 - g_\infty - p_\infty) = a_0 + g_\infty(1 - g_\infty) - p_\infty \ ,$$

and

$$\log(p_\infty/1 - g_\infty - p_\infty) = b_0 + p_\infty(1 - g_\infty) - g_\infty \ ,$$

where $b_0 > 0$ is the intercept term for the oil success ratio equation. Numerical estimates of a_0 and b_0 vary widely over districts. For example in district 1 $(a_0, b_0) = (0.078, 0.26)$, in district 7 $(a_0, b_0) = (0.677, 0.081)$, and in district 11 $(a_0, b_0) = (0.211, -0.289)$. This implies very large variation in long-run equilibrium values for success ratios across districts—and is empirically questionable.

Extensions and revisions are described in more detail by version 4 than in earlier versions. Development wells drilled per year is the turnkey physical variable driving measurement of extensions and revisions. The total number of development wells in a district in a given year is modeled as a log-linear function of oil and gas prices, direct drilling costs, and average amount of gas and of oil recoverable per development well. Equations for amount recoverable per gas extension well and for amount recoverable per oil extension well are specified as functions of oil and gas prices to "explain directionality" and drilling costs "which could induce operators to change drilling patterns and alter the size distribution of resulting extensions."[28] Revisions are regarded as independent of wells drilled but as dependent on the previous year's reserve and production levels, on changes in production, and on price.

Offshore Louisiana drilling, success ratios, discovery sizes, extensions, revisions, and production are modeled separately in the same fashion as version 3.[29]

Offshore well equations differ from their onshore counterparts. Most notably, the wildcat well equation includes total acreage leased as an explanatory variable and an equation for "field wells," wells other than wildcats, is made dependent on the number of offshore rigs.

Both versions 3 and 4 were simulated historically over each of two intervals, 1967–1972 and 1968–1974, in order to compare their predictive performances. Pindyck reports that:

- Both versions underpredict successful gas wells. Over the time interval 1967 to 1972, version 3 overpredicted average size of gas discovery resulting in approximately a 30 percent error of predicted versus actual new discovery wells.

- The underprediction of successful gas wells by version 4 is canceled by over-prediction of average gas discovery size.

- Whereas version 3's overprediction of extensions and revisions resulted in about a 40 percent error in addition to reserves, the corresponding errors in version 4's addition to reserves is near 0 percent.

- Version 4 underpredicts new oil discoveries.

With a policy of increasing new contract field prices in all districts by 25 cents/MCF in 1975 and by 7 cents/MCF per year thereafter,

- Version 3 predicts market clearing of excess demand by 1980, but version 4 predicts large and growing shortages! As the demand forecasts are very similar, large differences in production and in additions to reserves account for this.

- Version 3 projects discoveries to increase by a factor of seven by 1978, and additions to gas reserves to quadruple by 1978.

Estimates of parameters of some key explanatory variables in both models are sensitive to shifts in data. Over the interval 1964–1974 gas prices increased, and the increase in 1973 and 1974 was large in all regions. Even so the responses to these price increases of success ratios and discovery sizes within regions prescribed by the model exhibited no clear-cut upward pattern. Pindyck points out that the hypothesis that price increases lead to more wildcat drilling (extensive margin drilling) is not substantiated by the data. He reports that success ratios for these two years, not only do not respond in an orderly way to price increases, but also "fluctuate considerably around average values that differ widely across regions"; he adds that perhaps a much more finely disaggregated model may be necessary to capture accurately interactions between price, success ratio, and discovery size.

Price rises in 1973 and 1974 were accompanied by the specter of massive federal intervention in the domestic oil market, which undoubtedly dampened exploration company enthusiasm for a rapid increase in the rate of exploratory drilling. Moreover, large price increases appear in only the last two years of the rather short time series used to estimate model parameters. One would not expect the impact of expectations of longer term price increases on exploratory drilling rates to be readily decipherable from such a short time series.

Between July 1975 and June 1976 the FPC increased a uniform national area rate from 42 cents/MCF to $1.42/MCF. Version 4 was used to examine two future pricing policies, a small, yearly, cost-based price increase of 5 cents over 60 cents/MCF in 1976, and a 1976 price of $1.42/MCF with a 4 cent price rise in each succeeding year.

With identical assumptions for exogenous variables over the time span of the forecast,[a] and with the first of the aforementioned policies in force, demand is projected to rise to 36 TCF per year while additions to gas reserves decline and production remains approximately level. This regulation-induced shortage is manifested by excess demand of 40 percent of total demand, most of which occurs in the Northeast, North Central, and Southeast regions.

Forecasts generated by the second policy are very different. A national area price of $1.42/MCF in 1976 plus a 4 cent/year price rise lead to a growth in reserves and production of 25.6 TCF with 20 percent of this growth from offshore. By 1980, the average price per thousand cubic feet at the wellhead in the United States grows to $1/MCF. This more aggressive pricing policy reduces excess demand to 2 TCF by 1980, spread reasonably evenly across the country.

Variations of this policy are examined. Pindyck reports that enlarging the Bureau of Land Management's offshore leasing policy to 4 million acres per year and assuming that nine instead of four offshore rigs are added each year results in only a small increase in gas reserves and an increase of 0.5 TCF of production by 1980. A "recession forecast" in which real annual growth of the economy varies between 0 and 6 percent in the years 1977–1980 causes a negligible drop in demand. A third variation assumes higher prices for alternative fuels. Domestic crude oil prices grow to $8 per barrel in 1977 and at $1 per year between 1978 and 1980. Alternative fuels other than natural gas grow in price at 14.5 percent, 8.5 percent, 7 percent, and 6 percent in the four years 1977–1980. The effects are a decrease in gas reserves, an increase in oil reserves, and, by 1980, a decrease in gas production of 2 TCF. Excess demand grows to 7 TCF in 1980.

a. Pindyck's assumptions include a 50 cent/year increase from a 1975 base of $6.25/ barrel for crude oil; 10 percent yearly increase in drilling costs 10 percent (constant) interest ratio; 2 million offshore acres leased annually; five new offshore drilling rigs installed per year offshore Louisiana. Income, value added, and capital expenditures grow at 7 percent in real terms in 1976 and at 4 percent per year from 1977 to 1980. Inflation is set at 6 percent in 1976 and 4 percent in succeeding years. Beginning in 1976, alternative fuel prices rise at 2 percent, 3.5 percent, 3 percent, 2.5 percent, and 2 percent in successive years. Average intrastate gas price rises in nominal dollars: $1/MCF in 1975, $1.25/MCF in 1976, $1.35/ MCF in 1977, and $1.50 in 1978 through 1980. Canadian imports per year are 10^9 MCF at $1.60/MCF.[30]

Version 4 policy simulations suggest that over a forecast period of 1976–1980:

- Expanding the federal offshore leasing program only marginally reduces shortages.

- Projections by version 4 are not sensitive to modest variations in assumptions about economic growth variables.

but

- Projections are sensitive to moderate to extreme changes in oil prices, and large oil price increases must be accompanied by large gas price increases to avoid excess gas demand.

Although this last version of the MacAvoy–Pindyck model tracks the behavior of natural gas demand in field and wholesale markets quite well, impediments to accurate representation of supply variables are not overcome, as Pindyck notes.

> Although the revised model provides an improved description of the natural gas industry, it is still lacking in its representation of exploration and discovery of new gas and oil reserves. The economic relationships that one would expect to hold on the micro-level are not supported by the data in the new discoveries equations, and those equations do not fully capture the geological determinants of the distribution of discoveries. This is partly the result of the level of regional aggregation used in estimating the model, and we may learn that it is more fruitful to model exploration and discovery at the micro-level of individual pools and fields. It is hoped that future work will help us better understand the dynamic response of exploration and discovery to higher energy prices.[31]

ATTANASI'S INDIVIDUAL BASIN ECONOMETRIC MODEL

Recognizing that aggregation of oil and gas data across distinct geologic provinces obscures the effects of costs and prices on patterns of exploration activity—patterns that may be decipherable from data for a single such province—Attanasi[32] studies exploration and development in the Denver–Julesberg Basin using econometric methods. The Denver–Julesberg Basin is a fortuitous choice, as the deposits in it possess a degree of geologic homogeneity not present in most basins. (Over 80 percent of the traps are lower Cretaceous stratigraphic traps in a single horizon at depths of 4,000 to 7,000 feet; see Chapter 15.)

Expectations of drilling operators about the sizes of fields that drilling may discover are explicitly incorporated in the models Attanasi examines. His treatment is novel in its use of an updated version of the Arps and Roberts discovery process model to generate expectations. The discovery process model is assumed to be the mechanism by which operators project the number of undiscovered deposits within each of a discrete number of size intervals.

Exploration effort is defined by Attanasi to be exploratory wells drilled per time period or alternatively exploration expenditures per time period. The rate of application of exploratory effort in a given period is made functionally dependent on expectations about profits that incorporate current perceptions of the sizes and numbers of undiscovered deposits.

Rather than fix on one method for incorporating current expectations about future discoveries, Attanasi examines several: an updated Arps and Roberts model blended with an economic analysis of drilling and discovery and three distinct autoregressive (distributed lag) models. In accordance with economists' terminology, Attanasi calls analysis employing the Arps and Roberts discovery process model an analysis based on "rational expectations." By this he means that the model produces projections based on a description of both the physical processes at play and an explicit microeconomic theory of how exploration decisions are made. The alternatives that he studies—distributed lag models that employ particular functions of empirical data as surrogates for operators' expectations about future discoveries—are not as closely tied to specific assumptions about how exploration decisions are made.

The Arps and Roberts model is used by Attanasi in the following way: recall (from Chapter 10) that if W_t denotes the cumulative number of wildcat wells drilled at time t, A_i is the areal extent of deposits in the ith size class, c is an exploration efficiency parameter, B_o is the potentially productive areal extent of the basin, $N(A_i, W_t)$ is the number of discoveries of areal extent A_i by W_t wildcat, and $N(A_i)$ is the number of deposits in the basin, then

$$N(A_i, W_t)/[1 - e^{-(cA_i/B_o)W_t}] \equiv \hat{N}(A_i) \tag{14-30}$$

is an estimate of $N(A_i)$ as a function of $N(A_i, W_t)$ and W_t.[b]

With $B_o = 5{,}700{,}000$ acres, $c = 2$, knowledge of the number of wells drilled by time t and of the number of discoveries in each size class, Eq. (14-30) produces an estimate of the number of fields remaining to be discovered in each size class at the end of each of the years 1946-1973.

An estimate of the expectation of marginal value at time t of undiscovered deposits in each size class is computed as a function of the wellhead price at t and averaged drilling, development, and production costs.[c] Letting $v_i(t)$ denote an estimate of the marginal expected value of a generic undiscovered deposit in size class i as a function of costs and wellhead price at t, the quantity

$$\hat{V}(t) = \sum_{i=1}^{M} [N(A_i) - N(A_i, W_t)] \hat{V}_i(t)$$

b. In order to make properties of this estimator precise an explicit probability law for the quantities $\Delta \tilde{N}(A_i, W) = \tilde{N}(A_i, W) - \tilde{N}(A_i, W - 1)$ must be assumed (see Chapter 15).

c. Details of this calculation are not presented by Attanasi.

is an estimate of the total expected value at t of all undiscovered deposits. If t denotes *year*, $\hat{V}(1)$, $\hat{V}(2)$, ..., $\hat{V}(t)$ is a sequentially updated sequence of estimates of economic values for undiscovered deposits; that is, for $t = 1, 2, \ldots,$ $\hat{V}(t)$ takes into account the exploration history of the basin up to and including the tth year. The estimate $\hat{V}(t)$ of undiscovered deposit values is a proxy at year t for operators' rational expectations about the future.

Because the Arps and Roberts model directly accounts for depletion of the resource base, the estimates $\hat{V}(t)$ incorporate the effects of depletion on future value. Consequently no explicit proxy for depletion such as cumulative wells drilled up to and including year t appears in Attanasi's specification of equations for wells drilled per year and for drilling expenditures per year. The wells equation is

$$\tilde{W}_t = \beta_0 + \beta_1 \, \Pi_{t-2} + b_2 \, \hat{V}(t) + \tilde{\epsilon}_t \ . \tag{14-31}$$

The parameters β_0, β_1, and β_2 are not known with certainty and are to be estimated from data and the $\tilde{\epsilon}_t s$ are assumed to be mutually independent and identically distributed with zero expectation. The explanatory variable Π_{t-2} is an estimate of expected exploration profit in year $t - 2$. This variable is designed "to reflect the overall profit of exploration in the region in terms of adjustment in overall firm allocation of exploration expenditures"[32] and is defined to be the product of the success ratio for wells drilled during year $t - 2$ and the difference between the dollar value of oil found by a successful wildcat and well costs. The explanatory variables Π_{t-2} and $\hat{V}(t)$ are different in character, because the profitability estimate Π_{t-2} is a statistic summarizing a feature of data observed by year $t - 2$, whereas $\hat{V}(t)$ is a point projection of one possible future value of undiscovered deposits.

With well-drilling expenditures Z_t in year t used as an index of exploratory effort, Eq. (14-31) is replaced by the following equation:

$$\tilde{Z}_t = \alpha_0 + \alpha_1 \, \hat{\Pi}_{t-2} + \alpha_2 \, \hat{V}(t) + \tilde{\delta}_t \ . \tag{14-32}$$

In addition to two models that incorporate the effect of depletion of the resource base on the economics of future drilling in an explanatory variable whose value is derived from a discovery process model, Attanasi studies three distributed lag models in each of which a proxy for exhaustion of the resource base is used: the value D_t in year t of discoveries of 500,000 barrels or more in size.[d] The idea behind this choice is that operators may use values of past discoveries to forecast the value of future discoveries. The particular fashion in which D_t enters the equation can be interpreted as a specification of the manner in which operators respond to patterns of discovery.

---

d. Attanasi tests several minimum sizes.

One of the distributed lag models examined by Attanasi is of the form:

$$\widetilde{W}_t = \beta_0 + \beta_1 \hat{\Pi}_{t-2} + \gamma I_t + \alpha \sum_{j=0}^{\infty} \lambda^j D_{t-j-1} + \widetilde{\epsilon}_t \ , \tag{14-33}$$

where β_0, β_1, γ, α, and λ are parameters not known with certainty, the $\widetilde{\epsilon}_t$s are mutually independent random variables with common distribution and zero means, and $\hat{\Pi}_{t-2}$, I_t, D_t, $D_{t-1}, \ldots$, are observed values of explanatory variables. ($D_t \equiv 0$ for $t \leq 0$.) The variable I_t is an index of exhaustion of the resource base, a weighted average of cumulative wildcat wells drilled in previous periods.[e]

If L is defined as a function that maps D_t into D_{t-1}, $L(D_t) = D_{t-1}$, then for $|\lambda| < 1$, the sum $\sum_{j=0}^{\infty} \lambda^j D_{t-j-1}$ can be more simply written as

$[1/(1 - \lambda L)] D_{t-1}$ and (14-31) becomes

$$\widetilde{W}_t = \beta_0 + \beta_1 \hat{\Pi}_{t-2} + \gamma I_t + \alpha(1/1 - \lambda L)D_{t-1} + \widetilde{\epsilon}_t \ . \tag{14-34}$$

In addition to (14-34), a "second order rational lag" model,

$$\widetilde{W}_t = \beta_0 + \beta_1 \hat{\Pi}_{t-2} + \gamma I_t + \alpha(1/1 - \gamma_1 L + \lambda_2 L^2)D_{t-1} + \widetilde{\epsilon}_t \ , \tag{14-35}$$

and a "polynomial distributed lag" model,

$$\widetilde{W}_t = \beta_0 + \beta_1 \hat{\Pi}_{t-2} + \gamma I_t + \sum_{j=0}^{n} k_j D_{t-j} + \widetilde{\epsilon}_t \tag{14-36}$$

are estimated.

Given a reasonably large number N of observations one might expect a finite distributed lag model with $2 < n + 1 < N$ "free" lag parameters $k_0, k_1, \ldots, k_n$ to fit observed data "better" than first- and second-order distributed lag models with only two and three "free" lag parameters respectively. This is indeed the case.

PARAMETER ESTIMATION AND COMPARISON OF MODELS

Basin well data for 1949 to 1973 were drawn from the Petroleum Information's Well History Control File, cost data from the Joint Association Survey of the U.S. Oil and Gas Production Industry for Colorado, and quarterly average

e. Acreage owned by the Union Pacific Railroad was not available for exploration until 1969. Measurement of I_t are designed to account for an expansion of potential targets when this acreage was opened to exploration.

wellhead prices (deseasonalized) were estimated for these years. Wellhead prices for oil in the Basin fluctuated between $2.63 to $3.46 per barrel from 1949 to 1972 and then "new" oil prices jumped to $8.70 per barrel in 1973.

When estimated with quarterly data, both the geometric and second-order distributed lag models implied that discoveries made more than fourteen quarters earlier had a nonnegligible effect on a projection of current discovery effort, while for the finite polynomial lag model the impact of past discoveries on current exploratory effort does not extend past eleven or twelve preceding quarters. Attanasi concludes that "when compared with the finite lag model . . . the two infinite lag models tended to overstate the magnitude and duration of the influence of new discoveries as measured by the variable D_t."[33] Both the fit (R^2 = 0.914-0.916) and the level of significance of coefficient estimates are superior for the finite polynomial lag model.

A reestimation done with semiannual in place of annual data allowed comparison of the Denver–Julesberg model with a similar model applied to data from the Powder River Basin.[34] The (empirically determined) lag lengths for finite polynomial lag models applied to the two basins were very close—four periods for the Powder River Basin and five periods for the Denver–Julesberg Basin:

> Just as for the Denver Basin, the results obtained for the Powder River Basin suggested that during an exploration play favorable expectations are sustained by the quantity of oil found per time period rather than by discoveries of certain critical size deposits.[35]

The rational expectations models, one employing well drilling and the other expenditures as measures of exploratory effort, both yield coefficient estimates with signs in accordance with a priori theory and fit the data almost as well as the finite distributed lag model when traditional measures of fit are employed.[f] Numerical values of the fitted rational expectation equations for a MERS of 500,000 barrels are[g]:

$$E(\widetilde{W}_t \mid \hat{\Pi}_{t-2}, \hat{V}(t)) = 14.50 + .00934\,\hat{\Pi}_{t-2} + .10172\,\hat{V}(t)$$
$$\phantom{E(\widetilde{W}_t \mid)} (.4) \quad (1.4) \quad\quad\quad (3.3)$$

$$E(\widetilde{Z}_t \mid \hat{\Pi}_{t-2}, \hat{V}(t)) = -49.01 + .02559\,\hat{\Pi}_{t-2} + .58802\,\hat{V}(t) \ . \quad (14\text{-}37)$$
$$\phantom{E(\widetilde{Z}_t \mid)} (.4) \quad (1.7) \quad\quad\quad (3.6)$$

Attanasi's next step is an important one. In order to examine the structural validity of his models, he breaks the data used to estimate parameters into two

---

f. A finite lag model with n = 10 yielded an estimated coefficient of determination R^2 of approximately 0.92 for the wells equation and 0.91 for the expenditures equation, while the corresponding estimates for the rational expectations model are 0.88 and 0.90. Estimates for coefficients of the profit variable $\hat{\Pi}_{t-2}$ for the rational expectations model are very close to those for the finite lag model: with D_t > 500,000 barrels,

subsamples of equal size and reestimates parameters of each model using each of the two subsamples. Then he tests whether or not "significant differences" between subsample estimates of a generic parameter are manifested. He reports that in no case are significant differences observed.[h] Recognizing that such a test may not have great power as a device for testing changes in structure over time, Attanasi examines residual plots generated by the two subsamples using each model to see if "obvious" systematic differences are present. He reports "no systematic variation." (The appearance of systematic variation in the residuals signals a misspecification of the model.)

In both the well-drilling and expenditure models, the term $\sum_{j=0}^{n} k_j D_{t-j}$ appearing in the finite distributed lag model may be interpreted as a surrogate for $\hat{V}(t)$ in the rational expectations model. Using estimates of $k_0, \ldots, k_n$ computed by use of the data and values of D_t for the time covered by the data, Attanasi regressed $\sum_{j=0}^{n} k_j D_{t-j}$ on $\hat{V}(t)$ and found that the fraction of variation in $\hat{V}(t)$ given corresponding terms $\sum_{j=0}^{n} k_j D_{t-j}$ not attributable to knowledge of the latter values was only 0.006. This is not surprising given nearly identical parameter estimates for coefficients of $\hat{\Pi}_{t-2}$ in both the distributed lag model and the rational expectations model along with nearly identical estimates of coefficients of determination. One may tentatively conclude that as a proxy for $\hat{V}(t)$, $\sum_{j=0}^{n} k_j D_{t-j}$ contains almost as much descriptive information as $\hat{V}(t)$ itself, over the period spanned by the data used to estimate model parameters about well-drilling rates and well expenditures even though $\hat{V}(t)$ is closer in definition to variables considered by exploration operators when they make drilling decisions. Attanasi favors the rational expectations model for use in forecasting future supply because of its simplicity and because it is based on a model directly describing physical aspects of discovery.

In a later publication Attanasi, Drew, and Scheunemeyer studied the performance of the rational expectations model as a device for predicting supply from undiscovered deposits.[36] Although the discovery process model employed is independent of time (it unfolds on a scale of wells drilled) the well-drilling equa-

f. (continued)	Finite Lag		Rational Expectations	
	Coefficient	t-statistic[g]	Coefficient	t-statistic[g]
Wells	.0096	(1.7)	.00934	(1.4)
Expenditures	.0275	(.9)	.02559	(1.7)

g. Estimate of one standard deviation.

h. Given the null hypothesis that parameter values remain constant over the time period spanned by both subsamples, there were no cases where the level of significance associated with a parameter exceeded 5 percent.

tion is not, so the precision of predictions of future supply depend in part on the time interval chosen to measure wells drilled. More specifically, without access to future values of explanatory variables, the wells equation must be respecified as recursive in W_t, $\hat{V}(t)$, and $\hat{\Pi}_t$. Consequently forecasting errors cumulate and the precision of a forecast made for a fixed future time T decreases as the length of time from "now" until time T increases.

The model is recast like this: Cost per well C_t and crude oil price p_t are given a priori for $t = t_0$, $t_0 + 1, \ldots, T$, and the value V_{t_0} of undiscovered deposits at period t_0 is specified. Initial values are also given for well drilled $W_{t_0 - 1}$ in period $t_0 - 1$, and for $\Pi_{t_0 - 2}$, expected profit in period $t_0 - 2$. Since the number of drilling successes and failures in a given period is not specified in the original well-drilling equation, expected profit is redefined as:

$$\hat{\Pi}_t = (R_t P_t / W_t) - C_t \ , \tag{14-38}$$

where R_t denotes reserves found in period t. For $t > t_0 - 2$, both R_t and W_t are endogenous, or determined via the well-drilling and discovery equations. In addition, empirical estimates of the number of deposits remaining to be discovered in each size class are used to represent the true number remaining to be discovered.

In light of these changes, the well-drilling equation (14-38) was reestimated[i]:

$$E(\widetilde{W}_t \mid \hat{\Pi}_{t-2}, \hat{V}(t)) = 8.386 + .06339 \hat{\Pi}_{t-2} + .10358 \hat{V}(t) \ . \tag{14-39}$$
$$(.2) \quad (1.7) \quad (3.0)$$

Given forecast values for W_{t-1}, W_{t-2} and W_{t-3} (via (14-36) at $t - 2$ and $t - 3$) equation (14-37) produces a prediction for $\widetilde{W}_t$ and allows calculation of a prediction $\widetilde{R}_t$ of reserves found in period t.

The recursive equation used to generate predictions of wells drilled in period t is

$$E(\widetilde{W}_t \mid W_{t-1}, \hat{\Pi}_{t-2}, \hat{\Pi}_{t-3}, \hat{V}(t), \hat{V}(t-1))$$

$$= .814 W_{t-1} + (1 - .814) 8.836$$

$$+ .06339 [\hat{\Pi}_{t-2} - .814 \hat{\Pi}_{t-3}]$$

$$+ .10358 [\hat{V}(t) - .814 \hat{V}(t-1)] \tag{14-40}$$

Predicted reserves $\widetilde{R}_t$ found in period t are derived from the discovery process model by computing the difference between cumulative amounts discovered by all wells forecast to be drilled up to and including period t for predicted $\widetilde{W}_t$ and

---

i. With an estimated coefficient of determination of 0.881, a standard error of estimate 21.1, and a Durbin–Watson statistic of 1.53.

Table 14-2.   Forecast Performance of Integrated System.

Dependent Variable		Wells			Reserves		
Forecast Length[a]	Period	RMS Errors[b]	Mean Errors[c]	Actual Mean	RMS Errors	Mean Errors	Actual Mean
A	1	7.551	3.589	109.250	2.103	0.786	3.763
	2	12.795	-9.213	92.250	3.183	-.025	3.161
	3	13.183	.868	110.000	3.036	-.573	4.111
	4	8.509	5.915	90.500	2.969	.277	2.555
	5	5.227	3.679	86.250	1.928	.330	2.096
	6	20.195	14.769	89.250	1.927	1.346	1.142
	7	12.703	9.698	68.750	3.381	-.519	2.374
	8	9.783	6.882	54.250	1.298	-.526	2.069
	9	19.146	-10.847	70.750	.585	-.458	1.837
Mean values		12.121	2.816	85.694	2.268	.071	2.567
B	1	22.953	14.986	100.750	2.729	0.776	3.462
	2	19.434	12.399	100.250	2.987	.059	3.333
	3	13.424	5.803	87.750	1.915	.805	1.619
	4	28.126	23.671	61.500	2.549	-.275	2.221
Mean values		20.984	14.215	87.563	2.545	.341	2.659
C → B	1	20.438	13.099	108.833	2.860	0.341	3.678
↓	2	15.643	11.746	88.667	2.345	.753	1.931
	3	29.878	23.761	64.583	2.102	-.141	2.093
Mean values		21.986	16.202	85.694	2.436	.318	2.567

a. Forecast lengths A, B, and C are four, eight, and twelve quarters, respectively.

b. Root mean square value is given by

$$\left(\frac{1}{N} \sum_{i=1}^{N} (P_i - A_i)^2\right)^{0.5} ,$$

where P_i is the predicted value and A_i is the actual value.

c. Mean error is based on actual minus predicted value.

Source: E. D. Attanasi, L. J. Drew, and J. H. Scheunemeyer, "An Application to Supply Modeling," in Petroleum-Resource Appraisal and Discovery Rate Forecasting in Partially Explored Regions, U.S. Geological Survey Professional Paper 1138-A, B, C (Washington, D.C.: U.S. Government Printing Office, 1980).

what is predicted to have been found by all wells forecast to be drilled up to and including time period $t - 1$.

Forecasting accuracy is measured by root mean square error; that is, the square root of the average of the sum of squares of differences between actual and forecast values. Table 14-2, from Attanasi, Drew, and Schuenemeyer, displays the results of this exercise. Mean prediction error is positive for three choices of time interval (four, eight, and twelve quarters) with respective ratios of mean error per period to the mean of actual values of 0.033, 0.162, and 0.189. Predictions of wells drilled are systematically (positively) biased. While root mean square estimates for reserves are large relative to actual amounts discovered, mean prediction errors are small by comparison (0.028, 0.128, and

0.124 for four, eight, and twelve quarter predictions, respectively. Attanasi, Drew, and Scheunemeyer conjecture that high root mean square errors in predictions of reserve additions "appear to be more the result of the erratic or stochastic nature of the historical arrival of discoveries than systematic bias in the discovery process model."[37] The magnitude of prediction errors they report are of the same order as those reported by other econometric modelers of well drilling and reserves from new discoveries (see MacAvoy and Pindyck.[38])

Attanasi et al. conclude with a discussion of the limitations of the models they studied:

> Although the linking of the discovery process model with the behavioral well-drilling model produced an analytical means of translating the forecasts of reserves per unit exploration effort, that is, wells drilled to reserves per unit time, several limitations of the analysis should be kept in mind. First, the economic model is specified to describe operator field decisions that are short term in nature. That is, the nature of decisions that are modeled are marginal adjustments in the rate of exploration rather than decisions to enter or exit a geologic basin. Second, because of the short-run nature of the models, it would be misleading to attempt to draw general conclusions about the effects of a general price change on drilling activity within the basin. That is, a general price change would induce some long-run adjustments to take place in the firm's internal allocation of resources across several regions. The type of price change that the behavioral well-drilling model might more appropriately capture corresponds to a change in the relative price of oil, for instance, the price of oil found in the Denver basin as opposed to another basin. In order to predict the effects of a general change in the price of oil on drilling behavior for a particular basin, the behavioral well-drilling equation should be respecified to reflect the firm's long-run decisions and include a variable that would denote the firm's alternative exploration opportunities in other geologic basins or its alternative opportunities for obtaining additional reserves.[39]

REFERENCES

1. For a thorough critical review, see P. W. MacAvoy and R. S. Pindyck, *Price Controls and the National Gas Shortage* (Washington, D.C.: American Enterprise Institute, 1975).
2. J. Daniel Khazzoom, "The FPC Staff's Econometric Model of Natural Gas Supply in the United States," *Bell Journal of Economics and Management Science* 2, no. 1 (Spring 1971): 51–93.
3. E. Erickson and R. Spann, "Supply Response in the Regulated Industry: The Case of Natural Gas," *Bell Journal of Economics and Management Science* 2, no. 1 (Spring 1971): 94–121.
4. P. W. MacAvoy and R.S. Pindyck, *The Economics of the Natural Gas Shortage (1960-1980)* (Amsterdam: North–Holland, 1975).

5. F. M. Fisher, *Supply and Costs in the U.S. Petroleum Industry* (Baltimore: The Johns Hopkins University Press, 1964).

6. E. A. Hudson and D. W. Jorgenson, "U.S. Energy Policy and Economic Growth, 1975-2000," *Bell Journal of Economics and Management Science* 5, no. 2 (August 1974): 461-514. A. S. Manne, "ETA: A Method for Energy Technology Assessment," *Bell Journal of Economics and Management Science* 7, no. 2 (August 1976): 379-406.

7. E. D. Attanasi, "The Nature of Firm Expectations in Petroleum Exploration," *Land Economics* 55, no. 3 (1979): 301.

8. Ibid.

9. A comprehensive evaluation of the predictive performance of econometric petroleum supply models is beyond the scope of this monograph. Much research on econometric energy model validation and comparison is currently in progress; for example D. Wood, "Model Assessment and the Policy Research Process: Current Practice and Future Promise," presented at Workshop on Validation and Assessment of Energy Models, National Bureau of Standards, Gaithersburg, Md., January 10-11, 1979.

10. *Oil and Gas Journal.*

11. R. S. Pindyck, "The Regulatory Implications of Three Alternative Econometric Supply Models of Natural Gas," *Bell Journal of Economics and Management Science* 5, no. 2 (Autumn 1974): 633-45.

12. For further discussion of predictive validation, see Wood, "Model Assessment and Policy Research Process."

13. Khazzoom, "The FPC Staff's Econometric Model of Natural Gas Supply in the United States."

14. Khazzoom writes, "These include (1) non-associated new field discovery, (2) associated-dissolved new field discovery, (3) non-associated new reservoir discovery, (4) associated-dissolved new reservoir discovery, (5) non-associated extensions, (6) associated-dissolved extensions, (7) non-associated revisions, and (8) associated-dissolved revisions. (1) to (4) make up ND; (5) to (8) make up XR. These terms are used in the same sense as defined by the AGA and the AAPG." Ibid., p. 54.

15. Ibid., p. 59.

16. Erickson and Spann, "Supply Response in the Regulated Industry."

17. P. MacAvoy and R. Pindyck, "Alternative Regulatory Policies for Dealing with the Natural Gas Shortage," *Bell Journal of Economics and Management Science* 4, no. 2 (Spring 1973): 454-98.

18. Ibid., pp. 474-75.

19. See ibid. for a more detailed presentation of demand equations.

20. Ibid., p. 486.

21. Pindyck, "The Regulatory Implications of Three Alternative Econometric Supply Models of Natural Gas."

22. Ibid.

23. Ibid., p. 645.

24. MacAvoy and Pindyck, *Price Controls and the National Gas Shortage* and *The Economics of the Natural Gas Shortage.*

25. Pindyck, "High Energy Prices and the Supply of Natural Gas." Some of the estimated coefficients in Version 4 of MacAvoy and Pindyck are insignificant and according to Pindyck ("High Energy Prices," p. 181), simulations yielded success ratios approaching zero or one when large price increases were introduced. There was no compelling evidence of shifts to intensive or extensive drilling as a function of large price increases.

26. Pindyck, "High Energy Prices," Appendix 2.

27. Ibid.

28. Ibid., p. 181.

29. MacAvoy and Pindyck, *Price Controls and the National Gas Shortage* and *The Economics of the Natural Gas Shortage*.

30. Pindyck, "High Energy Prices."

31. Ibid., p. 198.

32. Attanasi, "The Nature of Firm Expectations in Petroleum Exploration," p. 300.

33. Ibid., p. 308.

34. E. D. Attanasi and L. J. Drew, "Field Expectations and the Determinants of Wildcat Drilling," *Southern Economic Journal* 44, no. 1 (1977): 53–67.

35. Ibid.

36. E. D. Attanasi, L. J. Drew, and J. H. Scheunemeyer, "An Application to Supply Modeling," in *Petroleum-Resource Appraisal and Discovery Rate Forecasting in Partially Explored Regions*, U.S. Geological Survey Professional Paper 1138-A, B, C (Washington, D.C.: U.S. Government Printing Office, 1980), pp. C1–C20.

37. Ibid., p. C18.

38. MacAvoy and Pindyck, *Price Controls and the National Gas Shortage* and *The Economics of the Natural Gas Shortage*.

39. Attanasi, Drew, and Scheunemeyer, "An Application to Supply Modeling."

15 DISCOVERY PROCESS MODELS

A discovery process model is one built from assumptions that directly describe both physical features of the deposition of individual pools and fields and the fashion in which they are discovered. When such a model describes in complete detail a probability law governing the generation of observable data, it may also be called an objective probability model. The parameters of this probability law may not be known with certainty, but it is generally assumed that the functional form of the class of distribution functions characterizing the law is known with certainty.

Although in principle models of this type might be applied to data from geographic regions of any size, their underlying assumptions lose relevance as the region or target area is expanded to include descriptively dissimilar deposits. This was recognized by the authors of the first application of a discovery process model to oil and gas deposits, Arps and Roberts, who chose a sample area of 5.7 million acres on the east flank of the Denver–Julesburg basin, stating:

> The production on the east flank of the Basin is almost entirely from the Dakota group of Lower Cretaceous age, commonly called the "Graneros Series." In this series the "D" and "J" sands are the most prominent, although some heavy oil has occasionally been produced from the lower, or "O", series. Locally, some production has also been obtained in the Basin from the Lyons sandstone of Permian age.
>
> Because almost all production is from producing formations of one geologic age, and because the accumulations appear to be similar in nature, a statistical study was made of the Lower Cretaceous fields which have been found to date in a 5.7 million-acre "sample area."[1]

In addition, nearly 70 percent of the deposits discovered in the sample area as of 1958 were stratigraphic, and less than 10 percent were purely structural. Arps and Roberts classify fields into three trap types:

1. Structural traps, with trapping conditions entirely dependent upon structure;
2. Structural-stratigraphic traps, formed by a combination of structural anomalies and lithologic changes, both of which must occur together to produce a trap;
3. Stratigraphic traps, which are entirely controlled by lithologic changes and which would be productive even if regional dip were the only structural component.

Most subsequent applications of discovery process models hew to the principle that the target population must consist of geologically similar deposits.[2]

A petroleum play is a natural unit for analysis, because it consists of a collection of descriptively similar prospects and deposits in a stratigraphic unit within a petroleum basin. In practice a precise definition of the notion *geologically similar* and consequently a precise definition of a play is often elusive. Nevertheless, the concept of a petroleum play is a descriptive template universally used by explorationists to design exploration strategies. Like pornography, a play may be difficult to define, but as Supreme Court Justice Potter Stewart remarked after rendering the Court's decision, "I cannot give an exact description of pornography, but I know it when I see it."

A play typically begins with an exploratory well test of an innovative geological idea. A significant deposit is discovered, the idea confirmed, and a burst of exploratory well drilling ensues. When returns to exploration effort diminish, the rate of exploratory drilling tapers off and the play enters maturity.

If sizes of discoveries in a petroleum basin are arranged in order of discovery and are not sequestered into discoveries within individual plays, the graph of ordered sizes of discoveries on a scale either of discovery number or of cumulative wells drilled usually appears saw-toothed, with intervals of mild to rapid decline in values between sharp rises. This is typical of graphs of discovery size data when discoveries from several plays, each beginning at a different real time and at a different value on a scale of exploratory wells drilled, are aggregated and viewed as coming from a single population of discoveries. When discoveries are classified into individual plays and discoveries within each play are arranged in order of occurrence, the saw-toothed pattern characteristic of ordered sizes when distinct plays are aggregated generally disappears.

Consider the discovery history of the Western Canada sedimentary basin, an important North American onshore province. J. T. Ryan[3] studies the rate of crude oil discovery in the basin and presents graphs of the rates of discovery of recoverable crude oil and of oil in place, first for the province as a whole and then for seven principal plays, each named after the particular time-stratigraphic unit in which the play deposits occur: D-2, D-3, Cardium, Beaverhill Lake, Gilwood, Viking, and Keg River. All other deposits are lumped into a single category. The sharp peaks in Figures 15-1 and 15-2, from Ryan, are directly connected with the introduction (on a scale of cumulative new-field wildcats drilled in the basin) of new plays: Viking, D-2, and D-3 begin in 1947 with the recognition that Paleozoic sediments contained large commercial deposits, followed by the Cardium in 1953, Beaverhill Lake in 1957, Gilwood in 1965, and Keg River shortly after.

The discovery history of the North Sea petroleum province is qualitatively similar. Figure 15-3, from Barouch and Kaufman, shows sizes of fields discovered in the North Sea measured in barrels of oil equivalent (BOE) in order of discovery with no separation of fields into individual plays.[4] The province is composed of four major plays: Chalk, Tertiary, Jurassic North, and Jurassic Central. All but four of the sixty discoveries made by the middle of 1975 belong to one of these four plays. The four largest are as follows:

Name	Discovery Number	Discovery Well Spud Date	$BOE \times 10^9$
Ekofisk	3	9/69	1.932
Forties	7	8/70	1.800
Brent	11	5/71	2.375
Statfjord	32	12/73	4.595

Ekofisk is the first discovery in the Chalk play, Forties is the fourth discovery in the Tertiary play, and Brent is the first discovery in the Jurassic North play. Statfjord, the largest field discovered in the North Sea to date, appears as the tenth of twenty-five discoveries in Jurassic North; that is, the second largest of discoveries in this play was discovered first and the largest discovered tenth. Jurassic Central is a "smaller" play in which the largest discovery to date (Piper, 11/72, 0.800×10^9 BOE) was discovered first.

Graphs of field sizes in order of discovery within individual plays are shown in Figure 15-4.[5] By appearing as the tenth discovery in Jurassic North rather than among the first two or three discoveries, Statfjord is an outlier relative to the generally monotonic decreasing trend of size of discovery with increasing discovery number characteristic of Chalk, Tertiary, and Jurassic Central.

Why Statfjord appears as an outlier in Jurassic North has an explanation, one that emphasizes the importance of understanding the institutional setting in which North Sea discoveries are made. When Brent, the first discovery in Jurassic North, was drilled and identified as a field, it was immediately apparent that a large structure adjacent to it was a prime prospect with very high probability of containing hydrocarbons in the same sedimentary unit. Had exploration drilling proceeded unrestricted by the Norwegian government's exclusion of acreage containing this structure until the second round of licensing in the Norwegian sector, this structure almost certainly would have been drilled immediately after Brent, and Statfjord would appear as the *second* rather than the tenth discovery

Figure 15-1.   Discovery Rate of Recoverable Crude Oil in Alberta.

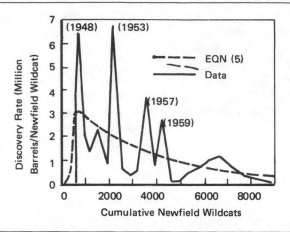

Source: J.M. Ryan, "An Analysis of Crude-Oil Discovery Rate in Alberta," *Bulletin of Canadian Petroleum Geology* 21, no. -2 (June 1973): 219–35, Figure 3. Reprinted with permission.

Figure 15-2.   Discovery Rate of Oil in Place in Alberta.

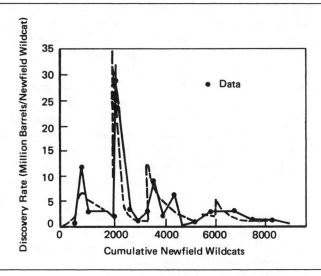

Source: Ryan, "Analysis of Crude-Oil Discovery Rate in Alberta," Figure 4. Reprinted with permission.

Figure 15-3.   Sizes of North Sea Oil Fields in Order of Discovery.

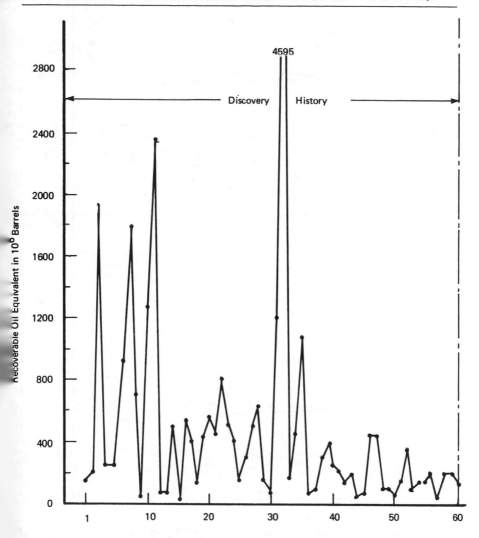

Source: E. Barouch and G.M. Kaufman, "Oil and Gas Modelled as Sampling Proportional to Random Size," Massachusetts Institute of Technology Sloan School of Management Working Paper 888-76, Cambridge, Mass., 1976. Reprinted with permission.

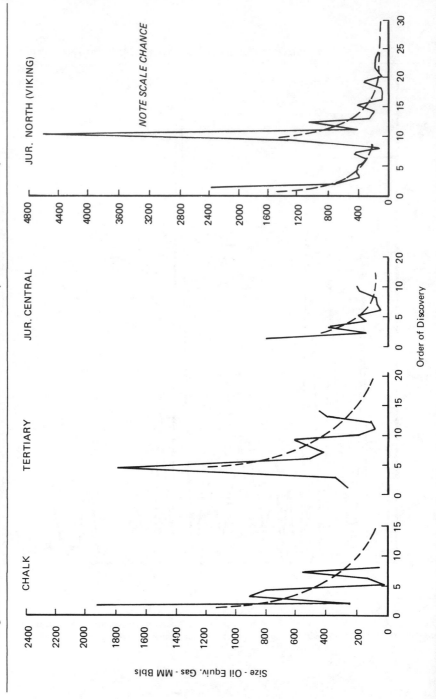

Figure 15–4. Sizes of North Sea Oil Fields in Order of Discovery within Individual Plays.

in Jurassic North. The Norwegian government has pursued a deliberate policy of excluding prime acreage from early licensing rounds in order to minimize the inflationary impact on their economy of a tremendous surge in drilling and discovery that would obtain if licensing were unrestricted in their sector. The fourth round of licensing will include four more very large and promising structures, all known to exist for several years.

Examination of these data and of sizes in order of occurrence within plays from other petroleum provinces[6] suggests an empirical proposition: The largest deposits tend to be found early in the evolution of a petroleum play. The nature of exploration technology and economic incentives join to make this proposition a component of exploration folklore, and there is evidence in its favor. It appears most intuitively acceptable in a province where exploration effort can be applied from the outset over the area encompassing a play, unfettered by leasing restrictions that withhold promising acreage. Most discovery process models incorporate this idea, either explicitly or implicitly. It is a version of a law of diminishing returns to increasing cumulative exploration effort.

The second empirical proposition underpinning discovery process models is that the relative frequency distribution of sizes of deposits in a petroleum play has a characteristic shape: Most relative frequency distributions are unimodal and positively skewed with a "fat" right tail.

Precise versions of these two propositions separately or conjoined are difficult to test with observed data. Many sources of measurement error distort reported deposit sizes however *size* is defined. When discoveries are generated by a process in which, within a finite collection of deposits, a large deposit is more likely to be found by a given exploratory well than a small one, individual discovery sizes are probabilistically dependent. These features of the data and of the process generating them complicate statistical testing of model assumptions.

DEPOSIT ATTRIBUTES, SIZE, AND MEASUREMENT ERROR

Attributes of individual deposits are the primitive elements of a discovery process model. Among the most important are the projective surface area of the deposit, its rock volume, the volume of pore space in its rock volume capable of containing hydrocarbons, the volume of hydrocarbons in place in it, and the volume of hydrocarbons ultimately recoverable from it by use of a particular mix of recovery methods. *Size* is often used as a label for any one of these attributes; for example, the size of a deposit is said to be 1,200 acres, or the size of the deposit is given as 100×10^6 ultimately recoverable BOE. Which particular attribute is labeled size is usually clear from the context. Volume of rock, volume of pore space, area, and average pay thickness of a deposit are functionally related to one another through constraints: In order for a deposit to be a deposit, it is constrained to conform to a restricted set of geometric configurations. Furthermore, if deposits in a petroleum basin are classified into descriptively homogeneous subpopulations and a probabilistic property is ascribed to one attribute of one of these subpopulations, then descriptive coherence requires that it be consistent with probabilistic properties assigned to other attributes. An absurd example illustrates: Suppose that the areas and rock volumes of N deposits in a sedimentary unit with maximum thickness of 2,000 feet are viewed as realizations of $2N$ mutually independent uncertain quantities; areas possess identical probability distributions with densities concentrated on $(0, \infty)$, each having a mean of 1 acre; and rock volumes possess identical probability distributions with densities concentrated on $(0, \infty)$, each having a mean of 12,000 acre-feet.

This assignment of probabilities is not descriptively coherent if it assigns positive probability to an event such as the area of a deposit being 1 acre $\pm$ 0.1 acre and its rock volume being 12,000 acre-feet $\pm$ 10 acre-feet, for this event implies a deposit configuration with about 12,000 vertical feet of rock, which exceeds the thickness of the sedimentary unit by 10,000 feet. The range of physically realizable values of any one deposit attribute is in practice a finite interval and choices of unbounded intervals, as ranges for uncertain quantities are usually dictated by mathematical convenience. Little harm results so long as incoherent values are achievable only with negligible probability.

Exploration technology determines which individual deposit attributes are *physically observable.* Deposit area, rock volume, and reservoir variables such as porosity and permeability are usually inferred from a limited set of measurements, at irregularly spaced well locations.

As noted in Part II, *reported* ultimately recoverable reserves and reported hydrocarbons in place in a field tend to grow as a function of the time elapsed from the date of the field's discovery. More particularly, the proportional growth in reported hydrocarbons in place and in ultimately recoverable reserves as a function of time elapsed from date of discovery is generally smaller for small deposits than for large ones. Pool reserve reporting is conservative in practice and closely tied to the number of development wells drilled. Compare a pool with small projective area requiring only one or two wells to develop it fully with a large pool requiring over 100 development wells to delineate and develop it. After two wells have been drilled in each, the reported reserves in the large pool may still change by an order of magnitude, whereas reported reserves for the small pool would be much less likely to experience as large a proportional change.

Observations of members of the finite collection of deposits set down by nature in the geographic area over which a play evolves is a sampling process, and rarely if ever are all of the deposits in a play both discovered and reported.

Economics and the character of exploration combine to exclude small deposits from tabulated deposit data. Geologic and geophysical search has limited resolving power: Even very intensive application of search effort applied to a play area may not detect all anomalies capable of containing hydrocarbons. Larger anomalies are more likely to be detected than smaller ones and many small anomalies that are deposits of hydrocarbons may be missed. If a discovery is too small to be profitable, it will not be developed and may not even appear in a tabulation of deposit data. Thus data sets describing sizes of deposits in a play are subject to economic truncation of sizes attributed to discoveries and to missing values of deposits that are profitable to produce but not yet discovered. A deposit data set rarely if ever includes all of the deposits set down by nature in a play. Ratios of reported to actual reserves for fully developed small pools requiring few development wells are likely to have greater dispersion than similar ratios for fully developed large pools. Very large pools are proved by delineation drilling, and the physical boundaries and productive area of such pools at maturity are generally determined with a reasonable percentage of accuracy. The productive area of very small pools is often guessed at and rounded off to the nearest well-spacing unit. Pool area is a potential equation variable, and it enters the equation linearly or with a power close to one. Hence, if reported reserves are computed using a reservoir potential equation, a given percentage error in reported versus actual area leads to a percentage error of the same magnitude in reported versus actual reserves. This holds true for reported versus actual hydrocarbons in place.

Although reserves attributed to a pool tend on the average to appreciate as development progresses, the rate at which changes take place in reported reserves is very much a function of the area and the geologic character of the pool. Ryan has studied the behavior of reported reserves for pools in individual plays located in the Western Canada Sedimentary Basin (WCSB).[7] The variation in time patterns of reported reserves varies markedly across plays. At one extreme are pinnacle reefs (Rainbow–Zama Lake), which are very prolific "one-well" pools fully developable in one or two years. At the other extreme are pools with a very large projective area that may require 100 or more development wells and a correspondingly longer time to be developed fully. A natural hypothesis is that reported reserves for a pool grow with the number of development wells drilled in it.

According to Ryan,

The most common development program consisted of intense drilling for a period of several years followed by a significant decrease in activity. During the initial phase, the estimate of reserves in the pool increased above the original estimate in direct proportion to the number of wells. Following the peak in drilling activity, the estimate usually tended to increase erratically until the current estimate was obtained. The development of the Bonnie Glen D-3A pool [see Figure 15-5] illustrates this pattern quite well.

The Provost Viking C pool exhibits a completely different pattern of development as shown in [Figure 15-6]. There are two conclusions to be drawn from this figure. First, the major development phase can take place

Figure 15-5.   Development of Bonnie Glen D-3A Pool.

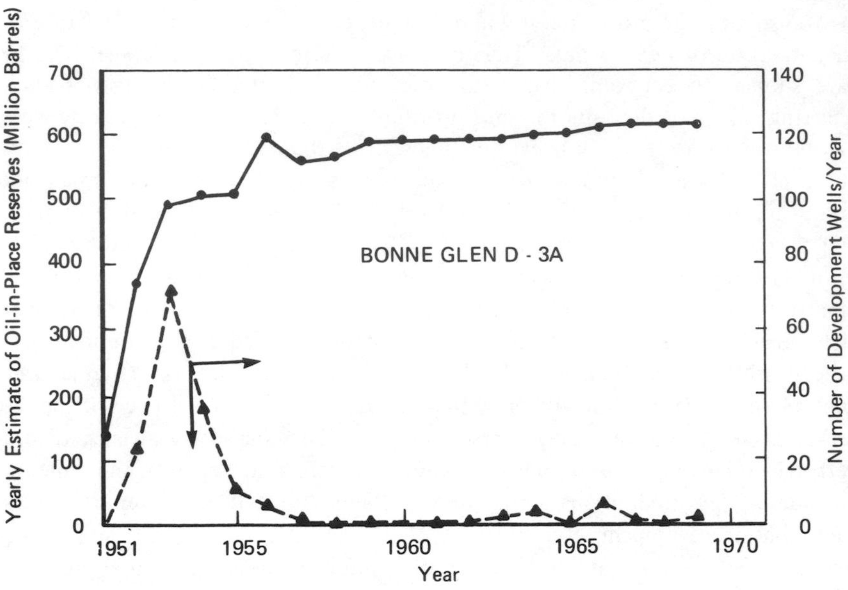

Source: Ryan, "Analysis of Crude-Oil Discovery Rate in Alberta," Figure 1. Reprinted with permission.

long after the pool is discovered. Second, the reserve estimate can be correlated with the number of wells drilled. Thus this pool exhibits an orderly though belated development sequence. In contrast, the correlation between the number of development wells and the reserve estimates of the Willesden Green Cardium A pool is weak and even negative. The data are shown in [Figure 15-7]. As can be seen, there is little relation between wells drilled and reserves estimated. The last two figures suggest that pools of both types are capable of further appreciation but that the degree of appreciation is not predictable.[8]

Hubbert, Pelto, and Arps et al. have each modeled reserve appreciation for a generic field as a function of time in years from date of discovery using aggregate U.S. data.[9] Root[10] has done so for fields in the Permian Basin and for fields in the offshore Gulf Coast province. A suitable model of reserve appreciation at the play level might be constructed by first stratifying pools or fields according to projective area, then projecting reserve appreciation as a function of time for each stratum.

Figure 15-6.   Development of Provost Viking C Pool.

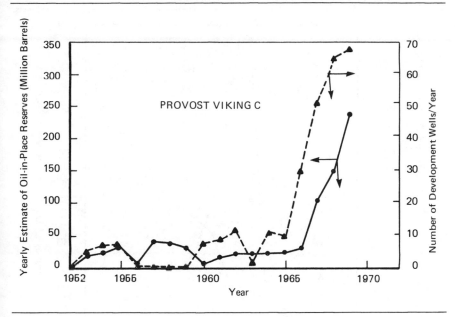

Source: Ryan, "Analysis of Crude-Oil Discovery Rate in Alberta," Figure 2. Reprinted with permission.

DEPOSIT SIZE DISTRIBUTIONS

Unimodel and positively skewed with a fat right tail are characteristic traits of the relative frequency histogram of sizes—areas, rock volumes, hydrocarbons in place, or recoverable hydrocarbons—of deposits discovered in a generic play that has reached maturity. (If geologically distinct types of deposits are mixed and interpreted as a single population, the corresponding relative frequency histograms may display two or more modes.) Economics and the limited resolving power of exploration technology combine to exclude very small deposits from observed data, so a histogram of the relative frequency of deposit sizes is truncated, although the point of truncation is seldom, if ever, precisely known. Since the rock volume of the largest deposit in a sedimentary unit cannot exceed the rock volume of the unit itself, the relative frequency histogram is also truncated from above.

The observation that relative frequency histograms of petroleum deposit sizes and of deposit sizes for minerals other than petroleum are roughly similar in shape—whether size is defined as deposit rock volume or as projective area or as volume of mineral in place—led researchers to ask a question. Suppose that

Figure 15-7.   Development of Willesden Green Cardium A Pool.

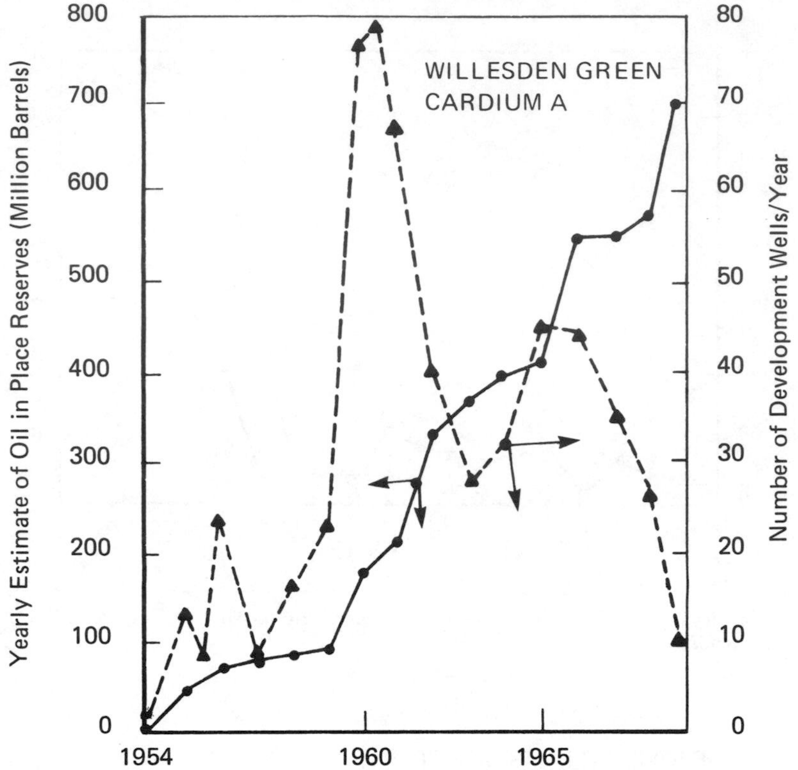

WILLESDEN GREEN
CARDIUM A

Source: Ryan, "Analysis of Crude-Oil Discovery Rate in Alberta," Figure 3. Reprinted with permission.

deposits in a given petroleum or metallogenic province are classified into descriptively homogenous subpopulations and the sizes of deposits within each subpopulation are regarded as coming from an idealized sampling process that generates values of mutually independent, uncertain quantities possessing a common probability distribution. Is the *functional form*[a] for this probability distribution the same from subpopulation to subpopulation for a given mineral? For each possible definition of size? For all minerals? If it is hypothesized that the same functional form characterizes subpopulations of deposit sizes for a given mineral and a particular definition of size, then what specific functional form is suggested by observational data?

a. A mathematical formula defining a family of probability distributions.

The hypothesis most frequently entertained is that mineral deposit size distributions are approximately lognormal; that is, that relative frequency histograms of the logarithms of deposit sizes for a prespecified population of deposits are closely fit by a normal or Gaussian density. The idea that, among an infinite number of possible functional forms, the lognormal distribution is a logical candidate for this role has its root in statistical studies of deposit and mine value, most notably those of Krige; Matheron, Blondel and Ventura; and Allais.[11] Since the problem of determining the distribution of values of sample blocks in a deposit is related to but not the same as the problem of determining the distribution of deposit sizes, this work is at most suggestive.[b] Krige and his students demonstrated that the gold values obtained in sampling reef areas in the Witwatersrand are closely approximated by lognormal density. Figures 15-8 and 15-9 afford a picture of two lognormal densities fit to Krige's data.

Other investigators also fit the lognormal density to relative frequency histograms of sedimentary rock attributes and to samples of hard-rock mineral attributes drawn from an individual deposit (see Thebault as well as Matheron[12]).

W. G. B. Phillips presents an interesting theoretical explanation for lognormality of the distribution of hard-rock minerals in the earth's crust. He views ore genesis as a random process driven by plate tectonic action, subduction, and erosion:

> The lithosphere, or outer shell, of the earth is made up of about a dozen rigid plates that move with respect to each other. New lithosphere is created at mid-ocean ridges by the upwelling and cooling of magma from the earth's interior. Simultaneously the old lithosphere is subducted, or pushed down, into the earth's mantle. As the formerly rigid plate descends it slowly heats up, and over a period of millions of years it is absorbed into the general circulation of the earth's mantle. Plate tectonics and the subduction of the lithosphere have helped to explain and unify a wide range of geologic phenomena that previously were puzzling or treated on an *ad hoc* basis. The implications for the theory of mineralization and the formation of ore bodies have yet to be worked out in detail. However, it seems clear that the new lithosphere can act as a vehicle to transfer elements of high atomic weight from the earth's deeper levels to the surface. It also seems probable that these elements (which include the base and precious metals) can be transferred to the continental crust during the process of subduction and remelting. The temperature within the earth increases rapidly with depth reaching about $1200°C$ at a depth of 100 km. This temperature is sufficient to melt any rock-forming minerals and, in the presence of water, heavy metal ions would be taken into solution. The mineralized solutions would then be injected under pressure into the continental crust along fissures and planes of weakness. The formation of the high-grade concentrations having economic importance are controlled by the

b. If in fact a complete sample of the mineral content in identically sized mineral blocks *within* a generic hard-rock mineral deposit were lognormally distributed, the aggregate mineral content of the deposit would not be precisely lognormal.

Figure 15-8.   Frequency Distribution of 28,334 inch DWT Values on Mine A by Means of a Histogram and the Lognormal Curve Fitted to These Observations.

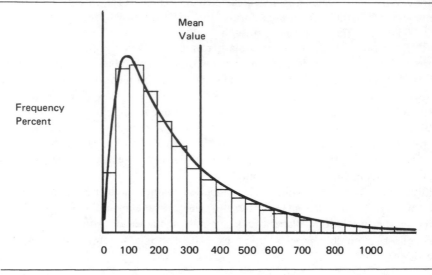

Source: D.G. Krige, "A Statistical Approach to Some Mine Valuation Problems on the Witwatersrand," *Journal of the Chemical Metallurgical and Mining Society of South Africa* 52, no. 6, (1951): 119-39.

Figure 15-9.   Frequency Distribution of inch DWT Values within an Average Ore Reserve Block on the Witwatersrand.

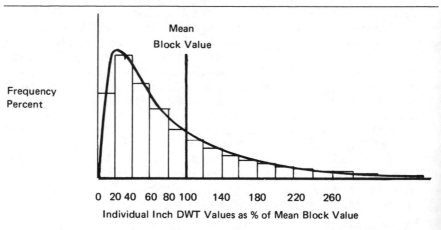

Source: Krige, "Statistical Approach to Some Mine Valuation Problems on the Witwatersrand," p. 126.

physical and chemical characteristics of the host rock. Folding, faulting metamorphism and saturation with mineralized solutions will tend to alter the
host rock in ways that facilitate metamorphism and mineralization at subsequent epochs. This is an example of a very general process whereby the existence of a property at some place in a system affects the probability that it
will exist later at the same adjacent places. Such a process leads directly to a
lognormal distribution of minerals, which can be shown empirically to be
characteristic of many ore bodies that have been studied in detail. The location and structure of the porphyritic copper deposits that are found around
the Pacific basin close to an active subduction zone are consistent with the
theory of mineralization described above . . .

Consider a positive variate which is the outcome of a discrete random process, and assume that the process takes place at successive points in time, as
may be the case for the process by which a mass of host rock is mineralized
by solutions containing metal ions of high atomic weight (e.g., copper, lead,
zinc, iron). If the variable is the concentration of metal in the rock, the
degree of concentration may be regarded as the joint effect of a large number
of mutually independent causes acting in an ordered sequence during successive epochs of mineralization.[13]

Defining X_t as the concentration at time t in a prespecified rock volume, Phillips
first describes the process as unfolding at discrete time points $t = 0, 1, 2, \ldots$
according to

$$\widetilde{X}_t = \widetilde{X}_0 \prod_{j=1}^{t} (1 + \widetilde{\epsilon}_j) , \qquad (15\text{-}1)$$

or equivalently to

$$\log \widetilde{X}_t = \log \widetilde{X}_0 + \sum_{j=1}^{t} \log (1 + \widetilde{\epsilon}_j) . \qquad (15\text{-}2)$$

Given

$$\widetilde{X}_{t-1} = X_{t-1} , \qquad \widetilde{X}_t = X_{t-1} (1 + \widetilde{\epsilon}_j) , \qquad (15\text{-}3)$$

so the concentration at time t is a random proportion of that at time $t-1$ in
accord with the idea that "the effect of future exposure to solutions containing
metal ions would depend on the degree of concentration that had already been
obtained by the host rock."[14] In order to ensure that the degree of concentration does not increase beyond all bounds as $t \to \infty$, behavior "contrary to the
geological evidence," Phillips posits the effect of erosion to be a countervailing
force and modifies the usual proportional effects model (in which it is assumed
that $\widetilde{X}_0, \widetilde{\epsilon}_1, \widetilde{\epsilon}_2, \ldots$ are mutually independent), to

$$\log (1 + \widetilde{\epsilon}_t) = -\alpha \log X_{t-1} + \widetilde{\delta}_t, \qquad t = 1, 2, \ldots, \qquad (15\text{-}4)$$

or equivalently to

$$\widetilde{X}_t = X_{t-1}^{1-\alpha} \, e^{\widetilde{\delta}_t}, \qquad t = 1, 2, \ldots, \qquad (15\text{-}5)$$

with $0 < \alpha < 1$, $\widetilde{\delta}_1$, $\widetilde{\delta}_2, \ldots$ mutually independent and identically distributed with mean δ, and $\widetilde{X}_0$ independent of $\widetilde{\delta}_t$. With these assumptions, $E(\log \widetilde{X}_t) = \delta/\alpha$ and $\mathrm{Var}(\log \widetilde{X}_t) = \mathrm{Var}(\widetilde{\delta}_1)/\alpha(2 - \alpha)$. Consequently, the distribution of concentration has finite variance so long as $0 < \alpha < 2$. This is the first orderly discussion tying together plate tectonic theory and lognormality of mineral concentration in the earth's crust. The formula for the variance of $\log \widetilde{X}_t$ is incorrectly stated as $\mathrm{Var}(\widetilde{\delta}_1)/\alpha^2$ by Phillips.[15]

According to Aitcheson and Brown, Kalecki (1945), in his study of the size distribution of incomes, was the first to suggest introducing negative correlation between $\log(1 + \widetilde{\epsilon}_j)$ and $\log \widetilde{X}_j$ in order to bound probabilistically the growth of $\widetilde{X}_j$.

Given data and hypotheses of this kind, it is natural to ask if the same functional form is a reasonable fit to relative frequency histograms of deposit sizes. Allais and his colleagues Blondel and Ventura gathered deposit data describing hard-rock mineral deposits of several types in France, North Africa, the western United States, and the world at large.[16] Blondel and Ventura classified deposits by region and by mineral type, computed the dollar (or franc) value of each deposit, and assumed that:

> The value of different deposits is supposed to be distributed according to a lognormal distribution. . . . This hypothesis is justified *a priori* by the fact that the major part of economic quantities has a lognormal distribution and *a posteriori* by the fact that in all cases which we have studied the distribution of the value of deposits or of mineral outputs fits the lognormal curve remarkably well. The fitting of the lognormal distribution has been made using the customary straight-line method proposed by Henry; the goodness of fit is remarkable.[17]

The definition of a deposit varies from source to source, for it may be a mineral district, a mine, or a collection of mines. Allais claims that this fuzziness in definition did not affect the results significantly and that the fits were good.

A graphical method was used to fit a lognormal density to the data: Individual deposit values were grouped into size classes and the number of deposits in each size class was used to compute a sample fractile corresponding to that class. The sample fractiles were then plotted on lognormal probability paper. Although a straight line fit to the grouped data appears good in almost all cases, grouping obscures features of the data. In particular, the behavior of the tails of the relative frequency histogram of values is not clearly visible. Figure 15-10 is an example. It shows 125 North African mineral deposits grouped into seventeen size

Figure 15-10. Frequency Distribution of Sizes of 128 North African Mineral Deposits.

G(x) Applications numeriques Graphique B6a

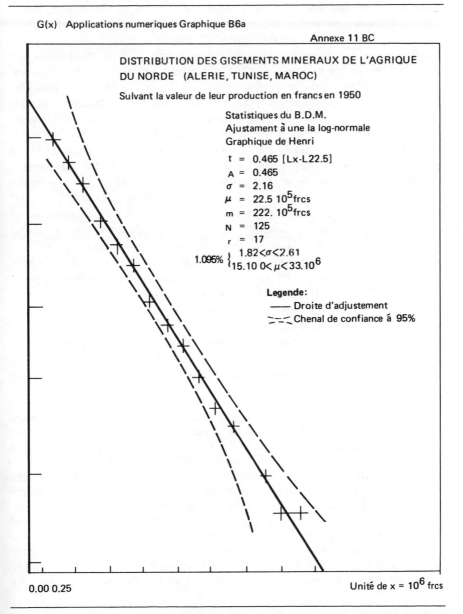

Annexe 11 BC

DISTRIBUTION DES GISEMENTS MINERAUX DE L'AGRIQUE DU NORDE   (ALERIE, TUNISE, MAROC)

Suivant la valeur de leur production en francs en 1950

Statistiques du B.D.M.
Ajustament à une la log-normale
Graphique de Henri

$t = 0.465 [Lx-L22.5]$
$A = 0.465$
$\sigma = 2.16$
$\mu = 22.5 \cdot 10^5 frcs$
$m = 222. \cdot 10^5 frcs$
$N = 125$
$r = 17$

$1.095\% \begin{cases} 1.82 < \sigma < 2.61 \\ 15.10 \, 0 < \mu < 33.10^6 \end{cases}$

Legende:
—— Droite d'adjustement
⊃⁼⁻⊂ Chenal de confiance á 95%

0.00 0.25

Unité de x = 10^6 frcs

Source: M. Allais, "Method of Appraising Economic Prospects of Mining Exploration over Large Territories," *Management Science* 3 (1957): 285–347. Reprinted with permission.

classes (only sixteen of which appear). Each size class represents a fractile. Although Allais recognizes that the data are most likely doubly truncated—from below by economics and from above by physical limits on the maximum tonnage of deposits—the fitting procedure does not directly account for it. A 95 percent confidence band is drawn about the fitted distribution.[c]

Allais states at the outset, "Our task was to supply the best possible solution within a given period of time using the information available . . . [and] to find as quickly as possible the essential orders of magnitude."[18] Petroleum resources were deliberately excluded from his published manuscript because Algerian hydrocarbon potential was then regarded as a politically explosive issue by the French government. Shortly it would not belong to France: The Algerian revolt blossomed and Algeria gained its independence. Allais's study raises a host of methodological questions, nevertheless it served as a springboard for attempts to fit the lognormal distribution to many other mineral deposit data sets, including oil and gas data.

Using publicly available data[19] Kaufman examined frequency histograms of oil and gas deposit sizes for several U.S. regions, with *size* defined as cumulative production, as oil (gas) in place, or as ultimately recoverable oil (or gas).[20] The data could not be carefully partitioned into definitionally homogeneous deposit types and are composed of mixtures of two or more geologically distinct populations in all but one case. No rigorous statistical testing was done. Fractile plots on lognormal probability paper were straight enough to suggest that the lognormal hypotheses deserves further study.

McCrossan separated oil and gas deposits in Canada's Western Sedimentary Basin into genetically related types and examined, type by type, properties of fractile plots of pool areas and of ultimate recoverable reserves on lognormal probability paper. He states,

> One of the justifications for the study was to answer the question: are giant pools unique geologically, with some peculiar reason for their existence that makes them freaks? The answer would appear to be that they are not, but are rare because of the improbability of the confluence of favorable factors necessary for their creation and preservation. An examination of most of the plots clearly indicates that large pools fall logically into the lognormal distribution, or that the probability of their occurrence is consistent with this type of distribution. Where occasionally there is a point or two significantly off the upper tail of the distribution, a logical explanation should be sought and may well lie in an incorrect estimation of the reserve size. Even in the cases of the heterogeneous distributions, the giant pool seems to fall in with a group and are not off by themselves. Properties of natural resources might be more

c. It is computed for a given value x and cdf $F(x)$ by viewing the sample of size n as a Bernoulli process with probability $F(x)$ of a value falling at or below x and $1 - F(x)$ of a value falling above x. A standard 95 percent confidence interval is then generated at each value of x.

precisely described by other types of distributions such as negative binomial, Poisson, etc. The data considered here, however, seem to approximate the lognormal distribution sufficiently closely, particularly the type of data involved, to justify the assumption.[21]

The number of deposits within each type varies from fifteen to fifty-three; that is, the number of observations range from small to moderate. Many of McCrossan's fractile plots display significant departures from a straight line; inflection points appear and, for pool area in particular, as many as five observations may have exactly equal value.

Table 15-1 shows for which deposit types a relative frequency histogram for discoveries has apparent inflection points when plotted on lognormal probability paper. McCrossan offers an explanation:

> Most of the reserve size-frequency distributions of Western Canada display some heterogeneity. At least three hypotheses to explain the heterogeneity can be entertained.
>
> The first is that it is the result of mixing two geologically unlike groups and that there is some fundamental difference in reservoir characteristics between the two, porosity type for instance.
>
> Second, it might result from an inaccuracy in the estimation of one or both groups of reserves, that is, the heterogeneity could be induced. The group of larger sizes could be over-estimated, which seems less likely, or the group of smaller sizes could be underestimated. One reservoir parameter that is often estimated for smaller pools is the area, and this would create a larger percentage error in a small pool than in a large one. Newly discovered pools would also, of course, be under-estimated.

Table 15-1.   Reserves Frequency Distribution Types.

Unimodal		Bimodal		Uncertain
Oil	*Gas*	*Oil*	*Gas*	*Oil and Gas*
Turner Valley	Charlie	Lower	Leduc	Granite Wash
and Pekisko	Glauconite?	Mannville	Wabamun	Gilwood
Cardium	Ellerslie	Viking	Turner Valley	Debolt
(truncated)	(truncated)	Keg River	(unconf.)	structure
Belly River	Permo–Penn	Upper	Slave Point?	
Leduc	Mississippian	Mannville	Nisku?	
	(NEBC0	Nisku	Pekisko	
	(truncated?)	Beaverhill	Jurassic	
	Turner Valley	Lake	Viking	
	(foothills)		Halfway?	
			Baldonnel	
			(truncated?)	

Source: R.G. McCrossan, "An Analysis of Size Frequency Distribution of Oil and Gas Reserves of Western Canada," *Canadian Journal of Earth Sciences* 6, no. 201 (1969): 209, Table 1. Reprinted with permission.

Third, the group of larger-sized pools with the larger mean might be the result of enhanced recovery as opposed to the group of small average-size pools, which might be those with only primary production. Three other possibilities have been suggested as follows: (1) it is possible that large pools are found early in an exploration play and so would be disproportionately represented in a sample; (2) perhaps the controlling processes forming accumulations might have accelerated at a certain size and thus made large pools more frequent than expected; and (3) possibly the non-linearity between lognormally distributed variables upon which the reserves are dependent could cause departures from normality. It is almost certain, however, that the estimations made in calculating the reserves is one of the most important factors in creating two modes. The fact that some of the distributions are unimodal makes the last two hypotheses and possibly the last three somewhat less tenable.[22]

In a commentary on the hypothesis that the distribution of mineral content in igneous rock is lognormal, Prokhorov[23] points out that lognormality cannot be either confirmed or disproved if data consist of a large number of distinguishable samples of "small" to "medium" size. (Two samples are "distinguishable" if they are generated by distributions with the same functional form but possibly different parameter values. One hundred observations is regarded as a medium-sized sample.) More generally the functional form of a distribution cannot be reconstructed from a very large number of distinguishable samples of small to medium size (see Petrov[24]). Thus, even if deposit size distributions for petroleum plays all have the same functional form, it would not be possible to identify it with precision using deposit size data from a large number of plays with at most a moderate number of discoveries in each play.

Prokhorov clarifies the nature of the formal argument by example, showing that when appropriately normalized, the logarithm of a gamma-distributed random variable is asymptotically lognormal. Consequently a very large sample is required to distinguish a lognormal from a gamma distribution. If $\tilde{x}$ has density $\alpha x^{\alpha-1} \exp\{-x\}/\Gamma(\alpha)$, $\alpha > 0$ and $x > 0$, then $\tilde{u} = [\log \tilde{x} - \log \alpha] \sqrt{\alpha}$ has density

$$\frac{1}{\sqrt{2\pi}} \exp\left\{-\frac{1}{2} u^2\right\} \left[R(\alpha) \exp\left\{-\alpha(e^{u/\sqrt{\alpha}} - 1 - \frac{u}{\sqrt{\alpha}} - \frac{u^2}{2\alpha})\right\}\right]$$

$$(15-6)$$

with $R(\alpha) = \alpha^{\alpha-1} \exp\{-\alpha\} \sqrt{2\pi\alpha}/\Gamma(\alpha)$. When $\alpha = 1$ and $\tilde{x}$ has an exponential distribution with density $\exp\{-x\}$ this density can be distinguished with type I and type II errors of 0.05 from a lognormal distribution with density $\exp\left\{-\frac{1}{2} \log^2 u\right\}/u \sqrt{2\pi}$, $u > 0$, only if the sample size is close to 100. A similar argument can be made for the one-sided stable distribution with density $\exp\{-\alpha^2/2x\}/x^{3/2} \sqrt{2\pi}$, $x > 0$: for α large, $\log(x/\alpha^2)$ is approximately lognormal.

ARPS AND ROBERTS' AND DREW, SCHEUNEMEYER, AND ROOT'S MODELS

The pioneering work of Arps and Roberts and extensions of it by Drew, Scheunemeyer, and Root portray exploration as a process by which wildcats are randomly placed at coordinate points within the boundary of a well-defined play or basinal area.[25] If the wildcat lies within the perimeter of the projective area of a field, a discovery is made; otherwise the wildcat is a dry hole. A view of drilling effort as "exploring area" converges logically with the search for stratigraphic trap-type deposits. Identification of specific drillable prospects by predrilling exploratory efforts is typically less effective for this type of deposit than for structural deposits.

Arps and Roberts postulate that the number of fields $\Delta N(A, W)$ found by the next increment ΔW of new-field wildcats drilled is proportional to the total area $A[N(A) - N(A, W)]$ of fields of size A remaining to be discovered:

$$\frac{\Delta N(A, W)}{\Delta W} \propto A[N(A) - N(A, W)] . \tag{15-7}$$

If the factor of proportionality rendering the left- and right-hand sides of (15-7) equal is assumed to be a constant c_o, and $N(A, W)$ is interpreted as a (deterministic) continuous function of W, then

$$N(A, W) = N(A)\left[1 - \exp\left\{-c_o A W\right\}\right] . \tag{15-8}$$

This model interrelates a specific field size A, number of wells drilled, and number of fields of size A, and in this respect must be distinguished from rate-of-effort models that do not incorporate such features. A further distinction is that the model and its successors are micromodels designed to portray the evolution of discovery in a single province, while other rate-of-effort models are designed to treat aggregates of provinces, such as the coterminous United States onshore. Consequently, the Arps–Roberts model is labeled here as a discovery process model.

The postulate (15-7) does not by itself constitute a discovery process model, that is, an objective probability model based on assumptions about the interaction of exploration technology and properties of deposits as deposed by nature. If the change $\Delta N(A, W)$ in the number of fields of size A discovered by an increment ΔW of wells is regarded as an uncertain quantity, then an appropriate interpretation is that the expectation $E(\Delta \tilde{N}(A, W) \mid H_W)$ of $\Delta N(\tilde{A}, W)$, given an exploration history H_W generated by wells $1, 2, \ldots, W$, is proportional to $A[N(A) - N(A, W)]$. Interpreted in this way, (15-7) describes conditional expectations for $\Delta \tilde{N}(A, W)$, $A > 0$, but not a joint probability law for

$$\left\{\Delta \tilde{N}(A, W) \mid A > 0, \Delta W, H_W\right\} .$$

The physics of exploration suggests that Eq. (15-7) can be reshaped to incorporate two observable features of the discovery process: *areal exhaustion* and *discovery efficiency*. As wildcat wells are drilled in a basinal or play area of extent B_o, the effective sample space or target area is reduced by discovery of new fields and by the elimination of unproductive area "explored" by dry holes. By definition, the areal extent of a discovered field no longer forms part of the sample space to be explored by future wildcats.[d] The idea that a dry hole condemns a positive portion of the sample space, eliminating it as a target for future drilling, has been studied by Singer and Drew.[26] It is also used by Drew, Schuenemeyer, and Root[27] in their analysis of exploration in the Denver-Julesberg Basin using (15-7) in a more sophisticated way than Arps and Roberts. Drew, Scheunemeyer and Root use the cumulative area exhausted by drilling of wildcats as a measure of exploratory effort rather than the number of wildcats drilled.

The Arps and Roberts study and that of Drew, Schuenemeyer, and Root assume the existence of a finite number $A_1, \ldots, A_m$ of target sizes (areal extent of fields), and that nature has deposed $N_i, i = 1, 2, \ldots, m$ fields of areal extent A_i in a play or basinal area B_o. The N_i's and B_o are fixed parameters, none of which are known with certainty.[e] The methods used to estimate parameters not known with certainty are strictly marginal; each size class is considered separately from all others.

A hierarchy of models can be designed by employing the idea portrayed in Eq. (15-7). Which is "best" to use depends on what observational data are available and on tolerance for analytical and computational complexity.

Among options for treating areal exhaustion (in increasing order of descriptive accuracy) are the following:

1. Ignore areal exhaustion due to elimination of the area of discovered fields from the target area or sample space. Regard a dry hole as a point that exhausts zero area.
2. Eliminate the areal extent of discovered fields from the target area or sample space. Regard a dry hole as a point that exhausts zero area.
3. Eliminate the areal extent of discovered fields from the target area or sample space *and* eliminate the area condemned by dry holes.

Arps and Roberts adopt (1) as a working assumption, whereas Drew et al. work with (3).

---

d. Uncertainty about the areal extent of a field prior to its discovery by extension and development wells may result in a well drilled with the intent of discovering a new field (new-field wildcat) penetrating a discovered field. Here such a well is not considered to be a new-field wildcat.

e. Arps and Roberts do not estimate B_o from a time series of observations of drilling outcomes; they estimate B_o directly and use this estimate as a certainty equivalent.

If wildcats are drilled in B_o in a manner assigning equal probability, a priori, to a wildcat being sited in a rectangle of area ΔR no matter where the area ΔR is located in B_o, then drilling is random, and when no areal exhaustion takes place, the probability of discovering a particular field or area A is exactly A/B_o.

In their Denver–Julesberg Basin study, Arps and Roberts assert:

> In hunting for stratigraphic traps . . . the larger potential traps, aside from being easier to hit with any one wildcat, will also be the more obvious when studying the regional sand distribution.
>
> As a first approximation it may be stated therefore that . . . when drilling on geological and geophysical leads, . . . the chance should be better than for random drilling, and the constant c_o [in Eq. (15–8)] is therefore larger than $[1/B_o]$.[28]

Hence they replace A/B_o with cA/B_o, $c > 1$. Reasoning as follows, they set $c = 2$:

> According to the annual A.A.P.G. statistics on exploratory drilling, the success ratio between wildcatting on technical advice such as geology and/or geophysics over the period 1944–1956 is 2.75 times as good as the success ratio for wildcats which were drilled for non-technical reasons. It is the opinion of the authors that this ratio for the Denver–Julesberg Basin was probably not as high as the 2.75 United States average because of the nature of the traps involved, and for the purposes of this estimate we have therefore used a ratio of two. Those who feel that a different number should be used can easily work up their own results with the formulas given.[29]

When efficiency is modeled by scaling up A/B_o in this fashion, values of c greater than one may be interpreted as reducing the size of the sample space (effective basin or play area). This interpretation does not hold if c is not a constant function of the areal extent of a field—that is, if c varies from size class to size class of fields.

Arps and Roberts' aim is to predict ultimate production from Lower Cretaceous fields in the basin. They discuss the economics of production in some detail. Drilling costs include, in addition to productive development wells, the cost of drilling dry holes to delineate the productive limits of a field. They model the number $D(A)$ of dry holes needed to delineate a field of area A as proportional to a power of A and, using dry hole data for each of the 338 fields in their sample, they estimate by least squares that $D(A) \cong .076(A)^{0.345}$. A similar relation between ultimate recovery $U(A)$ from a field of area A leads to $U(A) \cong 530(A)^{1.275}$. They explain why $U(A)$ is proportional to a power greater than one:

> The ultimate recovery U is proportional to the 1.275 power of the area A, and increases, therefore, faster than the areal extent, itself. This means that the average recovery in barrels per acre, U/A, apparently increase as the .275

power of the areal extent, and therefore improves as the fields grow larger. There is, of course, a logical geological explanation for this, since the larger fields will generally have a thicker oil column, and, in the case of stratigraphic traps, because of their sizes, will usually extend farther away from the pinch-out line, thus generally having a less shaley and a better-developed sand section.[30]

In order to project ultimate production from the same area in the basin, the number of fields remaining to be discovered in each size or areal extent class must be projected. Once this is done, the economics of future production (the cost per barrel of future discoveries) can be forecast.

As of January 1958, $W = 3,705$ wildcats had been drilled in a sample area $B_O = 5.7$ million acres, so that $3,705/5,700,000 = 0.00065$. With $W = 3,705$ and $c_O W = 2(0.00065) = 0.0013$, the number $N(A)$ of fields of areal extent A can be expressed as:

$$N(A) = N(A, 3705) \left[1 - \exp\{- .0013A\} \right] \quad . \tag{15-9}$$

Since the numbers $N(A, 3,705)$ of fields of area A discovered by the first 3,705 wildcats is known, the right-hand side of the preceding equation is an estimate of $N(A)$. Arps and Roberts use this relation to generate a projection of the number of fields in each size category (see Table 15-2).

As can be seen, their point estimates of $N(A)$ for each size category suggests that very few fields of large size remain to be discovered, while a very large number of small fields remains undiscovered. Figure 15-11 shows their projections of increments of cumulative ultimate recoverable oil in each size category as a function of cumulative wildcat wells drilled.[f] An inset records their forecast of cumulative reserves to be found as a function of wells drilled; for example, 9,705 wells will discover 407 million barrels (MB) of ultimately recoverable reserves.

The Arps and Roberts analysis is strictly marginal in that each size category is treated on its own as a function of cumulative wells drilled. There is no formal interaction between size classes built into their model. Probabilistic error bounds for their point estimates of the $N(A)$ are not cited. This requires a more complete description of probabilistic properties of their model than they offer. That is, their basic assumption is not a complete description of a probability law governing the generation of discoveries.

Drew et. al study a probabilistic model for each size category, analyze the model rigorously, and provide error bounds (cf. Drew et. al, Parts A and C).

RYAN'S MODEL

Ryan analyzes conventional crude oil discovery in the Western Sedimentary Basin of Canada using a model based on a descriptive postulate: "The rate of crude oil discovery is proportional to the accumulated geological knowledge and

---

f. The symbol F_O denotes $N(A)$ and the symbol F_W denotes $N(A, W)$ in our notation.

Table 15-2. Statistical Data on Denver–Julesberg Oil Fields within Sample Area versus Ultimate Recovery.

Group	Range of Estimated Ultimate Recovery, MB	Number of Fields after 3,705 Wildcats	Number of Fields (Smoothed Data)	Estimated Average Ultimate Recovery, MB	Average Productive Area, Acres	Estimated Ultimate Number of Fields
1	1–2	3	3.0	1.7	2.5	967.7
2	2–4	8	7.4	3.0	3.9	1,453.8
3	4–8	14	13.0	5.3	6.1	1,651.8
4	8–16	23	25.0	12.2	11.7	1,656.7
5	16–32	26	35.0	23.4	19.4	1,401.1
6	32–64	52	43.0	47.6	34.0	994.4
7	64–128	36	46.0	87.5	54.7	669.4
8	128–256	45	45.0	181.3	97.3	378.7
9	256–512	38	40.5	390.7	178.2	195.8
10	512–1,024	34	34.0	791.0	309.8	102.6
11	1,024–2,048	28	27.5	1,420.7	490.5	58.3
12	2,048–4,096	17	17.0	2,944.7	868.5	25.1
13	4,096–8,192	10	10.0	4,962.0	1,306.0	12.2
14	8,192–16,384	3	3.2	10,987.0	2,442.0	3.3
15[a]	16,384–65,536	1	.5	50,750.0	8,108.0	.5

a. Two binary cycles.

Source: J.J. Arps and T.G. Roberts, "Economics for Cretaceous Oil Production on the East Flank of the Denver–Julesberg Basin," AAPG Bulletin 42, no. 11 (1958), Table II. Reprinted with permission.

Figure 15-11. Projection of Increments of Cumulative Ultimate Recoverable Oil.

Cumulative Wildcats in Sample Area	Cumulative Drilling Cost MM$	Cumulative Reserves Found MM Bbls	Incremental Drilling Cost $ per Gr. Bbl
			---------0.92
3705	258.46	279.70	
4705	305.16	311.29	---------1.48
5705	348.32	336.49	---------1.71
7705	428.06	376.18	---------2.01
9705	501.46	406.82	---------2.40

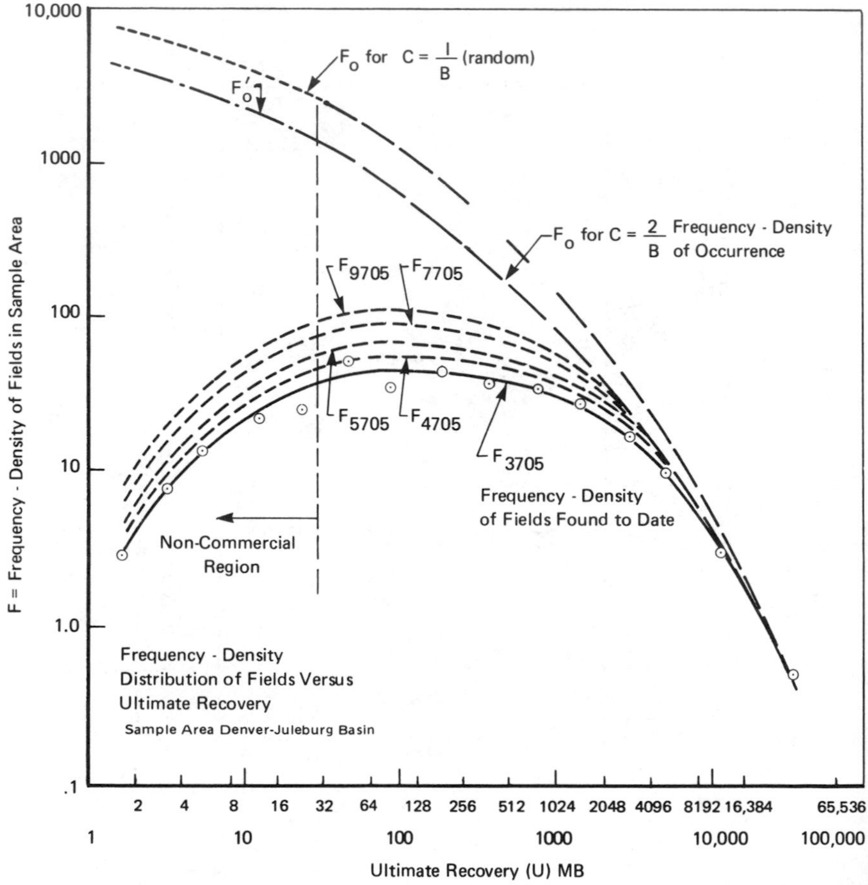

Source: J.J. Arps and T.G. Roberts, "Economics for Cretaceous Oil Production on the East Flank of the Denver–Julesburg Basin," *AAPG Bulletin* 42, no. 11 (1958), Figure 9. Reprinted with permission.

the amount of undiscovered oil."[31] He first treats the province as a single entity and then repeats his analysis, breaking the province up into seven major play categories and lumping all deposits not in one of these seven plays into a single omnibus play. Ryan says,

> The theory is useful for explaining the seemingly erratic growth of provincial reserves. In reality, additions to the total provincial reserve are simply the sum of the additions occurring in each major play. In contrast to the aggregate provincial reserve, discovery of reserves within a play follows an orderly pattern. Once the play has been discovered, the largest fields are found quickly. As exploration continues, all major pools are located. Any further drilling results in the discovery of small pools that contribute relatively insignificant amounts to either the play or the provincial reserve. All Alberta plays exhibit this pattern. The apparent randomness in the growth of provincial reserves results from the sporadic discovery of new plays coupled with the simultaneous activity in other plays in various stages of maturity. With most plays in the middle to late phase of development, any future major additions to the provincial reserve must come from some new horizon or environment not currently thought to be productive.[32]

Ryan's data source was the Alberta Oil and Gas Conservation Board's compilation of individual pool information and well data files, one of the better publicly available data bases as of 1970.[g]

What Ryan calls the "simple" theory is a simple model for the evolution of discovery in the entire basin of the following form: Let $t = 0, 1, 2, \ldots$ be discrete time periods (years), $W(t)$ be cumulative wildcats drilled up to and including year t, $R(t)$ be cumulative recoverable reserves (or oil in place) discovered up to and including year t, U_∞ be ultimately recoverable oil (or oil in place), β be a constant, and $K(t)$ a "function describing the knowledge of where the oil is to be found." Then, if $\Delta R(t) = R(t) - R(t-1)$ and $\Delta W(t) = W(t) - W(t-1)$,

$$\frac{\Delta R(t)}{\Delta W(t)} = \beta K(t) [U_\infty - R(t)] . \qquad (15\text{-}10)$$

The function $K(t)$ is assumed to depend directly on the cumulative amount $W(t)$ of exploratory drilling up to and including t. He argues,

> the knowledge of where the oil is to be found must be a constant or a function of the number of wildcat wells. The second supposition seems more reasonable and would imply that the geological knowledge of a region depends on how many wells are drilled in the region. At first glance, this may seem to indicate that exploratory drilling is the only way to develop geological knowl-

---

g. The Alberta pool file has been updated recently by the Economics Subdivision of the Institute of Sedimentary and Petroleum Geology of the Energy, Mines, and Resources Bureau of Canada.

edge. Obviously this is not so, since knowledge can be gained by a variety of other activities such as seismic, surface, gravity and magnetic surveys. Historical data indicate that the number of new field wildcats and the geophysical and geological activities are closely correlated. Therefore, the knowledge gained from deductive surveys and from exploratory drilling is treated as a single parameter. These arguments indicate that the variable in the term involving knowledge is exploratory wells, but do not establish the form of the function. At best, one can only hope to make some general statements about geological knowledge and its accumulation. It seems reasonable to assume that if the number of wells drilled is small, the petroleum geology of the region is not known in detail. As more and more wells are drilled, the geological information and the knowledge derived from this information increase significantly until, at some time in the future, they are for all practical purposes complete.[33]

Ryan chooses the (ubiquitous) logistic function and sets

$$K(t) = \left[1 + M \, \exp\left\{ -\alpha W(t) \right\} \right]^{-1} , \qquad (15\text{-}11)$$

M and α being parameters to be determined from data. Passing to the limit in (15-10) and integrating gives

$$R(t) = U_\infty \left[1 - e^{-\beta W(t)} \left\{ M/K(t) \right\}^{\beta/\alpha} \right] , \qquad (15\text{-}12)$$

with $K(t)$ as in (15-11). Using the data, the parameters α, β, and U_∞ were estimated using a nonlinear least-squares method with $M = 10,000$.[h] The results are displayed in Figures 15-12 and 15-13 and Table 15-3.

The dotted lines in Figures 15-12 and 15-13 showed that this model for the province as a whole does not explain the apparently erratic rate of discovery. There are distinct, distinguishable periods of high rates of discovery corresponding to the discovery of new plays. The "knowledge function" $K(t)$ increases so rapidly at $W = 600$ (in 1947) that it is almost artifactual and could easily be replaced by a simple step function. According to Ryan:

in 1947, the knowledge of the economic geology of Alberta increased from essentially zero to one. These values should not be taken literally; to do so would be misleading. The sudden increase in the value of knowledge function to its maximum does not mean that everything about the petroleum geology of the province was established at the 600th well. Rather, it implies that the most significant key to finding oil in Alberta had been established. What was learned in 1947 was that large commercial quantities of crude oil could be found in the Paleozoic sediments and, in particular, in the Devonian reefs.

h. Ryan found that the results were insensitive to changes in values of M and α when M is large and β is much smaller than α.

Figure 15-12.   Growth of Initial Recoverable Crude Oil Reserves in Alberta.

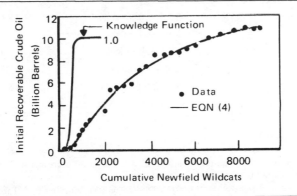

Source: Ryan, "Analysis of Crude-Oil Discovery Rate in Alberta," Figure 1. Reprinted with permission.

Figure 15-13.   Growth of Initial Oil-in-Place Reserves in Alberta.

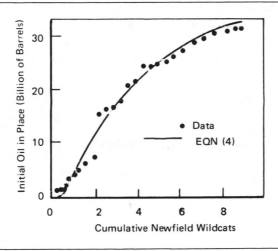

Source: Ryan, "Analysis of Crude-Oil Discovery Rate in Alberta," Figure 2. Reprinted with permission.

Table 15-3. Ryan's Estimates of Recoverable Oil, Oil in Place, and Model Parameters.

	Recoverable Oil	Oil in Place
M	10,000	10,000
α	18.6/1,000 wells	18.6/1,000 wells
$\Delta\alpha$[a]	± 2.1/1,000 wells	2.1/1,000 wells
β	.304/1,000 wells	.281/1,000 wells
$\Delta\beta$[a]	± .037/1,000 wells	.045/1,000 wells
U_∞	11.82 billion STB	35.27 billion STB
ΔU_∞[a]	± .669 billion STB	2.7 billion STB
Variance	.1270	1.63
Standard deviation	.356	1.27

a. $\Delta\alpha$, $\Delta\beta$, and ΔU are the error estimates of these parameters at the 95 percent confidence level. The value 10,000 for M was predetermined.

Source: Ryan, "Analysis of Crude-Oil Discovery Rate in Alberta," Table 1. Reprinted with permission.

Since most of the oil in the province is in these and older rocks, the knowledge of where oil is to be found was established.[34]

In order to match patterns in the data, Ryan recasts his original assumption, bringing it to bear on individual plays: "The rate of discovery of oil in a play is proportional to the undiscovered oil in the play and the knowledge of the existence of the play."[35]

Equation (15-10) is reinterpreted to hold for a generic play: Let $i = 1, 2, \ldots, I$ denote the ith play in order of birth. Then

$$\frac{\Delta R_i(t)}{\Delta W(t)} = \beta_i K_i(t)(U_i - R_i(t)) \ . \qquad (15\text{-}13)$$

If the ith play commences at the W_ith well, then

$$K_i(t) = \begin{matrix} 0 \\ 1 \end{matrix} \quad \text{if} \quad \begin{matrix} W(t) < W_i \\ W(t) \geq W_i \end{matrix} \ . \qquad (15\text{-}14)$$

With $R(0) = 0$, cumulative reserves for the province as a whole are

$$R(t) = \sum_{i=1}^{I} R_i(t) = \sum_{i=1}^{I} U_i \left[1 - \exp\left\{ -\beta_i K_i (W(t) - W_i) \right\} \right] \ . \quad (15\text{-}15)$$

Figures 15-14 and 15-15, from Ryan[36] show nonlinear least-squares fits for each of seven plays of (15-13), to cumulative oil in place and to cumulative recoverable oil reserves as functions of cumulative new-field wildcats. Cumula-

Figure 15-14.   Recoverable Oil Reserve Growth by Play.

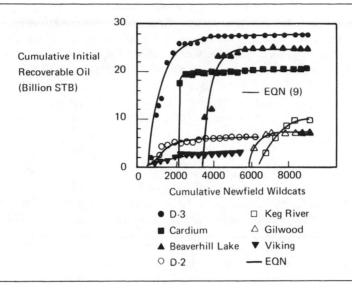

Source: Ryan, "Analysis of Crude-Oil Discovery Rate in Alberta," p. 233, Figure 5. Reprinted with permission.

Figure 15-15.   Oil-in-Place Reserve Growth by Play.

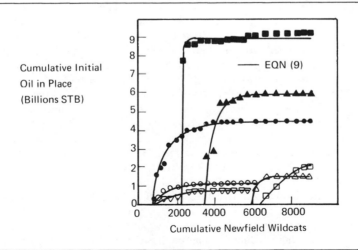

Source: Ryan, "Analysis of Crude-Oil Discovery Rate in Alberta," p. 233, Figure 6. Reprinted with permission.

tive initial recoverable oil and cumulative initial oil in place for the province as a whole fitted with (15-15) appear in Figures 15-16 and 15-17. Discovery rates fitted with (15-15) are displayed in Figures 15-18 and 15-19. Since the value W_i, the starting time on a scale of new-field wildcats of the ith play, is determined ex post, one would expect good fits to both cumulative in place and reserve amounts, as well as to discovery rates. This is the case.

There are discrepancies, however, between the actual discovery rate series and the fit of (15-15) to it. Ryan's discussion of why this happens highlights how distinctive one play in a given province can be from another and emphasizes the effects such differences may have on the ability of geological and geophysical search techniques to identify them:

[A] discrepancy appears on both figures at the 4500th well, and represents principally the Judy Creek and Swan Hills South fields. The theory implies that in any play, the largest fields or pools will be found in the initial stages of exploration, the middle range found next, the small pools last. Given a high level of knowledge, the rational man would drill the best prospect in the play first, and then the other prospects in order of their decreasing reserves. It would be easy to accept these two fields as exceptions to the theory or attribute them to the "randomness" of petroleum exploration. An examination of the question indicates that the reason for this particular difference lies in the knowledge function used in the extended theory. This theory assumed the knowledge was either zero or one. The one-year delay in finding these two major fields points out the weakness of this assumption. If the geological knowledge of the Beaverhill Lake reefs had been complete, these fields would have been found sooner. Obviously, the knowledge was not absolute as required by the mathematics of the theory. The word *knowledge* implies that there is something to be learned. Most of the plays considered in this analysis have had some characteristic that is observable from a surface survey, or could be deduced from geological evidence. The nature of this characteristic is much more definite in some than in others. The most distinctive geological feature in the province is that of the pinnacle reefs of the Keg River play. These reefs show extremely well on seismic records and the success ratio in this area was extraordinarily high in spite of the small areal extent of the pools. If the identifying characteristic of the Keg River reefs is sharp, the profile of the Beaverhill Lake and the D-2 plays shows vaguely on seismic records due to a lack of velocity contrast. The success ratio in these two plays was much less than that in the Keg River, and would account for the delay in finding the Judy Creek and Swan Hills South fields as well as some of the D-2 fields not associated with D-3 reefs. If the Keg River reefs represent one extreme, and the Beaverhill Lake and D-2 the middle ground, the other extremes are those plays or fields with an extremely vague or not identifying characteristic. These sediments may be those found in stratigraphic traps or in the foothills, where the seismic records can be very difficult to interpret.[37]

Ryan's model can be used to project future discoveries from existing plays, but, as he points out, it does not provide a way to predict the occurrence of as

Figure 15-16.   Growth of Initial Recoverable Crude Oil Reserves.

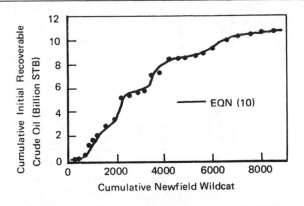

Source: Ryan, "Analysis of Crude-Oil Discovery Rate in Alberta," p. 233, Figure 7. Reprinted with permission.

Figure 15-17.   Growth of Initial Oil-in-Place Reserves.

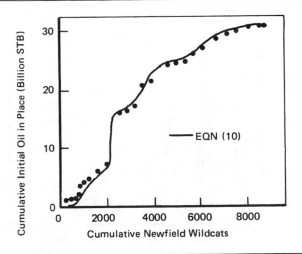

Source: Ryan, "Analysis of Crude-Oil Discovery Rate in Alberta," p. 233, Figure 8. Reprinted-with permission.

Figure 15-18.   Discovery Rate of Recoverable Crude Oil.

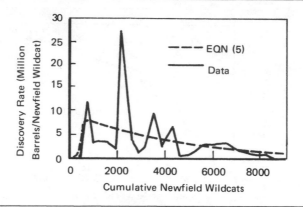

Source: Ryan, "Analysis of Crude-Oil Discovery Rate in Alberta," p. 233, Figure 9. Reprinted with permission.

Figure 15-19.   Discovery Rate of Oil in Place.

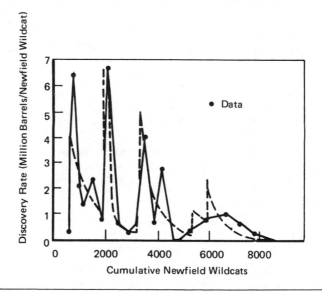

Source: Ryan, "Analysis of Crude-Oil Discovery Rate in Alberta," p. 233, Figure 10. Reprinted with permission.

Table 15-4.   Ryan's Projections.

Estimator	Recoverable	Oil in Place
Canadian Petroleum Association	25	75
Energy Resources Conservation Board	20	60
Van de Panne	17-18	51-54
Folinsbee	12-13	36-39
Ryan	12-13	35-39
McCrossan	16	48

Source: Ryan, "Analysis of Crude-Oil Discovery Rate in Alberta." Reprinted with permission.

yet undiscovered plays. If no new plays exist, then estimating ultimate oil in place or ultimate recoverable oil by summing estimates $\hat{U}_i$ of U_i, $i = 1, 2, \ldots, I$ is reasonable. If new plays remain to be found however, then such estimates will be too conservative. Ryan delegates the problem to the geologist:

> The analysis presented here shows it is permissible to extrapolate [(15-15)] into the future only if no new play is found. Whether such a play exists must be answered by the geologist. In the light of the past history of the discovery of oil in the province, the geologist who wishes to predict ultimate reserves need not concern himself with fields or barrels of oil. He need only examine the possibility of discovering a family of fields having a common characteristic and probably relatively close to one another. This remark is consistent with a statement made by L. G. Weeks (1952) twenty years ago: "It is not an exaggeration to state that fully 80 percent of the commercially recoverable hydrocarbons in the world are accumulated in pools that are scattered beneath 20 percent of the total sedimentary area." [38]

Ryan then projects recoverable oil and oil in place for known plays, as described previously, and compares his projections with others, as shown in Table 15-4. Since other estimates are for recoverable oil, Ryan divides them by an assumed recovery factor of 0.33 to obtain comparable estimates of oil in place.

When displayed on a scale of exploratory wells drilled, do patterns of occurrences of plays exhibit regularities that would allow such occurrences to be modeled meaningfully? This is an important unanswered research question that has not been addressed in the published literature on exploration analysis that focuses on plays.

BAROUCH AND KAUFMAN'S MODEL

The concept of a field size distribution and the observation that the largest deposits tend to be found early in the evolution of a petroleum play underlie a discovery process model investigated by Kaufman, Balcer, and Kruyt[39] and Barouch and Kaufman.[40]

Deposits in a play are presupposed to be generated by a random process whose outcome consists of values of deposit attributes, for example, field areas, volumes, and so forth. This process may be called a *superpopulation* process in accordance with standard statistical terminology. The superpopulation process describes features of deposits that nature put into the ground.

Discovery may be conceived of as sampling from a *finite* population of deposit attributes generated by this superpopulation process in accordance with a sampling scheme describing the order in which deposits are observed. To be compatible with the data, this sampling scheme should assign larger probabilities to orderings in which "large" deposits appear early than to those in which "large" deposits appear late.

We first present the finite population sampling scheme employed to model discovery, and contrast it with Arps and Roberts' model. The superpopulation process is then tied to it and applications presented in following sections.

Finite Population Sampling without Replacement
and Proportional to Size

In order to simplify analysis, Arps and Roberts (as discussed earlier) discretize possible field areas into a modest number of representative size (area) classes and model the number of discoveries in each class as a function of cumulative wildcats drilled. The model for each size class is *functionally independent of what is discovered in all other size classes.* Consequently, given M size classes $i = 1, 2, \ldots, M$, the number of discoveries made by W wildcats in size classes other than that corresponding to the ith does not influence inferences made about the number remaining to be discovered in the ith size class.

Drew, Scheunemeyer, and Root's modification of Arps and Roberts' model, while accounting for areal exhaustion (both dry holes and discoveries eliminate basinal area as a target for drilling, so as the number of wildcat wells being drilled increases, the remaining target area decreases), also treats discoveries in each size class as functionally independent of discoveries in other size classes. Both models are structured for marginal analysis of individual size classes. They exclude the contribution to inference about the number of fields in a given size class from observation of discoveries and dry holes in all other size classes. A *joint* analysis of observations from size classes is in principle desirable.

To this end imagine an areal sampling scheme that splits discovery sizes in order of occurrence from wildcat successes and dry holes: On a scale of wildcats drilled, $W = 1, 2, \ldots$, the first field discovered is (i_1, A_{i_1}) at $W = W_{j_1}$; the second discovered is (i_2, A_{i_2}) at wildcat $W_{j_2}, \ldots$; the nth discovered is (i_n, A_{i_n}) at wildcat W_{j_n}. Sizes of discoveries in order of occurrence, while embedded in a scale of wildcats drilled, can be modeled as functionally independent of drilling successes and failures and simultaneously incorporate the effect of *all* discovered fields on the probability that an as yet undiscovered field will be the next discovery.

In this spirit Kaufman, Balcer, and Kruyt model the discovery process as sampling without replacement and proportional to size from a finite population $A = \{(1, A_1), \ldots, (N, A_N)\}$ of fields: The probability that the N fields in A will be discovered in the order $1, 2, \ldots, N$ is

$$P\{(1, 2, \ldots, N) \mid A\} = \prod_{j=1}^{N} A_j / A_j + \ldots + A_N$$

(Model I)   (15-16)

In the absence of externalities restricting access to elements of A —lease blocking, sequestering of acreage —the larger a deposit is, the larger is the probability that it will be found early in the exploration history of a play. The sampling scheme represented by Model I incorporates this descriptive property of exploration.

Among the issues to be considered when using models like Model I as a sampling scheme are:

- What is the appropriate definition of size? Should it be area, rock volume, oil equivalent in place, or some other function of one or more of these variables?

- How should Model I be modified to accommodate delayed access to drilling in one or more basinal areas in which a play is located?

In their initial formulation Kaufman, Balcer, Kruyt used volume of recoverable oil (BOE) as their definition of size (cf. Barouch and Kaufman also). Arps and Roberts postulate the probability of discovery of a field to be proportional to its area.

Bloomfield et al.[41] extend Model I.[i] They hypothesize that, given any size measure, the probability of discovery of a field of size A may be proportional to some function other than A (the "discoverability function"). They choose A^α, with α a fixed number not necessarily equal to one, and replace Model I with:

$$P\{(1, 2, \ldots N) \mid A, \alpha\} = \prod_{j=1}^{N} A_j^\alpha / (A_j^\alpha + \ldots + A_N^\alpha).$$

(Model II)   (15-17)

---

i. The following discussion of discoverability is taken from "Model Mis-specification and the Princeton Study of Volume and Area of Oil Fields and Their Impact on Order of Discovery" by G. M. Kaufman and J. W. Wang, MIT Energy Laboratory Working Paper No. 80-003, January 1980, 27 pp.

If $\alpha = 0$, then this probability is $1/N!$ and all $N!$ permutations of $A_1, \ldots, A_N$ are equally likely, that is, size does not influence the discovery order. When $\alpha = 1$, Model II is identical to Model I. In general, the larger α is, the greater the influence of size on the order of discovery.

Model II allows testing of the hypothesis that the probability of discovery of a field is proportional to some particular measure of size.[j] To this end Bloomfield et al. examine relations between the volume and area of oil fields discovered in Kansas from 1900 to 1975. Maximum likelihood estimation of α and tests of significance of hypothesized α-values are performed, and they conclude that the assumption that discoverability of a field is proportional to its area $(\alpha = 1)$ is

> . . . untenable for the Kansas data. The discoverability was found, instead, to be proportional to a surprisingly low power of area $(\alpha = .33)$. The results of these analyses are of course limited to the Kansas data . . . They indicate that models assuming that discoverability is proportional to either area or volume should not be used on a regional basis without further study along the lines of this report.

The statistical techniques employed are innovative and useful. However, it is dangerous to extrapolate from the conclusions these authors draw from their analysis of the Kansas data. First, in their study, there is no recognition of the possibility that deposits discovered in Kansas come from several descriptively distinct deposit populations and that as discovery effort grew, so did the number of deposit populations recognized as targets for drilling. An explicit assumption on which their analysis is based is that all discovered fields in their sample constitute a target for drilling from the beginning of exploration.

Some deposit types are more easily discovered than others. Geologic knowledge gleaned from past drilling history, and geological, geochemical, and geophysical analyses often lead to a perception of previously unrecognized types of anomalies as viable prospects or targets for drilling. Exploration has been conducted in Kansas for about 80 years' during which time exploration technology has undergone dramatic changes. It is implausible to assume that all deposits discovered as of 1975 were perceived as prospects at the outset of exploration. Yet the model used by Bloomfield et al. to study the order of discovery of oil ields (Model II) effectively assumed that this is true.

Second, as the authors point out, the models they study were not designed to be applied to large regions containing distinguishable populations of deposits with widely varying characteristics (trap-type, time-stratigraphic unit, etc.). In addition to defining the setting in which sampling without replacement and pro-

---

j. The following discussion is drawn from Kaufman and Wang "Model Mis-specification and the Princeton Study of Volume and Area of Oil Fields and Their Impact on The Order of Discovery."

portional to size from *all members* of a deposit population might serve as a model for order of discovery, the proposers of this assumption asserted that:

> Assumptions about the physical nature of the discovery process are stated in a way which tacitly implies that economic variables may influence the temporal rate of drilling exploratory wells in a play, but they do not affect either the probability that a particular well will discover a pool or the size of a discovery *within a given play*. This assertion is patently false if applied to a population consisting of a mixture of subpopulations with widely varying geologic characteristics. For example, a large price rise may accelerate exploratory drilling in high risk (low probability of success) subpopulations with large average-pool sizes at a substantially different rate than in subpopulations with small pool sizes but high success probabilities. The overall probability of success for a generic well among the wells drilled in a mixture of these subpopulation types, as well as the size of discovery, will depend on the relative proportions of wells drilled in each subpopulation; and these proportions are influenced by prices and costs. By contrast, it is reasonable to assume that, within a given subpopulation, the precision of information-gathering devices and the quality of geologic knowledge of that subpopulation are the principal (perhaps sole) determinants of the probability of success of a generic well. A price rise may accelerate the temporal rate of drilling within that subpopulation, but it will not affect the quality of geologic knowledge at any given point on a scale of cumulative wells drilled into it. Exceptions can be found, of course, but this assumption is plausible as a broad descriptive principle.

Bloomfield et al. go on to say that:

> The context for which these models were developed . . . has been described as a *play* or as a *geologically homogeneous subpopulation of fields*. However, resource estimates are required for regions such as geological *basins*, and neither concept coincides with such a region.

On its face this statement is factually correct, but at the same time it is misleading. Resource estimates are often expressed as point estimates or as probability distributions for *aggregate* amounts of petroleum remaining to be discovered in a region containing populations of deposits with widely differing geologic and engineering characteristics. Resource projections are truly useful as an instrument for policy analysis only when expressed so that a projection can be made of future supply over time under alternative assumptions about costs, prices, fiscal regimes, regulatory policy, and so forth. For this reason models of exploration and discovery should ideally account for distinguishable characteristics of deposits that influence the *supply* of petroleum that can be drawn from them. A first step is to classify deposits in a basin into subsets, each composed of deposits from a single time-stratigraphic unit exhibiting lateral persistence of similar geologic and engineering attributes. This was not done for the Kansas data.

Distinguishable Deposit Populations and Model Misspecification

Consider a petroleum basin containing K distinguishable populations of deposits. As exploratory effort in it expands, so does the number of distinguishable populations that presently are, or in the past were, potential targets for drilling, until possibly all K populations have been so identified. At any given point on a scale of discovery numbers 1, 2, . . . , suppose that discovery proceeds as if sampling is proportional to a power α of size from a collection of undiscovered deposits, each of which is a member of one of the distinguishable populations whose deposits are targets for drilling at that point on the aforementioned scale. Consider two populations $A = \left\{ A_1, \ldots, A_N \right\}$ and $B = \left\{ B_1, \ldots, B_M \right\}$ where $A_i \epsilon A$ is a generic field size in A, and similarly for $B_j \epsilon B$. Suppose that the first m discoveries are $A_1, \ldots, A_m$ from A and that at the $(m + 1)$st discovery the population B is introduced. Define $C = \left\{ A_{m+1}, \ldots, A_N \right\} \cup B$, let C_k be a generic element of C, and label the elements so that C_k is the kth among $N + M$ discoveries, $m < k \leq N + M$. Then the probability of observing $A_1, \ldots, A_m, C_{m+1}, \ldots, C_{N+M}$ in that order is

$$\prod_{j=1}^{m} \frac{A_j^\alpha}{A_j^\alpha + \ldots + A_m^\alpha + U} \prod_{k=m+1}^{N+M} \frac{C_k^\alpha}{C_k^\alpha + \ldots + C_{N+M}^\alpha}$$

(Model III) (15-18)

where $U = A_{m+1}^\alpha + \ldots + A_N^\alpha$, $A_{m+1}, \ldots, A_N$ being sizes of the $N - m$ fields in population A discovered subsequent to the introduction of population B as a target. Model III can be easily extended to treat the case of more than two populations.

When staggered mixing of two populations as described is ignored, sampling proportional to the α power of size and without replacement implies a probability

$$\prod_{j=1}^{m} \frac{A_j^\alpha}{A_j^\alpha + \ldots + A_m^\alpha + C_{m+1}^\alpha + \ldots + C_{N+M}^\alpha} \prod_{k=m+1}^{N+M} \frac{C_k^\alpha}{C_k^\alpha + \ldots + C_{N+M}^\alpha}$$

(Model II) (15-19)

for the event "$A_1 \ldots, A_m, C_{m+1}, \ldots, C_{N+M}$ in that order." Use of Model II to compute an estimate of α when Model III is the correct model can seriously distort the estimated value of α. A MLE for α computed using (II) when the data are generated by (III) is negatively biased. That is, it is on the average smaller than the value of α specified for (III).

If, as is highly probable, a model of the form (III) represents the order of discovery of fields in a large region more accurately than (II), then estimation of α as done by Bloomfield et al. is a typical case of estimation based on a misspecified model. Table 15-5 presents results of a Monte Carlo simulation experiment designed to demonstrate the effects.

Simulated data were generated using Model III with $\alpha = 1.0$. For each of 100 replications a MLE $\hat{\alpha}_{III}$ for α was computed with Model III, and a MLE $\hat{\alpha}_{II}$ was computed using Model II—an incorrectly specified model since the data were generated by Model III. Throughout, finite population values $A_1, \ldots, A_N, B_1, \ldots, B_M$ were held fixed in order to focus attention on the behavior of $\hat{\alpha}_{II}$ and $\hat{\alpha}_{III}$ as functions of the order of discovery. The sample means and standard deviations of $\hat{\alpha}_{III}$ and of $\hat{\alpha}_{II}$ for these 100 replications are displayed in Table 15-5.

Two features of Table 15-5 are:

- MLEs for α computed using the model by which the data were generated exhibit slight positive bias (1 or 2 percent).

- MLEs for α computed using Model II exhibit very large negative bias. As the fraction m/N of fields discovered in A before B is introduced increases, so does the magnitude of the bias.

In addition:

- These attributes of sampling distributions for $\hat{\alpha}$ are neither sensitive to a choice of an underlying distribution for elements of A and of B nor to a choice of α.

Table 15-5. Properties of the Sampling Distributions of $\hat{\alpha}_{III}$ and $\hat{\alpha}_{II}$ when Data Are Generated by Model III.

N	M	M	$\hat{\alpha}_{III}$		$\hat{\alpha}_{II}$	
			Mean	Standard Deviation	Mean	Standard Deviation
50	50	0	1.013	.156	–	–
		10	1.025	.099	.919	.073
		20	1.020	.091	.745	.040
		30	1.016	.092	.586	.026
		40	1.016	.097	.440	.018
		50	1.017	.095	.271	.010
40	20	0	1.003	.128	–	–
		10	1.002	.129	.894	.088
		20	1.004	.127	.648	.040
		30	1.021	.121	.450	.018
		40	1.015	.125	.223	.008

The particular values chosen for $A_1, \ldots, A_N$ and for $B_1, \ldots, B_M$ in the calculations leading to Table 15-5 are lognormal fractiles. Essentially similar results are obtained when the A_is and the B_js are chosen to be fractiles of a uniform distribution and when $\alpha = 0.5$ in place of $\alpha = 1.0$.

As can be seen in Figure 15-20, the Monte Carloed sampling distribution of $\hat{\alpha}_{II}$ and of $\hat{\alpha}_{III}$ behave very differently as a function of the fraction m/N of the population A of fields discovered prior to introduction of population B. The sampling distribution for $\hat{\alpha}_{II}$ concentrates at smaller and smaller values of α as m/N increases, while that for $\hat{\alpha}_{III}$ does not appear to shift significantly. Figure 15-21 is a graphical comparison for $N = 80$, $M = 20$, $\alpha = 0.5$ and $\alpha = 1.0$, and for $N = M = 50$, $\alpha = 0.5$ of sample means of $\hat{\alpha}_{III}$ and $\hat{\alpha}_{II}$.

If three populations of deposits are mixed in the fashion described above, α can be even more strikingly negatively biased (see Table 15-6).

Incomplete Sampling of a Deposit Population

More often than not, the number n of discoveries made from a population $A = \{A_1, \ldots, A_N\}$ of deposit sizes is less than N. Model II says that the probability of discovering $A_1, \ldots, A_n$, $n \leq N$ in that order is $\prod_{j=1}^{n} A_j^\alpha / A_j^\alpha + \ldots + A_n^\alpha + U$, where $U = A_{n+1}^\alpha + \ldots + A_N^\alpha$. If (1) $n < N$ and $A_1, \ldots, A_n$ is interpreted as an exhaustive sample of A and (2) a MLE of α is computed by use of $\prod_{j=1}^{n} A_j^\alpha / A_j^\alpha + \ldots + A_n^\alpha$, this MLE is negatively biased.

Table 15-7 and Figure 15-22 show the effects on the sampling distribution of $\hat{\alpha}$ of interpreting $n < N$ observed field sizes from a population with N fields as if the n observed fields constitute a complete population.

Effects of Sequestering Acreage

Federal offshore leasing policy prevents unrestricted search by drilling over the full areal extent of potentially productive offshore sediments. Area is opened up to exploration piece by piece, as in the U.S. Offshore Gulf Coast oil and gas province. Significant restrictions on acreage available for exploration are still in effect in the North Sea. Exploration in a portion of the Denver-Julesberg basin was restricted in a somewhat similar fashion by Union Pacific Railroad: In 1949 the company released for exploration 4,400 square miles of virtually unexplored land in the basin.[42]

The effects on the order of discovery of deposits of releasing large blocks of acreage piece by piece are similar to those introduced by a staggered mixing of distinguishable populations of deposits. Consider a population of $N + M$ deposits of similar trap type within a given time-stratigraphic unit in a basin. The areal

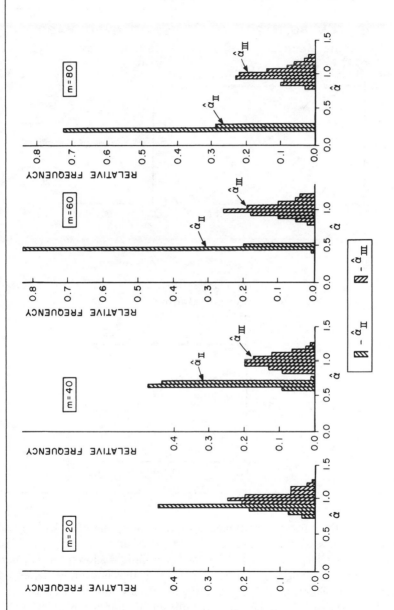

Figure 15-20. Monte Carloed Sampling Distributions of $\hat{\alpha}_{II}$ and $\hat{\alpha}_{III}$ when Data Are Generated by Model III.

Figure 15-21. Comparison of Sample Means of $\hat{\alpha}_{III}$ and $\hat{\alpha}_{III}$ when Data Are Generated by Model III.

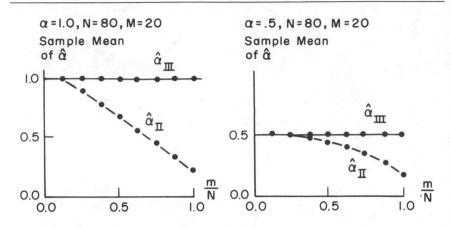

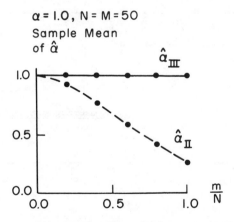

extent of the basin is partitioned into two parts, one of which contains N of these deposits, $A = \{A_1, \ldots, A_N\}$, the other of which contains M of these deposits, $B = \{B_1, \ldots, B_M\}$. Exploratory drilling is allowed in that portion of the basin containing B at the nth discovery in that portion containing A. Then (15-18) is a simple model for capturing the effects of introducing B.

In some cases, *both* the sequential introduction of new populations of deposits *and* sequential release of large blocks of acreage take place. The North Sea province is an example.

Table 15-6. Properties of Sampling Distribution of $\hat{\alpha}_{III}$ and $\hat{\alpha}_{II}$ when $\alpha = 1.0$, Three Populations Are Mixed, and Data Are Generated by Model III.[a]

		$\hat{\alpha}_{III}$		$\hat{\alpha}_{II}$	
m	q	Mean	Standard Deviation	Mean	Standard Deviation
10	20	1.002	.084	.814	.048
	30	1.005	.099	.690	.039
	40	1.003	.081	.574	.020
	50	1.013	.093	.466	.015
	60	1.005	.087	.355	.010
	70	1.019	.099	.216	.005
20	30	1,031	.095	.661	.032
	40	1.012	.091	.553	.024
	50	1.025	.083	.448	.015
	60	1.026	.097	.343	.011
	70	1.021	.094	.207	.005
30	40	1.003	.105	.497	.021
	50	1.014	.091	.421	.017
	60	1.014	.089	.324	.010
	70	1.029	.093	.191	.005
40	50	1.007	.100	.314	.012
	60	1.005	.106	.267	.011
	70	1.012	.094	.163	.006

a. The number of elements in the first, second, and third populations are 40, 30, and 30, respectively; m is the number of elements discovered prior to the introduction of the second population; and q is the total number of elements discovered prior to the introduction of the third population.

The Jurassic North play began with the discovery of the Brent field, the second largest of twenty-five discoveries made through 1975 in this play. Thirty kilometers away lies Statfjord, the largest field found in the North Sea to date. It is the tenth of twenty-five Jurassic North discoveries as of 1975 and appears to be an outlier with respect to a decreasing trend of discovery size with increasing discovery number.

The explanation deserves repetition, as it emphasizes the importance of understanding externalities that may influence both the sizes of fields that are discovered as well as their order of discovery:

When Brent, the first discovery in Jurassic North, was drilled and identified as a field, it was immediately apparent that a large structure adjacent to it was a prime prospect with very high probability of containing hydrocarbons in the same sedimentary unit. Had exploration drilling proceeded unrestricted by the Norwegian government's exclusion of acreage containing this structure until the third round of licensing in the Norwegian sector, this structure

Table 15-7. Properties of a MLE of α Computed by Use of
$\prod\limits_{j=1}^{n} A_j^{\alpha}/(A_j^{\alpha} + \ldots + A_n^{\alpha})$ when $n < N$ and Data Are Generated
by Model II with $\alpha = 1.0$.

| | | $\tilde{\alpha}$ | |
| | | Mean | Standard Deviation |
N	n		
40	10	0.5558	0.3695
	20	0.7168	0.2240
	30	0.8662	0.1681
100	10	0.3914	0.3444
	20	0.4557	0.2224
	30	0.5541	0.1761
	40	0.6269	0.1525
	50	0.6991	0.1312
	60	0.7711	0.1164
	70	0.8341	0.1072
	80	0.8916	0.0985
	90	0.9500	0.0920
	100	1.0109	0.0879

almost certainly would have been drilled immediately after Brent, and Stratf-jord would appear as the *second* rather than the tenth discovery in Jurassic North. The Norwegian government has pursued a deliberate policy of excluding prime acreage from early licensing rounds in order to minimize the inflationary impact on their economy of a tremendous surge in drilling and discovery which would obtain if licensing were unrestricted in their sector. Future rounds of licensing may include four more very large and promising structures, all known to exist for several years.[43]

The order of discovery of Jurassic North fields might be better modeled using a model of the form III than II. However, the data are incomplete. With high probability additional Jurassic North fields will be discovered. In addition field data remain to be classified by licensing round.

Short of an analysis that accounts for these features of the data, insight into the effect of a reordering of data points imposed by externalities affecting exploration patterns on estimates of α is afforded by "guessing" the order of field discovery in the absence of externalities, then comparing estimates of α that obtain with a given model for order of discovery both when the data are reordered and when they are not. To this end, Figure 15-23 displays likelihood functions for α when Statfjord is recorded as the tenth and as the second discovery respectively, with the following (descriptively incorrect) assumptions in force:

1. The twenty-five fields discovered in Jurassic North as of 1975 constitute the *complete* population of fields in the play; that is, no additional Jurassic North fields remain to be found.

Figure 15-22.   Relative Frequencies of Values of $\hat{\alpha}$ Maximizing

$$\prod_{j=1}^{n} A_j^{\alpha}/A_j^{\alpha} + \ldots + A_n^{\alpha} \text{ when } \alpha = 1.0 \text{ and } N = 100 \text{ (500 Replications)}.$$

y : RELATIVE FREQUENCY

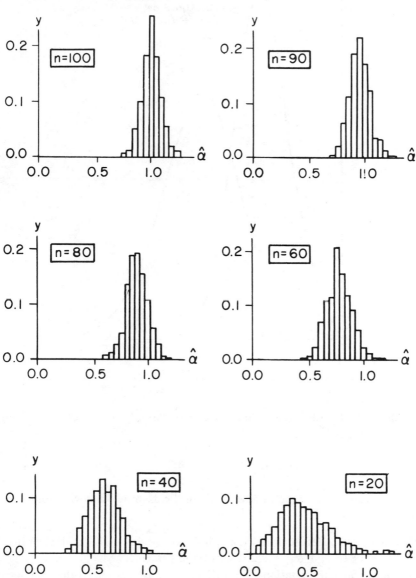

Figure 15-23. Posterior Distributions for $\tilde{\alpha}$.

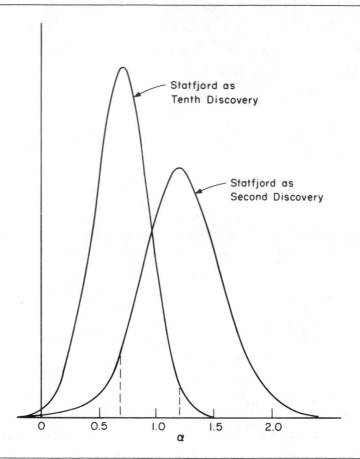

2. Model II describes the fashion in which the order of discovery of fields is generated and $N = 25$:

$$\prod_{j=1}^{25} A_j^\alpha / A_j^\alpha + \ldots + A_{25}^\alpha \ , \tag{15-20}$$

3. The relative order of discovery of all fields other than Statfjord is held fixed.

When Statfjord is recorded as the tenth discovery, $\hat{\alpha}_{II} = 0.7$, and when it is recorded as the second, $\hat{\alpha}_{II} = 1.2$, a shift of 58 percent.

A Bayesian interpretation: If a diffuse prior is assigned to the parameter α (the random variable $\tilde{\alpha}$ is assigned an (improper) prior density proportional to

$d\alpha$, $\alpha \epsilon (-\infty, \infty)$), then Figure 15-23 displays two posterior densities for $\tilde{\alpha}$, one for Statfjord as the tenth and one for Statfjord as the second discovery. The most likely value of α, a posteriori, when Statfjord is recorded as the second discovery is a 0.98 fractile of the posterior density for $\tilde{\alpha}$ when Statfjord is recorded as the tenth discovery. And the most likely value of α, a posteriori, when Statfjord is recorded as the tenth discovery is a 0.06 fractile of the *posteriori* density for $\tilde{\alpha}$ when Statfjord is recorded as the second discovery.

As explained in Chapter 12, if additional fields beyond the twenty-five fields in the data set used here remain to be discovered in Jurassic North, then it is possible that *both* of the posterior densities in Figure 15-23 are centered on smaller values of α than would be the case if undiscovered fields could be counted and their sizes measured.

In sum:

- The analysis of the Kansas data done by Bloomfield et al. only confirms that a statistical model designed for data from a *single* population of descriptively homogeneous deposits in a petroleum basin may not be an appropriate model for discovery data drawn from a mixture of distinguishable deposit populations.

- The "surprisingly low power" of the discoverability parameter α for the Kansas data as they estimate it ($\alpha = 0.33$) is *possibly* a result of use of an *incorrect* model: When data are generated by sampling without replacement and exactly proportional to size ($\alpha = 1.0$) from two or more distinguishable populations, each of which becomes a target for drilling at a different point in time, application of a model in which deposits from *all* populations are regarded as targets for drilling at the outset can yield an estimate of α much less than 1.0.

- For samples of moderate size (thirty observations or less) the maximum likelihood estimator for α as proposed by Bloomfield et al. can be *very sensitive* to the reported order in which fields are discovered. A shift of a single data point can cause large changes in a maximum likelihood estimate of α

Superpopulations

Field sizes $A_1, \ldots, A_N$ are primitive elements of the finite sampling scheme just described. A superpopulation process corresponding to an a priori choice of functional form for a size distribution can be introduced by assuming that $A_1, \ldots, A_N$ are realizations of a random process, rather than fixed but possibly unknown numbers:

Let A_i be the "size" of the ith deposit among N deposits in a petroleum play. The A_is are values of mutually independent identically distributed rvs $A_1, \ldots, A_N$, each with density $f(\cdot \mid \theta)$ concentrated on $(0, \infty)$ and indexed by a parameter $\theta \epsilon \Theta$.

Our discussion of size distributions suggests that a lognormal density is a reasonable candidate for f when "size" is interpreted as hydrocarbons in place or recoverable hydrocarbons in a generic deposit.

This assumption coupled with any one of the finite population models just discussed, Model I for example, defines a probability law for sizes of discovery in order of occurrence: Let Y_j be the size of the jth discovery and $Y = (Y_1, \ldots, Y_n)$ be a vector of discovery sizes in a sample of size $n \le N$. Given θ, N, and infinitesimal intervals $dY_1, \ldots, dY_n$, the probability of observing $\widetilde{Y}_1 \epsilon dY_1, \ldots, \widetilde{Y}_n \epsilon dY_n$ in that order is, letting $b_j = Y_j + \ldots + Y_n$,

$$P\{\widetilde{Y} \epsilon dY \mid \theta, N\} = \frac{\Gamma(N+1)}{\Gamma(N-n+1)} \prod_{j=1}^{n} Y_j f(Y_j \mid \theta) dY_j$$

(15-21)

$$x \int_0^\infty \cdots \int_0^\infty \prod_{j=1}^{n} [b_j + A_{n+1} + \ldots + A_N]^{-1} \prod_{k=n+1}^{N} f(A_k \mid \theta) dA_k .$$

The expectation of $\widetilde{Y}_n^k$, the kth moment of the nth discovery, possesses a representation

$$E(\widetilde{Y}_n^k \mid \theta, N)$$

(15-22)

$$= (N - n)\left(\tfrac{N}{n}\right) \int_0^\infty L^{(k+1)}(\lambda \mid \theta)[L(\lambda \mid \theta)]^{n-1} [1 - L(\lambda \mid \theta)]^{N-n} d\lambda$$

where L is the LaPlace transform of $f(\cdot \mid \theta)$ and $L^{(k)}$ is its kth derivative. Given θ and N, $E(\widetilde{Y}_n^k \mid \theta, N)$ is easily computed by standard numerical methods. Figure 15-24 displays graphs of the first moments of first, second, ..., nth discoveries when f is lognormal for $\mu = 6.0, \sigma^2 = 3.0$, and $N = 150, 300, 600$. The behavior of these moments as a function of n strongly depends on the choice of f. If, for example, f is exponential with mean $1/\theta$, then $E(\widetilde{Y}_n) = \frac{2}{\theta}\left[1 - \frac{n}{N+1}\right]$ and the first moment of $\widetilde{Y}_n$ declines linearly with increasing n.

After some manipulation[44] (15-21) can be reexpressed as

$$\prod_{j=1}^{n} Y_i (f(Y_j \mid \theta) dY_j \int_0^\infty Z(\lambda) [L(\lambda \mid \theta)]^{N-n} d\lambda$$

(15-23)

where

$$Z(\lambda) = \sum_{j=1}^{n} p_j e^{-\lambda_j} \text{ with } p_j = \prod_{\substack{k=1 \\ k \ne j}}^{n} [b_k - b_j]^{-1} ,$$

a function solely dependent on observed data.

Figure 15-24. Graph of $E(\tilde{Y}_n)$ for Lognormal Superpopulation with Mean = 1808.04, Variance 62.3910 × 10^6 ($\mu = 6.0$, $\sigma^2 = 3.0$).

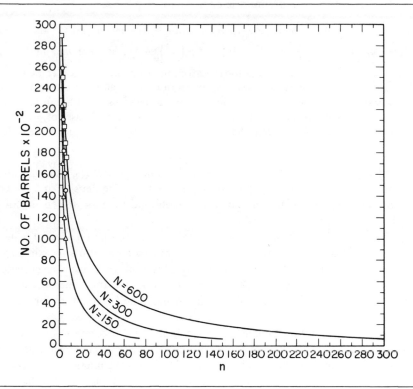

Maximum Likelihood Estimation

The sampling density for Y is indexed by parameters $\theta \in \Theta$ and $N \geq n$, so a likelihood function for θ and N given the data follows directly from (15-23). While the assumptions on which (15-23) are based are simple and easy to state, accurate computation of (15-23) is difficult. Coefficients of the partial fraction expansion defining $Z(\lambda)$ vary in typical cases by as much as 10^{20} from largest to smallest and significant cancellations occur for λ in the vicinity of zero.

Two approaches to calculating (15-23) have been studied: direct numerical computation[45] and asymptotic expansions. Barouch and Kaufman (1977) present an asymptotic expansion for the density of $\tilde{Y}$, valid for $p = N - n$ large and $b_j = Y_j + \ldots + Y_n$, $j = 1, 2, \ldots, n$ fixed, from which it follows that as $p \to \infty$ the density of $\tilde{Y}$ approaches $\prod_{j=1}^{n} Y_j f(Y_j \mid \theta)/M_1$, with M_1 the mean of $f(\cdot \mid \theta)$.

That is, in the limit $p \to \infty$ with $Y_1, \ldots, Y_n$ and n fixed, sampling is "length-biased,"[46] for example). Unfortunately, most cases of practical interest are not covered effectively by this asymptotic expansion:

When f is lognormal with parameters μ and σ^2, [its mean and variance are] $M_1 = \exp\left\{\mu + \frac{1}{2}\sigma^2\right\}$ and $V = M_1^2 [\exp\{\sigma^2\} - 1]$, so that $n \exp\{\sigma^2\}/p$ must be smaller than order one for the aforementioned expansion to be valid. A crude statistical analysis of plays in western Canada shows σ^2 to vary from .8 to 8.0 between plays, while n varies from 2 to 3 to several hundred. Consequently, [such] an expansion is not sufficient to cover all cases of practical interest; i.e., $n \exp\{\sigma^2\}/p$ order one, and $n \exp\{\sigma^2\}/p$ larger than order one.[47]

In order to treat a wider range of possible parameter values, Barouch and Kaufman compute a *uniform* asymptotic expansion for the density of $\widetilde{Y}$ when f is lognormal, and use it to approximate a likelihood function for (μ, σ^2, N) given data Y.

For fixed Y and n and each value of $N \geq n$, there is a unique pair of values $(\hat{\mu}(N), \hat{\sigma}^2(N))$ of the parameter (μ, σ^2) maximizing a uniform approximation to the sampling density (15-23). The uniform asymptotic expansion provides conditional (on N) maximum likelihood estimators for the superpopulation parameters, but not unconditional maximum likelihood estimators for μ, σ^2, *and N.*

In order to extent the domain of applicability and to compute joint maximum likelihood estimates for all three parameters, Nelligan (1980) has devised a scheme for evaluation of integrals similar to (15-23) by numerical quadrature. Nelligan's formulation differs from (15-23) in one respect. He assumed that the number N of fields is a Poisson distributed rv:

$$P\{\widetilde{N} = N \mid z\} = e^{-z} z^n/n! \ . \tag{15-24}$$

Upon randomizing (15-24) with respect to $\widetilde{N}$ and renormalizing (given n discoveries, $\widetilde{N} \geq n$ with probability one), the sampling density becomes

$$P\{Y \mid \theta, z\} = [z^n/P\{\widetilde{N} \geq n \mid z\}] \tag{15-25}$$

$$\times \ \prod_{j=1}^{n} Y_j \ f(Y_j \mid \theta)] \int_0^\infty Z(\alpha) \exp\{-z(1 - L(\lambda \mid \theta))\} d\lambda \ .$$

Some results of his calculations are shown in Table 15-8. A priori, the expectation of N is z, so his estimate of z is a surrogate for an estimate of N when the number of fields is assumed to be a fixed number not known with certainty, as in (15-23).

If z is held fixed, then (15-24) regarded as a function of μ, and σ^2 has a unique maximum but varies slowly in a neighborhood of this maximum, that is, the likelihood function is relatively "flat" in μ and σ^2. However, if μ and σ^2 are

Table 15-8. Nelligan's Joint MLE for μ, σ^2, and z (LeDuc Play, Western Sedimentary Basin).

Number of Discoveries	$\hat{z}$	$\hat{\mu}$	σ^2	$\hat{M}^{a,b}$	$\hat{z} \times \hat{M}_1$	$\log_e L$
15	228	2.3	2.7	38	8,789	-128
20	192	2.6	2.4	45	8,582	-168
25	119	3.0	2.1	57	6,830	-205
30	98	3.1	2.0	60	5,931	-242
31	104	3.0	2.0	55	5,700	-248

a. In units of 10^6 BOE in place.

b. $\hat{M}_1 = \exp\left\{\hat{\mu} + \dfrac{1}{2}\,\hat{\sigma}^2\right\}$.

held fixed and z is varied, (15-25) is sharply peaked in (n, ∞). The same behavior is exhibited by (15-23) with N as a surrogate for z.

For these data, the uniform expansion yields different maximum likelihood estimates (MLEs) for μ and σ^2 at $n = 24$ and similar MLEs at $n = 25$ (see Table 15-9). At $n = 24$ discoveries and $N = 100$ and 200, conditional MLEs for μ and σ^2 computed by use of the uniform expansion are (4.05, 1.67) and (3.97, 1.98), while at $n = 25$ and $N = 100$ and 200 these estimates are (3.0, 2.0) and (2.5, 2.4), respectively. Corresponding estimates of μ and σ^2 for $z = 119$ in Table 15-9 are (3.0, 2.1) (see Table 15-8).

Estimates of the (lognormal) superpopulation mean M_1 vary markedly with one additional observation: The twenty-fifth discovery is the fourth smallest among forty-three observations. The smallest is 2 million BBL; the largest 1,300 million BBL. For $n = 24$ the uniform expansion produces estimates of M_1 more than twice as large as those for $n = 25$. At $n = 25$ and $N = 100$, the estimate produced by this expansion is 55 million BBL, in close correspondence with Nelligan's estimate of 57 million BBL at $z = 119$.

It appears that MLEs computed by use of the uniform expansion are sensitive to moderately extreme observations. Possible explanations are that

- The uniform expansion is being applied outside its range of validity, $ne^{\sigma^2}/p = 0(1)$,[k]

- The flatness of the likelihood function in the μ - σ^2 half-plane allows a single moderately extreme observation to induce a large shift in the maximum,

or that

- Extreme numerical accuracy is required, and not provided.

k. For the range of parameter values discussed here, ne^{σ^2}/p values are close to (and in one case almost exactly) order one, so this explanation is less plausible than the others cited.

Table 15-9.   MLE for μ and σ^2 Given N via the Uniform Expansion.

Number of Discoveries	N	$\hat{\mu}$ (N)	$\hat{\sigma}^2$ (N)	$\hat{M}_1$	$N \times \hat{M}_1$	ne^{σ^2}/p
24	100	4.05	1.67	132	13,200	1.68
	200	3.97	1.98	143	28,519	.99
25	100	3.0	2.0	54	5,500	2.46
	200	2.5	2.4	41	8,090	1.57

In principle the correct explanation can be obtained by analysis of Monte Carlo experiments performed to this end. In practice such experiments are not practical because computation of MLEs for μ and σ^2 are computationally expensive.

Computational expense and the need for expertise in techniques of numerical analysis in order to implement these estimators properly are practical barriers to their widespread use.[1] Smith and Ward[49] partially overcome these barriers by discretizing the superpopulation size distribution and dropping the assumption that this size distribution must have a particular functional form. They incorporate a discoverability parameter α and use Model II as a model for the order of discovery. In addition, they "test" lognormality by imposing a priori restrictions on the number of fields deposed by nature in each discrete size class.

The data they use consist of ninety-nine North Sea discoveries made prior to 1977. Inasmuch as there are at least five distinct play sequences in the North Sea and sequestering of acreage has been and is practiced by host governments, the order of discovery of all North Sea fields is not well represented by use of Model II. Although Smith and Ward recognize that staggered mixing of play sequences introduces a downward bias in the estimate of the discoverability parameter α, their apparent rationale for use of Model II is that there are too few discoveries in individual North Sea plays to allow stable MLE of the numbers of fields in each of seven size classes ranging in size from 25 million BOE to 2.4 billion BOE.

MLEs are computed first for sampling exactly proportional to size (the discoverability parameter $\alpha = 1$) and four distinct cases: an arbitrary number of fields allowed in each size class, and constraints on the number of fields allowable in each size class compatible with three distinct functional forms: lognormal, Weibull, and exponential. The minimum economic field size in the North Sea is large enough to introduce nonnegligible trunctation of small sizes in the population of all North Sea fields. This, coupled with Smith and Ward's choice of rather large intervals for each size class so as to avoid sparsely occupied size

---

1. The model represented by (15-21) is naturally suited to MLE via application of the Expectation–Maximization algorithm with *marginal* expectations of future discoveries serving as predictive estimates of unobserved finite population sizes. This approach is currently being studied.[48]

classes, is one possible explanation of why the discretized relative frequency distribution of discoveries appears J-shaped. Smith and Ward's classification scheme is shown in Table 15-9 and MLEs for the number in each size class in Table 15-10.

Given their seven size classes, the lognormal and Weibull alternatives produce estimates of the number of deposits in place and of the total volume of recoverable North Sea reserves within 1 percent of estimates in the unconstrained case: 320 fields and 43-44 BBL.[m] Corporate estimates according to Smith (1980) range from 36 to 67 BBL. Approximate confidence intervals for total reserve volume varies little among these alternatives. For these three alternatives they compute the following 95 percent confidence intervals in billions of barrels of recoverable reserves:[n]

- Unconstrained   (38.2, 78.5)

- Lognormal   (38.3, 59.2)

- Weibull   (37.6, 55.4)

Whereas predictions of reserves are very close under alternative distributional assumptions, properties of the underlying size distribution are very different. Smith and Ward report an estimate of the standard deviation for a lognormal superpopulation three times that of a Weibull superpopulation and twice that of unconstrained discrete superpopulation. This large difference is explained by the authors as due to the J-shaped character of the discretized data, as in Table 15-10. Even though a proper lognormal density cannot be J-shaped, parameter values can be selected so that the mode of the density is arbitrarily close to zero. Truncation of the density at any preassigned field size greater than the mode will yield a J-shaped density. North Sea data as discretized in Table 15-10 suggest that the mode $\exp\{\mu - \sigma^2\}$ is less than 50 million barrels of reserves (Smith and Ward estimate is as 0.97 million barrels). In units of millions of barrels, this restriction implies that $\mu - \sigma^2 \leq 3.91$. Smith and Ward's estimate of $(\mu, \sigma^2) = (3.30, 1.825)$ clearly fulfills this restriction, but for $\sigma^2 = 1.825$ so does any value of $\mu \leq 5.74$. A difference of $5.74 - 3.30 = 2.44$ in μ implies a scale factor difference of $e^{2.44} = 11.47$ in the mean 67.53 million barrels of the underlying size distribution as estimated by Smith and Ward. This heuristic says that for $\sigma^2 = 1.825$ a superpopulation mean as large as $11.47 \times 67.53 = 774.5$ million barrels is consistent with the discretized data. Barouch and Kaufman[50] use data gathered by Beall[51] on fifty-nine North Sea discoveries as of 1974 to esti-

m. The assumption of an exponential superpopulation produces estimates " . . . of demonstrably poorer quality" and is dropped.

n. From Table 3 of Smith and Ward (1980): "Confidence intervals are derived by identifying total reserve volumes which, if introduced as a constraint during estimation, reduce the maximized log-likelihood by more than 1.92 units. Such reserve volumes would be rejected at the 95 percent confidence level on the basis of a likelihood ratio test."

Table 15-10. Reserves Estimates Based on Discoverability Proportional to Size ($\rho = 1$) (Reserve Volume Measured in Million Barrels).

Generating Process	Estimated Size Distribution of Deposits							Number of Deposits	Volume of Deposits	Estimated Process Parameters		Maximized Log-Likelihood[d]
	C_1	C_2	C_3	C_4	C_5	C_6	C_7			Mean	Standard Deviation	
unconstrained lognormal[a] ($\hat{\mu} = 3.30$, $\hat{\sigma} = 1.825$)	203	44	26	23	16	4	4	320	43,175	131	311	61.67
Weibull[b] ($\hat{\alpha} = .433$, $\hat{\beta} = 44.9$)	205	43	29	23	16	4	4	324	43,600	143	744	47.13
exponential[c] ($\hat{\beta} = 255.7$)	205	37	28	23	16	4	4	317	43,000	93	253	46.93
	40	20	17	19	16	4	4	120	34,750	256	256	13.69

a. $f(x \mid \mu, \sigma) = \dfrac{1}{x\sigma\sqrt{2\pi}} \exp\left[\dfrac{1}{2\sigma^2}(\ln x - \mu)^2\right]$.

b. $f(x \mid \alpha, \beta) = \dfrac{\alpha}{x}\left(\dfrac{x}{\beta}\right)^{\alpha} \exp\left[-(x/\beta)^{\alpha}\right]$.

c. $f(x \mid \beta) = \dfrac{1}{\beta} \exp\left[-x/\beta\right]$.

d. Values are reported up to an additive constant.

Source: Smith and Ward, 1980.

mate lognormal superpopulation parameters with their discovery process model assuming N = 300 fields in the province. Their estimate, $(\hat{\mu}, \hat{\sigma}^2)$ = (5.77, 1.38), yields an approximation to the mean of 645 million barrels of recoverable reserves and a mode of 80.6 million barrels, both of which are an order of magnitude greater than Smith and Ward's estimates. The particular choice of size classes employed appears to have a significant impact on estimates of superpopulation parameters.

In addition to examining alternative functional forms for the superpopulation, Smith and Ward study the effect of varying the discoverability parameter α in concert with specification of lognormal, Weibull, and unconstrained functional forms for the superpopulation distribution. They conclude that estimates of total reserves are insensitive to choice among these three functional forms, but very sensitive to the choice of α:

> The estimated number of deposits and total reserve volume fall precipitately as the value of α is reduced. This is a consequence of the structure of the discovery model. For high α values (which imply an efficient search), the historical discovery of small- and medium-sized deposits early in the sequence can only be rationalized by the presence of a large remaining number of such deposits. As the value of α declines, the early discovery of small deposits is more easily attributed to the randomness of the discovery process.[52]

Smith and Ward examine three values of α: $\alpha = \frac{1}{3}$, $\frac{2}{3}$, and 1.

> Rather inexplicably, the data strongly favor relatively low values of α, especially when a distributional form has been imposed a priori. The reductions in log-likelihood shown in Table 15–10 are sufficient to reject at the 95 percent level the null hypothesis $\alpha = 1$ in favor of $\alpha = \frac{1}{3}$ under the lognormal and Weibull specifications. The hypothesis of pure randomness ($\alpha = 0$) is also rejected by this criterion. Under the unconstrained distributional specification, more of the data is consumed in the estimation of the deposition itself, and we are left with less information and greater uncertainty regarding the value of α. Consequently we are not able to reject any values between 0 and 1 on the basis of likelihood ratios, although the value $\alpha = \frac{1}{3}$ still dominates.

In addition to negative bias in estimates of α induced by staggered mixing of plays, Smith and Ward conjecture that "small" MLEs for the model alternatives they consider are negatively biased even when the model is correctly specified. To test this conjecture they perform a Monte Carlo experiment. The finite population for the experiment consists of 320 fields distributed among size classes as shown in Table 15–10 for MLEs of the number in each size class in the "unconstrained" case. The discoverability parameter α was set equal to 1. Fifteen sequences of ninety-nine discoveries were generated. A likelihood ratio test of the null hypothesis that $\alpha = 1$ against the alternative $\alpha = 1/3$ resulted in rejection of the null hypothesis at a 95 percent confidence level in favor of $\alpha = \dfrac{1}{3}$ in seven of fifteen sequences. They assert that if maximum likelihood were unbiased, the probability of seven of fifteen rejections of the null hypothesis has an a priori probability of occurring less than 1/10,000! Fixing α at $\dfrac{1}{3}$ and then estimating the number of fields remaining to be discovered via maximum likelihood yields an estimate of one or two fields when in fact over 200 fields remain undiscovered. Estimation of the volume of reserves remaining to be discovered with α fixed at 1 is also negatively biased by an average of 30 percent over thirty simulated sequences.

Smith and Ward draw several conclusions about maximum likelihood procedures:

- When the sample size is "small," likelihood ratio hypothesis tests using the particular class of models they examine are misleading.

- Simultaneous MLEs of the discoverability parameter α and superpopulation parameters are negatively biased.

- While the hypothesis of a lognormal superpopulation cannot be formally rejected, the North Sea data as they use them do not support this hypothesis.

We would argue that an effective study of alternative size distributions cannot be done with sample data consisting of only ninety-nine observations drawn from four or five plays and then allocated to seven discrete size categories.

Mature plays in which many hundreds or even several thousand discoveries have been made allow sharper discrimination among alternative functional forms—provided that the data are carefully measured. The Lloydminster play in Western Canada is an example. More than 2,500 pools have been discovered in the Alberta and Saskatchewan components of this play, and attributes of each individual pool are measured by a uniform method.

OIL SUPPLY FORECASTING USING DISAGGREGATED DISCOVERY PROCESS MODELS

Eckbo, Jacoby, and Smith,[53] and Kaufman, Runngaldier, and Livne[54] describe intertemporal models of exploration of an individual play or basin designed to produce probabilistic projections of the time rate of additions to supply from future discoveries. They differ in several ways from econometric models designed for the same purpose. Salient geologic and engineering features of exploration, drilling, and production are explicitly incorporated in more detail than in most econometric models. Future expectations about discoveries are shaped by using the observed outcome of past drilling as input to discovery process models like those described in earlier chapters. Since both number and size of discoveries are conceptualized as rvs, supply additions over time are random super-positionings on a time scale of production profiles from discovered fields. Hence the supply function may be viewed as a probabilistic description of the future rate of production at discrete points in the future. Its properties depend jointly on such physical attributes of the play as:

- The size distribution of fields as deposed by nature,
- The number of prospects and the number of fields among these prospects,

and how:

- Drilling successes and failures occur,
- Sizes of discoveries unfold,
- Production profiles for discoveries are determined,

and on economic attributes, among which are:

- A projection of (future) prices per barrel,
- Exploratory drilling, development, and production costs,
- The fiscal regime in force (taxes, amortization, debt service, royalties, etc.),
- A normative criterion for making exploratory-well drilling decisions and development and production decisions.

Although Eckbo, Jacoby, and Smith's model is very similar in structure to Kaufman, Runngaldier, and Livne's model, the methods of analysis employed differ in an essential way. The former authors perform a static analysis of future supply. The time rate of drilling exploratory wells is specified a priori for each future time period and is held fixed. The latter authors perform an intertempo-rally dynamic analysis of the time rate of drilling. Exploratory drilling rates are rvs with a probability law determined by stochastic dynamic optimization of drilling behavior over time.

The schematic representation of discovery, development, and production pre-sented by Eckbo, Jacoby, and Smith (Figure 15-25) looks superficially similar to similar descriptive schema for econometric models ((cf.) discussion of Khaz-zoom and MacAvoy-Pindyck econometric models). However, microeconomic exploration, development, and production decisionmaking are incorporated in much finer detail. Their analysis unfolds in three stages. First, an estimate of the future time rate of exploratory drilling is made. Then drilling successes and fail-ures and discoveries are assumed to be generated in each of a preassigned number of discrete time periods by a discovery process model. Second, a microeconomic analysis of individual deposits is done for each future time period. The norma-tive criterion employed is maximization of expected net present value at a pre-

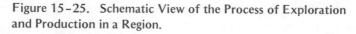

Figure 15–25. Schematic View of the Process of Exploration
and Production in a Region.

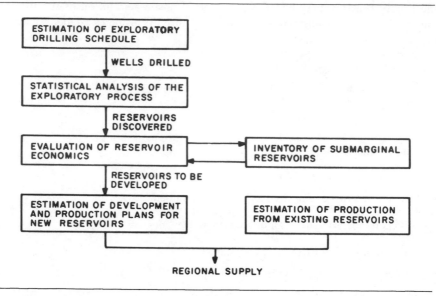

designated discount rate. Given estimates of development costs, production profiles as a function of deposit size, and a projection of future prices over time, a minimum economic reservoir size (MERS) is computed for each future time period. The MERS at a given point in time is that deposit size for which net present value is exactly equal to zero. It is an economic gate. If the size of a discovery exceeds the MERS, development begins. If not, then it goes into an inventory of current and past discoveries that are presently uneconomic to develop, but which may be developed in the future if price, cost, and/or tax changes reduce the MERS.

Their microeconomic analysis begins with the assumption that the production profile for a generic reservoir is known a priori and is a function of the size of the reservoir in barrels of recoverable oil. The planning period spans a time interval $[0, T]$ and is partitioned into T intervals of length one. Labeling $(t-1, t)$ as the tth time period, decisions to drill and to produce may be made at $t = 1$, $2, \ldots, T$ but not beyond. Define $\tau(s) + 1$ as the number of time periods over which a field size s can produce. The *production profile* for such a field is assumed to be a fixed vector $\delta(s) = (\delta_0(s), \ldots, \delta_{\tau(s)}(s))$ of numbers independent of the period at which production starts and such that $0 \leq \delta_j(s) \leq 1$ and $\sum_{j=0}^{\tau(s)} \delta_j(s) \leq 1$. The amount produced at time period j in the production life of a field of size s is $s\delta_j(s)$. Table 15-11, from Eckbo[55] shows typical production profiles and corresponding investment schedules for North Sea fields.

Given a present value discount rate β, cost, tax, and investment outflows $C_t(s)$, $C_{t+1}(s)$, ... at time periods t, $t+1$, ... for a field of size s discovered at t, and a projection of future prices P_t, P_{t+1}, ... per barrel, a MERS at t is a value s_t^* of s such that:

$$\sum_{j=0}^{\tau(s)} [\, s_t^* P_{t+j} \delta_j(s_t^*) - C_{t+j}(s_t^*)\,]/(1+\beta)^j = 0 \ . \qquad (15\text{-}26)$$

A precise mathematical accounting for fiscal variables such as royalty payments, petroleum revenue taxes, special taxes, corporate taxes, deduction and depreciation rules, oil allowance, withholding taxes on dividends and capital, and minimum liability provisions requires a more elaborate treatment of outflows than that represented by (15-26). Since we wish to focus on main principles and not on accounting details, these details are suppressed.

Figure 15-26 shows their computation of minimum economic reservoir size as a function of price.

Individual field economics is coupled to a drilling outcome model and to Barouch and Kaufman's discovery process model to produce a probabilistic description of future production. This is done as follows: Let

$$x_i = \begin{cases} 1 & \text{if the } i\text{th well is a discovery} \\ 0 & \text{otherwise} \end{cases} \qquad (15\text{-}27)$$

and

$$Z_i = \begin{cases} \text{size of discovery by } i\text{th well if } x_i = 1 \\ 0 \ \text{if } x_i = 0 \end{cases} \qquad (15\text{-}28)$$

Define $\omega(t)$ as the cumulative number of exploratory wells drilled up to and including period t. A complete history of drilling successes and failures and of discovery sizes in order of occurrence at period t is represented by the vector

$$H_t = [(x_1, Z_1), \ldots, (x_{\omega(t)}, Z_{\omega(t)})] \qquad (15\text{-}29)$$

Eckbo, Jacoby, and Smith[56] assume that $\tilde{x}_i, \ldots, \tilde{x}_i, \ldots$ are rvs with hypergeometric probability law: Given $N + M$ prospects, N are fields, M are dry and

$$P\{\tilde{x}_1 + \ldots + \tilde{x}_n = r \mid N, M\} = \binom{N}{r}\binom{M}{n-r}/\binom{N+M}{n} \ \text{for } r = 0, 1, \ldots, n. \qquad (15\text{-}30)$$

In addition, the sequence $\tilde{Z}_j | \tilde{Z}_j > 0, j = 1, 2, \ldots, N$ of discovery sizes is assumed to be conditionally independent of drilling outcomes $\tilde{x}_1, \ldots, \tilde{x}_i, \ldots$ and gen-

Table 15-11. Fraction of Total Exploration/Delineation Expenditures, Investment Expenditures, and Recoverable Reserves Occurring in Each Year Following Discovery.[a]

Year	Exploration/ Delineation Profile, all Fields	Fields < 300 Investment Profile	Fields < 300 Production Profile	300 ≤ Fields ≤ 1500 Investment Profile	300 ≤ Fields ≤ 1500 Production Profile	1500 < Fields Investment Profile	1500 < Fields Production Profile
1	0.1	0	0	0	0	0	0
2	0.2	0	0	0	0	0	0
3	0.2	.04	0	.01	0	.04	0
4	0.2	.44	0	.12	0	.12	0
5	0.2	.27	.09	.20	.03	.20	.04
6	0.1	.11	.13	.24	.08	.24	.04
7		.08	.15	.16	.10	.16	.06
8		.06	.13	.07	.10	.07	.09
9			.13	.06	.10	.06	.10
10			.11	.06	.10	.06	.10
11			.08	.05	.10	.05	.10
12			.07		.10		.10
13			.06		.08		.10
14			.05		.06		.08
15					.05		.07
16					.04		.05
17					.03		.03
18					.03		.03
19							.02
20							.01
21							.01

Source: P. Eckbo, "Estimating Offshore Exploration Development and Production Costs," MIT Energy Laboratory Working Paper, Cambridge, Mass., September 1977.

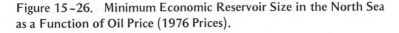

Figure 15-26.   Minimum Economic Reservoir Size in the North Sea as a Function of Oil Price (1976 Prices).

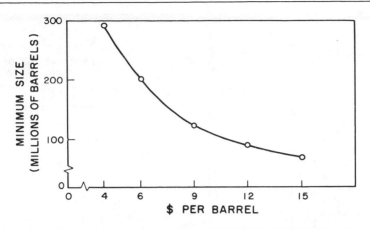

erated by the discovery process model represented by (15-21). More informally, sizes of discoveries in order of occurrence do not depend on when discoveries are made on a scale of exploratory wells drilled.

Given this characterization of drilling and discovery and a history H_t, the joint predictive distribution of drilling successes and failures and discovery sizes in order of occurrence at $t + 1, t + 2, \ldots$ can be computed. The assumption that the rate of exploratory well drilling is fixed and known a priori allows drilling outcomes to be allocated to time periods in a straightforward way. The discovery history in the North Sea as of 1975 is shown in Figure 15-27.

Sixty discoveries had been made. Figure 15-28 displays illustrative marginal predictive distributions for sizes of the sixty-first, sixty-second, and sixty-third discoveries. At $15/BBL the MERs is about 75×10^6 BBL and this appears as the dotted line in the graph. A MERS of 75×10^6 BOE is small enough so the probability of any one of these three discoveries falling below the MERs is also small.

Numerical assumptions underlying Eckbo, Jacoby, and Smith's computations are:

- Forty-four exploratory wells will be drilled in each of the years 1976, 1977, and 1978.

- The number of prospects and fields remaining in the North Sea is large enough so drilling successes and failures can reasonably be regarded as a Bernoulli process over a three-year drilling horizon.[o]

o. In other words, a hypergeometric probability law is replaced by a binomial probability law in their calculations.

Figure 15-27.   North Sea Discovery History and Forecast.

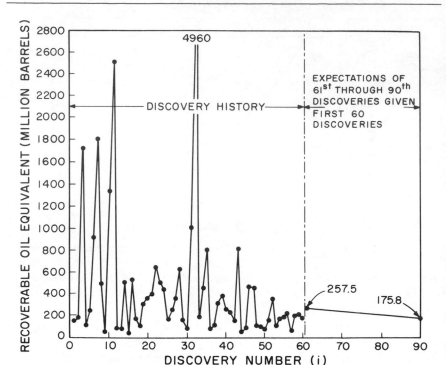

- The probability that an exploratory well will be a dry hole is 0.75, a number derived from historical experience and expert judgment.

- The size distribution of deposits is lognormal with parameter $(\mu, \sigma^2) =$ (5.78, 1.17) corresponding to a mean field size of 558×10^6 barrels.

- Two hundred deposits remain to be discovered.

The size distribution parameter values adopted are MLEs computed using the uniform approximation to the sampling density for the first sixty North Sea discoveries at $N = 200$.

A discretized version of predictive distributions for the sixty-first through sixty-fifth discoveries appears in Table 15-12.

A projection of future supply from both existing discoveries and further discoveries is now possible. Existing North Sea discoveries for which development plans have been announced are assumed to contribute to supply in accordance with these plans. For known discoveries for which development plans have not been announced but which are larger than the MERS of 75×10^6 BOE, future production is assumed to follow the profiles shown in Table 15-11. Dis-

Figure 15-28. The Sequence of Predictive Discovery Distributions.

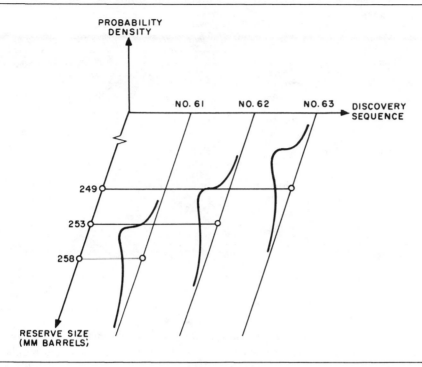

Source: P. Eckbo, H. Jacoby, and J. Smith, "Oil Supply Forecasting: A Disaggregated Approach," *Bell Journal of Economics and Management Science* 9, no. 1 (Spring 1978): 218–35.

coveries made in 1976–1978 that exceed the MERS produce according to this schedule of profiles. The particular projections reported are partial expectations. Conditional upon being given μ, σ^2, and N and having observed the first sixty discovery sizes $Y = (Y_1, \ldots, Y_{60})$, the $(60 + k)$th discovery has density $f(Y_{60+k} \mid Y; \mu, \sigma^2, N)$, $Y_{60+k} \in (0, \infty)$.[p] If the $(60 + k)$th discovery is made during period t and the MERS is s_t^*, development of this discovery will begin if $Y_{60+k} \geq s_t^*$ and will not begin if $Y_{60+k} < s_t^*$. If $Y_{60+k} > s_t^*$, amounts produced from it are

$$E(\widetilde{Y}_{60+k} \underline{\delta}(\widetilde{Y}_{60+k}) \mid \widetilde{Y}, s_t^*; \mu, \sigma^2, N) = \int_{s_t^*}^{\infty} y \underline{\delta}(y) f(y \mid Y; \mu, \sigma^2, N) dy.$$

$$(15-31)$$

p. Integral representations of the density of $\widetilde{Y}_{n+k} \mid (Y_1, \ldots, Y_n)$ for $k = 1, 2, \ldots$, $N-n$, and of the correlation structure of these $N-n$ rvs are given in Barouch and Kaufman (1976).

Table 15-12.   Predictive, Discovery Distribution, Millions of Barrels Oil Equivalent.

Size Category, k	Limits	Partial Expectation, P_{ik}, for Discovery Number				
		61	62	63	64	65
1	0-125	7	7	7	6	6
2	125-250	18	18	18	17	17
3	250-375	26	25	25	25	24
4	375-500	31	31	30	30	30
5	Over 500	176	173	169	166	163
Expected value, $E(\widetilde{Y}_i)$		258	253	249	244	240

Table 15-13.   North Sea Supply Estimates, Million Barrels Per Day.

	1980		1985	
	$9	$12	$9	$12
Existing reservoirs	2.82	2.82	2.50	2.50
Recent discoveries	1.68	1.82	1.57	1.66
1977-78 discoveries	0.47	0.48	1.68	1.72
Total	4.97	5.13	5.75	5.88

As the probability that a generic exploratory well will make a discovery is 0.24 and as 44 wells are assumed to be drilled per year, the expected number of discoveries per year is $(0.24)(44) = 10.56$, or, rounding off, eleven discoveries per year. Assuming that eleven discoveries are made in 1976, the expectation of amounts produced in future years from these discoveries is

$$\sum_{k=1}^{11} E(\widetilde{Y}_{60+k} \underline{\delta}(\widetilde{Y}_{60+k}) \mid Y, \ s_t^*; \mu, \sigma^2, N).$$ A similar calculation is per-

formed for discoveries in 1977 and 1978.

Table 15-13 from Eckbo, Jacoby, and Smith shows expectations of future supply in 1980 and 1985 at $9 and at $12 per barrel in 1976 dollars. A salient feature of these projections is that they are *static* expectations. The rate of drilling is not responsive to either price or to drilling outcomes, subsequent to 1975:

> Exploratory activity is not price sensitive . . . , and this is reasonable as long as the forecast is not extended beyond the period that will be influenced by currently planned (and therefore relatively inflexible) exploratory activity. Over longer periods, exploratory activity itself will be influenced by price (through expected returns) and this limitation in the current model becomes more serious.[56]

Exploration, discovery, and production constitute a sequence of interdependent intertemporal choices in the face of uncertainty. Projections made for more than a few years into the future should account for the feedback of both drilling outcomes and discovery sizes on operators' expectations of future returns to exploratory effort. Modification of operators' expectations in light of information generated as drilling progresses requires that the analysis be recast.

One possible generalization of the static expectations model just described is to view a generic operator as facing a sequential decisionmaking problem under uncertainty. The operator wishes to follow a drilling strategy over the planning period $t = 1, 2, \ldots, T$ that maximizes a normative criterion, net expected present value, for example. How the problem is formulated depends critically on what assumptions are made about information available to him at each point in time when the operator must make a decision. Does he have knowledge of

1. The *structure* of the drilling model and of the discovery sizes model?
2. The *parameters* of each of these models?
3. A *complete state history* H_t when at t?

In practice, most operators have no clear-cut knowledge of model structure, since few operators employ formal modeling as a guide to action. Even if they did, it would be more realistic to regard model parameters as uncertain a priori and estimates of them as subject to modification as drilling and discovery unfold.

In order to keep the amount of computation necessary to determine an optimal drilling strategy within reasonable bounds and yet retain the dynamic character of drilling and discovery, Kaufman, Runngaldier, and Livne assume that structure and model parameters are known with certainty, then formulate the operator's problem as a stochastic dynamic programming problem. An optimal drilling policy is a priori uncertain, so the solution consists of $d(0)$ and a joint probability law describing the number of wells $\tilde{d}(t)$, $t = 1, 2, \ldots, T$ to drill in future time period after the first. Interfacing $d(0)$, $\tilde{d}(1), \ldots, \tilde{d}(T)$ with a model for drilling successes and failures and a model for sizes of discoveries produces a probability law for discoveries in each time period. For example, if $\tilde{\omega}(t)$ denotes the cumulative number of wells drilled up to and including period t, then discovery sizes in period $t + 1$ are $\tilde{Z}_{\tilde{\omega}(t)+1}, \tilde{Z}_{\tilde{\omega}(t)+2}, \ldots, \tilde{Z}_{\tilde{\omega}(t+1)}$, a random number of rvs. In this setting, the number of exploratory wells drilled, the number of discoveries, sizes of discoveries, and the number of dry holes are all rvs. As with Eckbo, Jacoby, and Smith's model, each discovery must pass through a MERS gate in order to begin development.

Figure 15-29 is a schematic description of how components of the model of physical activities are related to process cost functions and wellhead prices. Figure 15-30 shows the major blocks of computation necessary to generate a probabilistic supply function and ancillary projections.

A numerical example best illustrates model output. Consider a play with M prospects of which N are fields and a lognormal size distribution whose parameter values correspond to MLEs for the Jurassic Central play in the North Sea.

Figure 15-29.   Intertemporal Model of Exploration in a Petroleum Play.

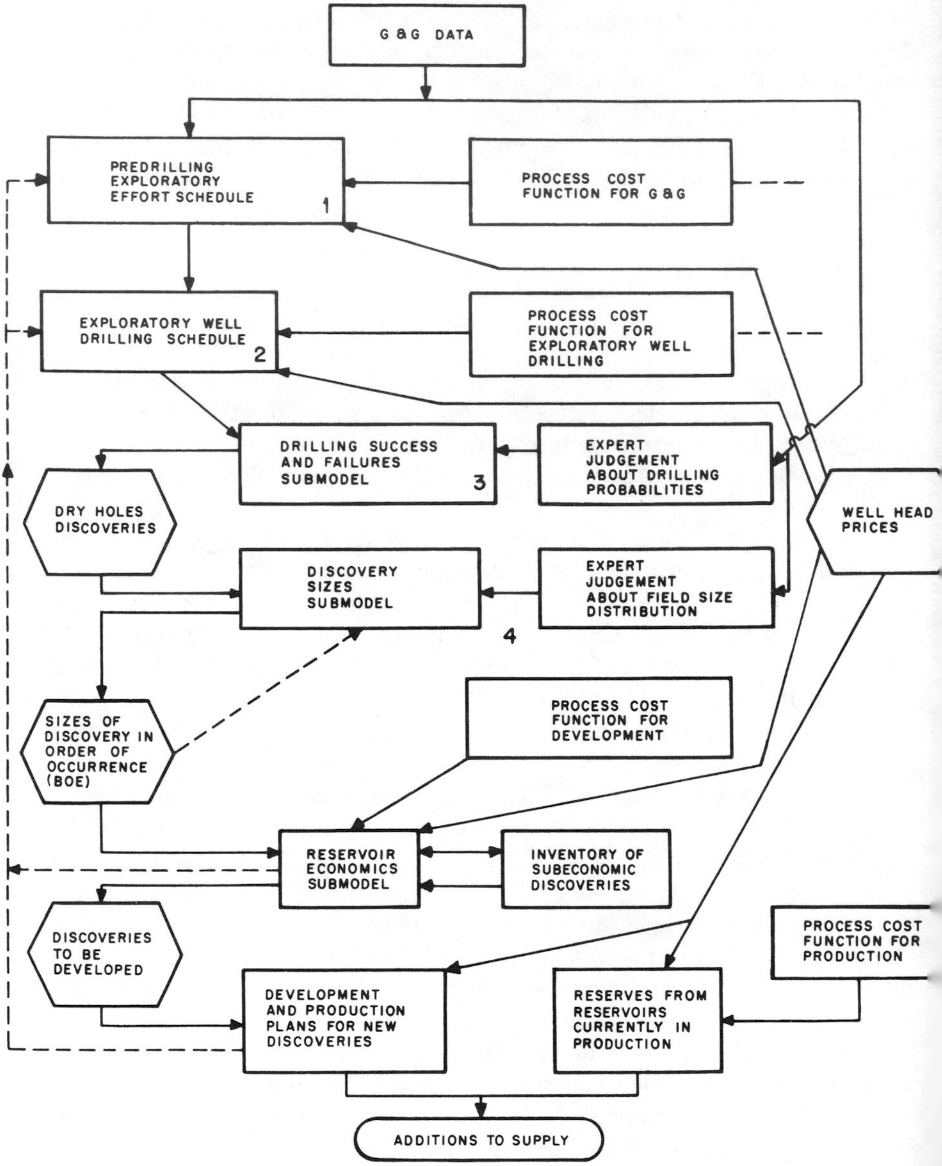

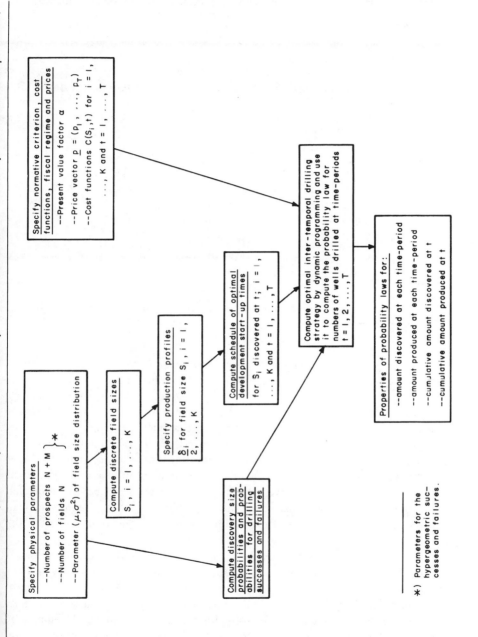

Parameters of Probability Laws Associated with Intertemporal Model of Exploration.

Specify physical parameters
--Number of prospects $N + M$ }*
--Number of fields N
--Parameter (μ, σ^2) of field size distribution

Compute discrete field sizes
S_i, $i = 1, \ldots, K$

Specify production profiles
$\underline{\delta}_i$ for field size S_i, $i = 1, 2, \ldots, K$

Compute schedule of optimal development start-up times for S_i discovered at t; $i = 1, \ldots, K$ and $t = 1, \ldots, T$

Compute discovery size probabilities and probabilities for drilling successes and failures

Specify normative criterion, cost functions, fiscal regime and prices
--Present value factor α
--Price vector $\underline{p} = (p_1, \ldots, p_T)$
--Cost functions $C(S_i, t)$ for $i = 1, \ldots, K$ and $t = 1, \ldots, T$

Compute optimal inter-temporal drilling strategy by dynamic programming and use it to compute the probability law for numbers of wells drilled at time-periods $t = 1, 2, \ldots, T$

Properties of probability laws for:
--amount discovered at each time-period
--amount produced at each time-period
--cumulative amount discovered at t
--cumulative amount produced at t

*) Parameters for the hypergeometric successes and failures.

Table 15-14.   Optimal Drilling Strategies for Various Parameters.

Model	Parameters	Parameter Values
Drilling successes and failures	Hypergeometric sampling parameters N and M	(N,M) = (5,10), (12,4) (20,4)
Size distribution	Lognormal parameters μ and σ^2	(μ, σ^2) = (5.78, 6.38)
Order of discovery	Sampling proportional to size and without replacement	–
Production profiles	As in table with time horizon of T = 20 years	–
Price path	Initial price grows at a constant proportional rate	Initial price P_0 = $5, $12 Growth rate γ = 6.6%/year and 4.5%/year
Process cost functions[a]	Exploratory drilling cost = $5 × 10^6	
	Total operating costs = $(18.87 × 10^6) + $0.04 (field size)	
	Total investment costs = $(296.1 × 10^6) + $1.12 (field size)	

a. These are individual components of a total cost function in simplified form. The cost growth rate is 6.6%/year.

Kaufman, Runngaldier, and Livne compute optimal drilling strategies for several choices of N, M, and economic parameters as shown in Table 15-14.

In order to conform to investment and production profiles as reported by Eckbo, the size distribution for fields was discretized into three size intervals of equal probability and a representative size (the geometric mean of each interval) assigned: 100 million barrels, 450 million barrels, and 1,500 million barrels.

Figures 15-31 and 15-32 describe features of the probability law at t = 0 of the number of exploratory wells to drill per period if an optimal drilling policy is followed from the outset, given that at most four wells can be drilled in each time period. With an optimal policy in force, $\tilde{d}(1), \ldots, \tilde{d}(\tau)$ are *not* independent; Figures 15-31 and 15-32 show only means and standard deviations of the $\tilde{d}(t)$s.

Coupling $d(0)$, and the rvs $\tilde{d}(1), \ldots, \tilde{d}(T)$ with models for drilling successes and failures, discovery sizes, and production profiles, the probability law for amounts produced per period can be computed. Figure 15-33 shows typical time profiles for expectations of amounts produced per period if an optimal drilling policy is pursued. A time profile for expected amounts produced may be interpreted as an intertemporal representation of expectations about future supply from a play, conditioned on the particular assumptions about prices, costs, and physical parameters employed.

Figure 15–31.   Mean and Standard Deviation of the Number of Wells Drilled in Runs 1 and 2.

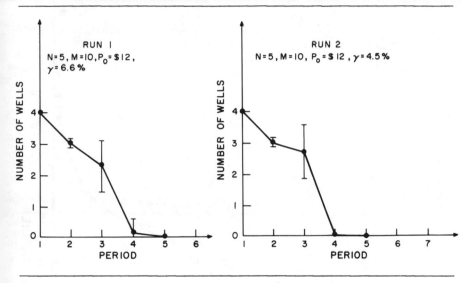

Figure 15–32.   Mean and Standard Deviation of the Number of Wells Drilled in Runs 3 and 4.

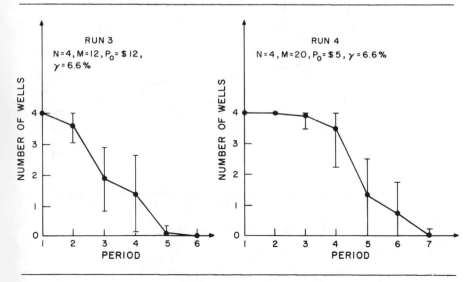

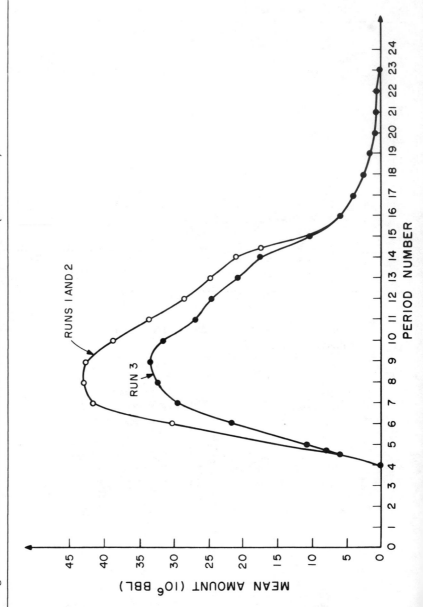

Figure 15-33.   Mean Amount Produced in Each Period of Runs 1-3 (in 10^9 BBL).

By comparison with most econometric models of oil and gas supply and demand or with static disaggregated supply models, numerical computation of supply projections via dynamic optimization is time consuming and expensive. A search is warranted for simpler alternatives that sacrifice some of the dynamic features of this particular model without major distortion of the character of the projections produced. To this end, Wang[57] assumes away uncertainty about discovery sizes and replaces Barouch and Kaufman's model with certainty equivalents for discovery sizes in order of occurrence. The certainty equivalents adopted are *expectations* of future discoveries in order of occurrence. The only remaining uncertainties are at which exploratory well a particular (expected) field size will be discovered and in what time period this will occur. The character of Wang's projections are very similar to those displayed in Figure 15-33. This modeling tactic reduces computation time by at least one order of magnitude and so allows the analyst to treat a much wider range of values of physical parameters (number of prospects and number of fields) at nominal computational cost.

While aspects of that portion of these dynamic optimization models describing discovery sizes in order of occurrence have been examined for structural and predictive validity, no such studies have yet been done for completely specified models. Alternatives to specification of drilling successes and failures as a hypergeometric sampling process are suggested by Kaufman, Runngaldier, and Livne and by Wang. Meisner and Demiren's[58] suggestion that drilling successes and failures be modeled as a logistic process deserves particular attention because they study the fit of this model to data describing a North Sea province in an orderly way and compare predicted and actual drilling outcomes. This model fits the data quite well and generates reasonable predictions.

REFERENCES

1. J. J. Arps and T. G. Roberts, "Economics for Cretaceous Oil Production on the East Flank of the Denver-Julesberg Basin," *AAPG Bulletin* 42, no. 11 (1958): 2549-66.
2. J. M. Ryan, "An Analysis of Crude-Oil Discovery Rate in Alberta," *Bulletin of Canadian Petroleum Geology* 21, no. 2 (June 1973): 219-35. E. Barouch and G. M. Kaufman, "Oil and Gas Discovery Modelled as Sampling Proportional to Random Size," Massachusetts Institute of Technology, Sloan School of Management Working Paper 888-76, Cambridge, Mass., 1976. L. J. Drew, J. H. Schuenemeyer, and D. H. Root, "Petroleum Resource Appraisal and Discovery Rate Forecasting in Partially Explored Regions: Part A," and J. H. Schuenemeyer and D. H. Root, "Petroleum Resource Appraisal and Discovery Rate Forecasting in Partially Explored Regions: Part C" in U.S. Geological Survey Professional Paper 1138 (Washington, D.C.: U.S. Government Printing Office, 1980).

3. Ryan, "Analysis of Crude-Oil Discovery Rate in Alberta."
4. Barouch and Kaufman, "Oil and Gas Discovery Modelled."
5. Private communication, M. King Hubbert, November 1975.
6. Arps and Roberts, "Economics for Cretaceous Oil Production on the East Flank of the Denver–Julesberg Bason; Drew, Scheunemeyer, and Root, "Petroleum-Resource Appraisal and Discovery Rate Forecasting in Partially Explored Regions, Part A"; G. M. Kaufman, Y. Balcer, and D. Kruyt, "A Probabilistic Model of Oil and Gas Discovery," *Studies in Geology, No. 1 – Methods of Estimating the Volume of Undiscovered Oil and Gas Resources* (The American Association of Petroleum Geologists, 1975), pp. 113–42.
7. Ryan, "Analysis of Crude-Oil Discovery Rate in Alberta."
8. Ibid.
9. M. King Hubbert, "U.S. Energy Resources: A Review as of 1972," in *A National Fuel and Energy Policy Study, Part I*, Committee on Interior and Insular Affairs, U.S. Senate, Serial no. 93–40 (Washington, D.C.: U.S. Government Printing Office, 1974). C. R. Pelto, "Forecasting Ultimate Oil Recovery," SPE Paper 4261, American Institute of Mining, Metallurgical, and Petroleum Engineers, 1o73. J. J. Arps, M. Mortada, and A. E. Smith, "Relationships between Proved Reserves and Exploratory Effort," *Journal of Petroleum Geology* (June 1971): 671–75.
10. See discussion of additional supplies from unknown fields in D. H. Root, *Future Supply of Oil and Gas from the Permian Basin of West Texas and Southeastern New Mexico*," U.S. Geological Survey Circular 828, U.S. Department of the Interior, 1980.
11. D. G. Krige, "A Statistical Approach to Some Basic Mine Valuation Problems on the Witwatersrand," *South Africa* 52, no. 6 (1951): 119–39; G. Matheron, "Application des méthodes statistiques à l'évaluation des gisements," Annueles des Mine (December 1955). F. Blondel and E. Ventura, *Le Structure de la Distribution des Produits Mineraux dans le Monde* (Paris, 1956). M. Allais, "Method of Appraising Economic Prospects of Mining Exploration over Large Territories," *Management Science* 3 (1957): 285–347.
12. Matheron.
13. W. G. B. Phillips, "Statistical Estimation of Global Mineral Resources," *Resources Policy* (December 1977): 268–80.
14. Ibid.
15. Ibid.
16. Allais, "Method of Appraising Economic Prospects of Mining Exploration over Large Territories." Blondel and Ventura, *Le Structure de la Distribution des Produits Mineraux dans le Monde.*
17. Blondel and Ventura, *Le Structure de la Distribution des Produits Mineraux dans le Monde*, p. 294.
18. Allais, "Method of Appraising Economic Prospects of Mining Exploration over Large Territories.
19. For example, *Oil and Gas Journal* 44, Jan. 1946, *Oil and Gas Journal Annual Review* vol. 58, Jan. 1960. Arps and Roberts, "Economics for

Cretaceous Oil Production on the East Flank of the Denver–Julesberg Basin"; and *The Financial Post* 15 (1959).

20.   G. M. Kaufman, *Statistical Decision and Related Techniques in Oil and Gas Exploration* (Englewood Cliffs, N.J.: Prentice-Hall 1962).

21.   Quoted from p. 201 of R. G. McCrossan, "An Analysis of Size Frequency Distribution of Oil and Gas Reserves of Western Canada," *Canadian Journal of Earth Sciences* 6, no. 201 (1969): 201–11.

22.   Ibid.

23.   A. A. Petrov, "Testing Statistical Hypotheses on the Type of a Distribution on the Basis of Small Samples," *Theory of Probability and Its Applications* 1 (1956): 223–45.

25.   Arps and Roberts, "Economics for Cretaceous Oil Production on the East Flank of the Denver–Julesberg Basin," *AAPG Bulletin* 42, 11 (1958): 2549–66. L. J. Drew, J. H. Scheunemeyer, and D. H. Root, "Petroleum-Resource Appraisal and Discovery Rate Forecasting in Partially Explored Regions," U. S. Geological Survey Professional Paper 1138, Parts A, B, C, U. S. Department of the Interior (Washington, D.C.: U.S. Government Printing Office, 1980).

26.   D. A. Singer and L. J. Drew, "The Area of Influence of an Exploratory Hole," *Economic Geology* 71, no. 3 (1976): 643–47.

27.   Drew, Scheunemeyer, and Root, "Petroleum-Resource Appraisal and Discovery Rate Forecasting."

28.   Arps and Roberts, "Economics for Cretaceous Oil Production on the East Flank of the Denver–Julesberg Basin."

29.   Ibid.

30.   Ibid.

31.   Ryan, "Analysis of Crude-Oil Discovery Rate in Alberta."

32.   Ibid., 219.

33.   Ibid., p. 222.

34.   Ibid., p. 226.

35.   Ibid.

36.   Ibid.

37.   Ibid., p. 232.

38.   Ibid.

39.   G. M. Kaufman, Y. Balcer, and D. Kruyt. "A Probabilistic Model of Oil and Gas Discovery." In *Studies in Geology, No. 1 – Methods of Estimating the Volume of Undiscovered Oil and Gas Resources* (American Association of Petroleum Geologists, 1975), pp. 113–42.

40.   Barouch and Kaufman, "Oil and Gas Discovery Modelled as Sampling without Replacement and Proportional to Size" and E. Barouch and G.M. Kaufman, "Estimation of Undiscovered Oil and Gas," in *Proceedings of the Symposia in Applied Mathematics*, vol. 21 (Providence, R.I.: American Mathematical Society, 1977), pp. 77–91.

41.   P. Bloomfield et al., "Volume and Area of Oil Fields and Their Impact on the Order of Discovery," Resource Estimation and Validation Project, Department of Statistics and Geology, Princeton University, Princeton, N.J., February 1979.

42. Drew, Scheunemeyer, and Root, "Petroleum Resource Appraisal and Discovery Rate Forecasting in Partially Explored Areas."

43. Bloomfield et al., "Volume and Area of Oil Fields."

44. Barouch and Kaufman, "Estimation of Undiscovered Oil and Gas."

45. P.Y. Fan, "Computational Problems in Modeling the Oil and Gas Discovery Process." MS thesis, MIT Sloan School of Management, Cambridge, Mass., 1976, and J.D. Nelligan, "Petroleum Resources Analysis within Geologically Homogeneous Classes," Unpublished PhD thesis, Department of Mathematics and Computer Sciences, Clarkson College of Technology, Clarkson, N.Y., June 1980.

46. D. Cox. "Further Results on Tests of Separate Families of Hypotheses," *Journal of the Royal Statistical Society*, Series B, 24, no. 2 (1962): 406–23.

47. E. Barouch and G.M. Kaufman, "Mathematical Aspects of Production and Distribution of Energy," in *Proceedings of the Symposia in Applied Mathematics*, vol. 21 (Providence, R.I.: American Mathematical Society, 1977).

48. A.P. Dempster, N.M Laird, and D.R. Rubin, "Maximum Likelihood from Incomplete Data via the EM Algorithm," *Journal of the Royal Statistical Society* Series B 39, no. 1 (1977): 1–38.

49. J.L. Smith and G.L. Ward. "Maximum Likehood Estimates of the Size Distribution of North Sea Oil Deposits." MIT Energy Laboratory Working Paper no. MIT–EL 80–027WP, Cambridge, Mass., 1980.

50. Barouch and Kaufman, "Estimation of Undiscovered Oil and Gas."

51. A.D. Beall. "Dynamics of Petroleum Industry Investment in the North Sea." MIT Energy Laboratory Working Paper no. MIT–EL 76–007WP, June 1976.

52. Smith and Ward, "Maximum Likelihood Estimates."

53. P. Eckbo, J. Jacoby, and J. Smith. "Oil Supply Forecasting: A Disaggregated Approach," *Bell Journal of Economics and Management Science 9*, no. 1 (Spring 1978): 218–35.

54. G.M. Kaufman, W. Runngaldier, and Z. Livne, "Predicting the Time Rate of Supply from a Petroleum Play," in *Economics of Exploration for Energy Resources, vol. 26 of Contemporary Studies in Economics and Financial Analysis*, ed. J. Ramsey, pp. 69–102.

55. P. Eckbo. "Estimating Offshore Exploration Development and Production Costs," MIT Energy Laboratory Working Paper, Cambridge, Mass., September 1977.

56. Eckbo, Jacoby, and Smith, "Oil Supply Forecasting," p. 224.

57. J. Wang, "Adaptive Optimal Control Applied to Discovery and Supply of Petroleum," Unpublished PhD thesis, MIT Department of Aeronautics and Astronautics, Cambridge, Mass., 1980.

58. J. Meisner and F. Demiren, "The Creaming Method: A Bayesian Procedure to Forecast Future Oil and Gas Discoveries in Mature Exploration Provinces," *Journal of the Royal Statistical Society*, Series A 143 (1980).

16 INDUSTRY APPROACHES TO FORECASTING FUTURE DISCOVERIES

The focus of this part has been on the principal approaches to forecasting amounts of undiscovered oil and gas, with little emphasis on the conceptual links that bind them together. Private sector projections of future supply and those made by the major oil companies in particular are tied together by threads of reasoning common to all companies, even though each major oil company has a unique analytical and procedural style. The purpose of the present chapter is to give the reader an idea of how methods as different in character as life-cycle, subjective probability, geologic-volumetric, rate-of-effort, and econometric methods may be employed by a single organization. A much more thorough documentation of methods used in industry is given in a report to the Energy Information Administration by ICF Incorporated,[1] and Holaday and Houghton present an insightful comparison of private sector supply forecasting and decisionmaking methods in *Energy Modeling Forum.*[2] As almost all private sector supply modeling is done in house and is generally regarded as proprietary, these two accounts are incomplete in some respects but nevertheless do an excellent job of conveying the flavor of approaches adopted by large oil companies to projecting future supply from U.S. petroleum provinces.

In their survey of private sector supply forecasting methods, Holaday and Houghton found that within a company, supply projections may be used for several purposes, each of which places different demands on basic geology, engineering and economic data, on the structure of an analysis of it, and on the form of the projection. Supply projections are used:

- To render explicit common assumptions decisionmakers use as a backdrop to allocate capital and personnel to specific projects,

289

- As input to overall exploration allocation decisions, and

- As a policy tool for interaction with federal government agencies and with the Congress.

Some executives Holaday and Houghton interviewed thought that the full potential of currently available methods for projecting future supply from basins and larger geographic regions had not yet been realized and that in the future a disaggregated approach to supply forecasting properly woven into a firm's decisionmaking fabric would provide a useful supplement to intuition, judgment, and methods currently employed:

- To allocate capital to individual basins and, more broadly, to allocate capital to petroleum exploration, development, and production as against allocation to alternative primary energy mineral exploration,[2] and

- To evaluate the impact on future supply of "rare" events such as the discovery of a supergiant field, a sudden change in tax or regulatory regime, or the occurrence of an embargo.

According to Holaday and Houghton, only Exxon and Shell are committed to highly disaggregated oil and gas supply modeling of basins in the United States. Some do no domestic supply modeling, and others do aggregate projections of future U.S. supply by piggybacking on more disaggregated forecasts made and published by their competitors.

Exxon is also actively engaged in research on modeling at a very disaggregated level. An indication of how the company uses a combination of geologic-volumetric methods, subjective probability, and rate-of-effort modeling to project future production from individual time stratigraphic units is given by White et al.[3]

Mobil's approach to projecting from new discoveries is an attempt to model individual plays, both known and potential within each basin. According to ICF,

- Future discoveries in plays in the lower forty-eight states are *probabilistically modeled*: "For each play, factors such as the expected recovery factors, the expected net feet of pay, and other geologic characteristics are input into the model with probabilities which estimate the subjectivity of these values. This information is collected from the field by the regional offices and transmitted to the main office. These parameters are expressed as cumulative probability curves, or risk profiles. A Monte Carlo computer simulation, involving the random sampling of these distributions, then produces curves which represent

---

a. As mentioned previously allocation problem may be likened to an investor's common stock portfolio problem: Given a capital budget to allocate among stocks with uncertain returns, what proportion of the budget should be invested in each stock available for investment?

future resource potential. Combining these curves gives an overall probability distribution of future resource potential in a given region." [4]

- The timing and amounts of future production from new discoveries in a partially explored basin are *judgmentally assessed* by the Exploration and Production Planning Group in light of currently available information about industry exploration development and production plays: "The resource potential numbers used by the Exploration and Production Planning Division generally reflect a 50 percent or greater probability that the estimated amount of resource is likely to exist." The Exploration and Production Planning Group then uses experience and judgment to determine the economic viability of such resources as well as the timing of discoveries and production. "Again, relevant information governing corporate plans for exploration, development, and production is assessed." [5]

- *Geologic-volumetric methods* are applied to frontier areas (the Beaufort Sea and onshore Alaska, for example):

    For frontier areas (e.g., offshore regions and onshore Alaska) where the data needed for the probabilistic model are not available, geologic analysis is used to estimate the potential resource base. Geologic information, on structural characteristics, rock formations, and basin geometry, is used to relate the frontier area to known fields with similar characteristics. The timing for drilling, developing, and producing from these areas is highly speculative and estimates rely heavily on expert judgment. [6]

ICF reports that Exxon's "Inland 48 Oil and Gas Production Methodology" is designed to aid in developing corporate policy positions with respect to regulation, taxes, and acquisition and development of primary energy mineral resources other than petroleum. The lower forty-eight states are divided into twenty-five regions, each containing a petroleum basin or group of basins sharing common geologic and engineering traits.

- *A model of growth of reported field reserves* for each region is used to project additions to currently reported reserves generated by revisions and extension. Projections of new discoveries are similarly treated.

- The historical series of amounts discovered as a function of cumulative footage drilled in each region is extrapolated with a *rate-of-effort model*. Historical ratios of "successful gas footage" are evaluated and total dry holes footage is directionally allocated based on these evaluations.

    Cumulative oil (gas) in place is modeled as an exponential function of cumulative exploratory footage drilled and the parameters of the function input from the company's current estimate of ultimate in-place amounts.

- *An in-house econometric model* with real resource prices for "new" oil and "new" gas as exploratory variables and exploratory drilling footage as a dependent variable is used to project the future amount of exploratory drilling footage per year. *Predicted amounts are judgmentally allocated to oil and gas*, region by region. Where Exxon has detailed information based on its own exploration activity in a region, this information is used in conjunction with predictions of the econometric model to forecast additions to revenues from new discoveries.

- The time rate of supply from each region is projected by application of a reserves-to-production ratio for each category of reserves. These ratios are *judgmentally determined.*

- Projections of future production in each region are *judgmentally adjusted* to conform to a variety of constraints exogenous to the models employed at earlier stages of analysis: capital requirements, limited rig and material availability, and so on.

There is no formal *overall* mechanism for projection of future production from a region. Expert judgment and opinion can intrude at any stage.

REFERENCES

1. ICF Incorporated, "A Review of the Methodology of Selected U.S. Oil and Gas Supply Models," 1979.
2. B. Holaday and J. Houghton, "Private Sector Supply Forecasting and Decision Making," *Energy Modeling Forum* 5 (1979).
3. D. C. White et al., "Assessing Regional Oil and Gas Potential," in *AAPG Studies in Geology, No. 1: Methods of Estimating the Volume of Underground Oil and Gas Reserves*, American Association of Petroleum Geologists, Tulsa, Okla.: 1975.
4. ICF, "Review of the Methodology of Selected U.S. Oil and Gas Supply Models."
5. Ibid.
6. Ibid.

SUMMARY

This part has treated, one by one, examples of each of the principal approaches to projecting future petroleum discoveries and supply. No single overarching concept links these approaches to one another. They are like ships passing in the night, headed for different ports. In the hold of one is a cargo of nongeologic methods of projecting petroleum supply designed to extrapolate exploitation history, while the other carries a cargo of methods based on assumptions about geologic and statistical attributes of deposits, how they are discovered, and economic decision making by oil and gas operators.

A remark borrowed from J. B. S. Haldane's 1928 *Possible Worlds* succinctly states my perception of the current state of oil and gas supply modeling "For every type of animal there is a most convenient size, and a large change in size inevitably carries with it a change in form." I quoted Haldane's remark in "Issues Past and Present in Modeling Oil and Gas Supply,"[1] where I expanded on this analogy as follows: Physical principles dictate the size of animals. Haldane points out that the human thigh bone breaks under about ten times the weight of a normal human. The strength of bone is in proportion to its cross-sectional area, while the weight of an animal is proportional to its volume. This lessened his respect for Jack the Giant Killer, for the Giant, a scaled-up human 60 feet tall, would have broken his legs with his first few steps toward Jack.

If we replace the words "animal" and "size" with "policy problem" and "model," respectively, then models of oil and gas exploration, discovery, and production currently in vogue loosely adhere to Haldane's observation about the animal kingdom. At one extreme are disaggregated approaches to modeling that focus on individual deposits as the basic unit for analysis; at the other extreme are models that treat time series of data at the national level. As with the giant, a

293

direct scaling up of models for supply from individual petroleum plays or petroleum basins breaks the model's legs in several ways. Yet there must be a logical
connection between micromodels of these physical entities and a countrywide
aggregation, for what happens at the national level to discovery and production
of oil and gas is, after all, determined by what happens in over 100 individual
petroleum basins.

The territory that lies between aggregated and unit-specific disaggregated
approaches is as yet uncharted. No logically tight methodology for aggregating
projections of supply over different time frames from individual geologic units in
varying stages of exploratory maturity is in sight. Ideally, we wish to have at our
disposal a system of logically interrelated methods and models sufficiently flexible to allow economic supply functions to be computed for mature, partially
explored, and frontier regions under a wide range of fiscal, regulatory, and technological alternatives at reasonable cost in time, human effort, and money. The
current state of the art is very far from this ideal.

A sketch of aspects of the current modeling environment has these salient
features:

- *No conceptual reconciliation of models* of individual plays, stratigraphic
  units, and petroleum basins that explicitly incorporate individual deposits as
  components with models that do not;

- *A movement toward process-oriented models*, which reflect key features of
  the physical processes of exploration, discovery, and production of petroleum.

- *Few publicly available data sets that allow meaningful structural validation* of
  highly disaggregated models;

- *An increasing use of personal (or subjective) probabilities* as a vehicle for
  representing expert judgments about uncertain quantities without a serious
  matching effort to train assessors to avoid cognitive biases that distort assessments; and

- *Policy issues as moving targets.* An often rapid change in what policy analysts
  view as "important" policy problems places a heavy burden on modelers
  working with models, most of which are difficult to reconfigure rapidly.

If allowed only one prescriptive conclusion, it would be that a massive effort to
collect and measure data accurately is needed. Real progress awaits data acquired
according to a design that permits meaningful structural and predictive evaluation of highly disaggregated models of deposition, discovery, and production.

REFERENCE

1. Appearing in *Oil and Gas Supply Modeling* NBS Special Publication 631.

IV COAL An Economic Interpretation of Reserve Estimates

Martin B. Zimmerman

INTRODUCTION

The analysis of production of natural resources is fundamentally different from the analysis of production in conventional industries. The difference is due to the phenomenon of depletion. In traditional supply analysis, cost is treated as dependent on the rate of output: As more units are produced per time period, costs rise. The relation between cost and output is reversible: As output declines, so does cost. In the production of natural resources, the nature of the production process is systematically changing. When cheaper deposits are exhausted, production moves to higher cost deposits. The vagaries of new discoveries or technological change might reverse the order for a while, but in a stable environment lower cost reserves are used and exhausted first.[a] This means that costs rise as output cumulates over time. The cost of any given rate of output increases with cumulative output. Because of this added relation, predicting future costs is more difficult than in traditional analysis. We want to know how the costs of mining will change as cumulative output increases. To answer this question, the analyst turns to reserve data. The present part of this book discusses how coal reserve data have been and can be used to analyze the impact of cumulative output on the future cost of coal.

---

a. It is well known in the theory of natural resources that in a competitive or monopolistic industry it will always pay to exploit the cheapest deposits first. In an oligopolistic industry, production of higher cost deposits can occur before a dominant firm, for example, produces all its low-cost deposits. Our interest here is with the American coal industry, which is a workably competitive industry.

The task involved is simpler than for oil, gas, or uranium, for a great deal more is known about the location and extent of coal deposits than about the location of those other energy sources. Past exploration and production of coal and even the exploration for oil, gas, and water have yielded a great deal of information about where coal is located. The same information also aids in delineating the extent of coal deposits because of the relative homogeneity of coal deposits. Coal seams lie relatively close to the surface and extend for miles underground. Extrapolation is not nearly as hazardous as for uranium, for example. What the coal industry calls exploration is called development activity in the oil industry. Exploration is aimed at determining the dimensions of the deposit, where fault lines lie, and so forth.

Because information is better about the location of coal deposits, this part focuses on already discovered, but undeveloped deposits. At issue primarily is the extraction cost of these deposits. Thus, the chapters of this part examine what reserve and resource data reveal about the future evolution of extraction costs.

Chapter 17 presents various resource classification schemes of the United States Geological Survey (USGS) and of the United States Bureau of Mines (BOM). Chapter 18 introduces some basic economic concepts essential to the economic interpretation of reserve data, then uses them to interpret the various classification schemes. Chapter 19 treats what has been left out of the reserve concepts: inferred reserves, the recoverability of coal, and technological change.

17 THE RESOURCE CLASSIFICATION SYSTEM

CERTAINTY CATEGORIES

The most basic distinction in resource classification is between identified and undiscovered resources. The latter category consists of hypothetical and speculative resources. Hypothetical resources are surmised to exist in known mining districts. Speculative resources, which are even less certain, are possible resources in areas where coal has not been previously discovered. Speculative resources are so uncertain, and the identified portion of coal resources is so large, that attention has focused on the identified portion.[1] In light of the large amounts of identified resources and the relatively higher costs associated with undiscovered resources, it is unlikely that speculative and hypothetical resources will play any significant role in the next twenty years or more. Consequently, the focus here is on identified resources.

Identified resources are classified along two dimensions: the certainty with which deposits are known and the physical characteristics of the deposits. Certainty categories are defined as measured, indicated, and inferred. The U.S. Geological Survey (USGS) define these terms as follows:

- *Measured*: Measured resources are resources for which tonnage is computed from dimensions revealed in outcrops, trenches, mine workings, and drill holes. The points of observation and measurement are so closely spaced, and the thickness and extent of the coal are so well defined, that the computed tonnage is judged to be accurate within 20 percent of the true tonnage. Although the spacing of the points of observation necessary to demonstrate continuity of coal differs from region to region according to the character of the coal beds, the points of observation are, in general, about half a mile apart.

- *Indicated*: Indicated resources are resources for which tonnage is computed partly from specific measurements and partly from projection of visible data for a reasonable distance on the basis of geological evidence. In general, the points of observation are about 1 mile apart from beds of known continuity. In several states, particularly Alabama, Colorado, Iowa, Montana, and Washington (where the amount of measured resources is very small) the measured and indicated categories have been combined.

- *Inferred*: Inferred resources are resources for which quantitative estimates are based largely on broad knowledge of the geologic character of the bed or region and for which few measurements of bed thickness are available. The estimates are based primarily on an assumed continuity in areas remote from outcrops of beds, which in areas near outcrops were used to calculate tonnage classes as measured or indicated. In the interest of conservatism, the areas in which the coal is classed as inferred are restricted as described under the heading, "Areal, Extent of Beds." In general, inferred coal lies farther than 2 miles from the outcrop in areas for which mining or drilling information is available.

- *Unclassified*: For a few states, particularly Georgia, Maryland, Pennsylvania, Utah, and West Virginia, the calculated resources have not been divided into the measured, indicated, and inferred categories.[2]

PHYSICAL DIMENSIONS

The identified portion of resources is further classified according to physical dimensions, the depth from the surface and the thickness of the coal seam. Resources are broken into three thickness intervals: 14–28 inches, 28–42 inches, and more than 42 inches. Depth is divided into three intervals: less than 1,000 feet from the surface, 1,000–2,000 feet deep, and 2,000–3,000 feet deep. Figure 17–1 summarizes this classification system.

RESERVE BASE

The foregoing description summarizes the contribution of the U.S. Geological Survey. The U.S. Bureau of Mines has gone further in attempting to define a more restricted measure of reserves, called the *Demonstrated Reserve Base*.[3] This aggregate consists of measured and indicated reserves in bituminous seams greater than 28 inches and up to 1,000 feet below the surface. (Where mining in a region is already taking place at less than 28 inches or deeper than 1,000 feet, local exceptions are made in these criteria.) To be included in the reserve base,

Figure 17-1. Distribution of Coal Resources by Physical Characteristics and Certainty Categories.

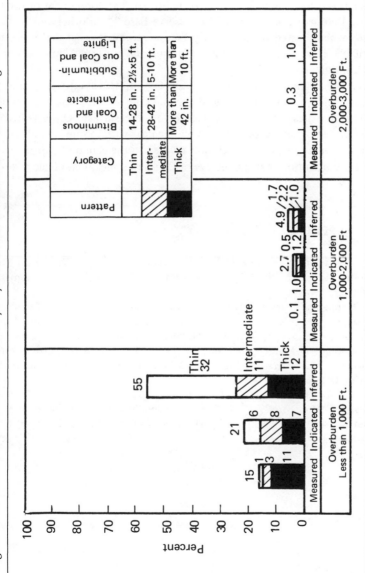

Source: Paul Averitt, Coal Resources of the United States, U.S. Geological Survey Bulletin 1412, January 1974 (Washington, D.C.: U.S. Government Printing Office, 1975), p. 40.

subbituminous seams and lignite seams must be greater than 5 feet. The reserve base is shown as the outlined area on the left-hand side of Figure 17-1.

The claim is made that a reserve is "that portion of the Identified Coal Resource that can be economically mined at the time of determination. The reserve is derived by applying a recovery factor to that component of the Identified Coal Resource designated as the Reserve Base."[4] In other words, the reserve base, once adjusted for a recovery factor, represents coal that is economically available according to the joint definitions of the U.S. Geological Survey and the U.S. Bureau of Mines.

Finally, the demonstrated reserve base is subdivided into strippable and underground reserves. In general, the strippable reserve base is the portion of the reserve base that lies less than 120 feet below the surface. Other restrictions sometimes placed on strippable reserves will be deal-with in some detail later.

The next several chapters, which are devoted to the economics of the reserve base test the contention that the reserve base is economically available at today's prices.

REFERENCES

1. Paul Averitt, *Coal Resources of the United States*, U.S. Geological Survey Bulletin 1412, January 1, 1974 (Washington, D.C.: U.S. Government Printing Office, 1975).
2. U.S. Geological Survey, *Principles of the Mineral Resources Classification System of the U.S. Bureau of Mines and U.S. Geological Survey*, Bulletin 1540-A. (Washington, D.C.: U.S. Government Printing Office, 1976).
3. U.S. Bureau of Mines, *The Reserve Base of U.S. Coals by Sulfur Content*. Information Circulars 8680 and 8693 (Washington, D.C.: U.S. Government Printing Office, 1975).
4. U.S. Geological Survey, *Coal Resource Classification System of the U.S. Bureau of Mines and the U.S. Geological Survey.* U.S. Geological Survey Bulletin 1450-B (Washington, D.C.: U.S. Government Printing Office, 1976).

18 AN ECONOMIC INTERPRETATION OF RESERVES

THE ECONOMIC CONCEPTS

Given the goal of determining the behavior of cost as output cumulates, the basic problem is that a measure of the cost of a ton of coal is meaningful only at a given *rate* of output. Economists speak of a cost function; cost depends on rates of output. The cost of exploiting any reserve depends on how fast the coal is taken out.

Long-Run Marginal Cost

Economic theory emphasizes the marginal cost of production—that is, the cost of the last unit produced. In competitive long-run equilibrium, this is equal to the minimum average cost of production. A firm that is producing at a level minimizing average cost and that is receiving a price equal to minimum average cost is earning a rate of return just large enough to keep it in business but not large enough to encourage entry of new firms.

In a mineral industry, the concept of equilibrium needs some modification, because all firms are not identical. Mines in better deposits (thicker seams and generally more favorable mining conditions) coexist with less productive mines. As mining proceeds from more to less favorable deposits, costs rise. Mines that opened under more favorable conditions earn a high rate of return at prices just high enough to keep the less favored mines in business. The last mine opened in order to satisfy demand can be thought of as the *incremental mine.* This mine would just break even under current prices, by producing at minimum average

301

cost. The minimum average cost of the incremental mine is the long-run marginal cost.[a]

If it is known how the incremental mine changes as output cumulates over time, the behavior of long-run marginal cost is known. Reserve data are useful to the extent they reveal the rate at which better deposits will be exhausted and show how the nature of the deposits exploited will change. The economic interpretation of this geological information translates the deterioration of geological conditions into a rate of cost increase.

THE DEMONSTRATED RESERVE BASE

The concept of the *Demonstrated Reserve Base* evolved as an attempt to tighten resource and reserve classification. The goal was to estimate the total amount of coal that the USGS felt confident was present and could be mined by current techniques. It has been widely interpreted as the *economically* available portion of the resource stock. As can be seen in Table 18-1, it is a large amount of coal. The numbers in Table 18-1 underlie the great optimism with respect to coal in this country. Let us examine these numbers in light of the concepts introduced earlier to see whether they measure what they claim to measure.

Figure 17-1 presented the classification system. The following questions must be answered: (1) How accurately do the chosen characteristics define a cost function? (2) Is the measure of reserves used by the Bureau of Mines the portion of the stock available at current costs? (3) If it is not, what information do reserve data convey about future costs?

Certainty: Measured and Indicated Reserves

A cross-section of the coal seam and other strata, or core, is obtained by drilling. Measured reserves are drilled before development commences. Indicated reserves are also drilled but not as much, so they are known with less certainty. The uncertainty includes the possibility of geological faults, which make mining difficult, or diminished seak thickness, which makes mining more costly. The extreme is, of course, that the seam thins to unminable thickness or has been removed by subsequent geologic events, such as channeling by ancient rivers across the coal swamps.

A mining company has the choice of mining a measured seam that is thinner or drilling an indicated seam with a greater expected thickness. At the margin, the two alternatives are equated. Therefore, the maximum the company is will-

---

a. Because of user cost it is possible that no mine is actually producing at minimum average cost. In the coal industry, where user costs are generally very small, the incremental mine produces close to that point. In any case, we can still measure long-run marginal cost as what it would cost to produce at that minimum average cost point.

Table 18-1.   Demonstrated Reserve Base[a] of Coals in the United
States on January 1, 1976, According to Rank,[b] Million Short Tons.

State	Anthracite	Bituminous	Subbituminous	Lignite	Total
Alabama	...	2,008.7	...	1,083.0	3,091.7
Alaska	...	697.5	5,446.6	14.0	6,158.2
Arizona	...	325.5	...	...	325.5
Arkansas	96.4	270.1	...	25.7	392.2
Colorado	25.5	9,144.0	4,121.3	2,965.7	16,256.4
Georgia	...	0.9	...	...	0.9
Idaho	...	4.4	...	...	4.4
Illinois	...	67,969.3	...	...	67,969.3
Indiana	...	10,714.4	...	...	10,714.4
Iowa	...	2,202.2	...	...	2,202.2
Kansas	...	998.2	...	...	998.2
Kentucky, East	...	13,540.1	...	...	13,540.1
Kentucky, West	...	12,460.8	...	...	12,460.8
Louisiana	...	...	...	c	c
Maryland	...	1,048.3	...	...	1,048.3
Michigan	...	126.8	...	...	126.8
Missouri	...	5,014.0	...	...	5,014.0
Montana	...	1,385.4	103,416.7	15,766.8	120,568.9
New Mexico	2.3	1,859.9	2,735.8	...	4,598.0
North Carolina	...	31.7	...	...	31.7
North Dakota	...	...	...	10,145.3	10,145.3
Ohio	...	19,230.2	...	...	19,230.2
Oklahoma	...	1,618.0	...	...	1,618.0
Oregon	...	c	17.5	...	17.5
Pennsylvania	7,109.4	23,727.7	...	...	30,837.1
South Dakota	...	...	...	426.1	426.1
Tennessee	...	965.1	...	...	965.1
Texas	...	...	...	3,181.9	3,181.9
Utah	...	6,551.7	1.1	...	6,552.8
Virginia	137.5	4,165.5	...	...	4,302.9
Washington	...	255.3	1,316.7	8.1	1,580.1
West Virginia	...	38,606.5	...	...	38,606.5
Wyoming	...	4,002.5	51,369.4	...	55,371.9
Total	7,371.1	228,924.6	168,425.0	33,616.6	438,337.3

a. Includes measured and indicated resource categories as defined by the USGS and
represents 100 percent of the coal in place.

b. Data may not add to totals shown due to rounding.

c. Quantity undetermined (basic resource data do not provide the detail required for
delineation of the reserve base).

Source: U.S. Bureau of Mines, as reprinted in Keystone Coal Industry Manual (New
York: McGraw-Hill, 1978), p. 694.

ing to pay for drilling is given by the increment in costs in the measured portion.
If it is cheap to drill and if the probability of finding a thicker seam is great,
there will be no move to much thinner seams in the measured reserve.

In fact, drilling is relatively cheap. The indicated reserves are estimated from
observations more widely spaced than is the case for measured reserves. Mea-

sured reserves are said by the USGS to be accurate within 20 percent. No estimate of accuracy is given for indicated or inferred reserves. The difference in
knowledge is represented by five core-holes per square mile. Observations are
0.5 mile apart in the measured stock and 1 mile apart in the indicated portion.
The individual mine operator faces a greater risk when dealing with an individual
parcel of indicated reserves. The risk is quite limited, however. After the drilling,
the mine operator can choose not to develop if the results indicate an unprofitable parcel. The most at risk is the cost of drilling.

Industrywide, even this risk disappears, for it can be "diversified away" by
drilling enough separate parcels. If all the indicated reserves were drilled, the
individual parcel results would vary, but unless the estimation process were
biased, the "expected" amount of coal would be found. The large number of
individual parcels and the law of large numbers assure this.

Assuming that all the holes are drilled, what cost is added to the cost of development? The cost of drilling using diamond bits is about $45 per vertical
meter to a depth of 300 meters. Therefore, the cost of drilling five holes in a
1-mile-square parcel is approximately $69,000.[1] This capital expenditure must
be amortized over the life of the investment, which depends upon total reserves
and annual output. To be conservative, assume the initial outlay is $100,000.
The amortized cost per ton of coal is

$$C = \frac{\$100,000.00}{\dfrac{q(1 - e^{-rt})}{r}} \tag{18-1}$$

where r = discount rate = 0.12; q = annual output rate; and t = life of investment = R/q. The life of the investment is simply R/q, where R is reserves
"proved." R, in turn, is equal to

$$R = 640(1,800)(Th) , \tag{18-2}$$

where Th = thickness of seam in feet. The number 640 refers to acres per square
mile, and 1,800 is the tons of coal per acre-foot.

A simple example illustrates that this cost is trivial. Assume the reserves are in
a seam with expected thickness of 42 inches, and output will be 100,000 tons
per year:

$$C = \frac{(100,000)(0.12)}{100,000(1 - e^{-rt})} = \$0.09 \text{ per ton} \tag{18-3}$$

The importance of this exercise is that for all intents and purposes the difference
between indicated and measured reserves can be ignored. Concerning the portion
of the stock being mined, it is not significant whether a new mine was developed
from the measured or from the indicated portion of the stock. More impor-

tantly, the two categories can be aggregated when analyzing the effect of increased output on cost.

Certainty: Inferred Reserves

Uncertainty with regard to inferred reserves is much greater, because these estimates are based not on core samples but on broad geological information over wide areas. Instead of data on a large number of individual parcels, estimates are based on broad extrapolation. The cost of drilling over such a large area cannot be estimated. The uncertainty in the estimates is reflected in the large adjustments made in them as new information becomes available. A good example is the massive reevaluation of coal reserves in Western states. The original estimates were based on a few observations and extrapolation. Now that drilling is taking place the estimates are changing.[2] By excluding resources from the Reserve Base, the U.S. Bureau of Mines highlights this large inferred uncertainty. We shall return to inferred reserves in Chapter 20; at this juncture note simply that they are substantially more uncertain than the Reserve Base.

Physical Characteristics and Costs

Besides certainty categories of reserve estimation, other yardsticks are provided by the reserve classification scheme. They relate to two main physical characteristics, thickness and depth. This section ignores problems of quality of information and assumes that the numbers provided are accurate. That is, the estimates of the amount of coal in each category of the Reserve Base are accepted. The question addressed is what this information means in the context of a cumulative cost function.

Seam Thickness. Two issues arise with respect to physical characteristics and costs. The first: How important are the characteristics used in defining the Reserve Base? This boils down to two related questions: Does thickness affect cost, and is it the only factor affecting cost? Do we need to consider the impact of other deposit characteristics on cost? The second issue is: How relevant are the physical limits used in classifying reserves? For example, how much information is conveyed by the fact that there is so much coal in seams greater than 28 inches in thickness? Would a cutoff different than 28 inches provide more information?

The effects of observed and unobserved characteristics must be considered. The analyst wants to know how important in determining cost are the observed characteristics for which data exists. But just as important is knowing what is left out. How important are cost-determining factors not considered in the classification?

Without question seam thickness is an important determinant of underground mining costs; this fact is extensively documented in the mining literature. However, the mining literature is also replete with articles about the importance of factors other than those recorded in the USGS classification:

> Natural conditions involve roof, floor, grades, water, methane and the height of the seam. . . . In addition to these normal conditions, there are, in some mines, rolls in the roof or floor, and clay veins of generally short horizontal distance that intersect the coal seam. All these must be taken into account.
>
> It is possible for an experienced engineer to examine previous conditions of the sections and the immediate area of the section and assess proper penalties. As an example, if the roof is poor, production is reduced by as much as 15 percent of the available face working time. If the floor is soft, and fine clay and water are present, the production handicap could be as much as 15 percent. If a great deal of methane is being liberated, so that it is necessary to stop the equipment until the gas has been bled off, this delay could run as high as 10 percent. Fortunately, only a few mines in the United States have such severe conditions. The same remarks apply to all the other natural conditions.[3]

Statistical analysis of coal-mining productivity and costs helps to sort out these factors. In what follows, some results of previous studies are summarized.[4] The goal is to develop a long-run cost relation that expresses cost as a function of observed and unobserved characteristics. The cost function will then be used to interpret the Reserve Base. However, to understand costs, we must first understand the technology.

Underground Mining Technology. There are three types of deep mine: drift, shaft, and slope. When the coal outcrops on a hill, a drift opening is constructed that provides access at the level of the seam. When the seam does not outcrop on a hillside, shafts or slopes are constructed. Once the seam is reached mining can begin.

The dominant technique for mining coal underground in the United States and the one upon which the following estimation is based is continuous mining. In continuous mining a large machine rips the coal from the seam and loads it onto shuttle cars in one continuous operation. The shuttle cars transfer the coal to a central transport network for removal to the surface. A mining machine, two shuttle cars, and a complement of miners constitute a mining section. A mine consists of a number of sections, each working independently, but sharing a common haulage system, ventilation system, and a set of openings to the seam that provides access for miners, for supplies, and a means for removing the coal.

Noncontinuous techniques are also used, but at current factor prices the large new mine—the one that determines the cost of coal at the margin—typically uses the continuous mining method. This is true for a wide variation in factor prices.

Statistical Analysis. The passage just quoted, about factors other than seam thickness, suggests a method for assessing the impact of observed and unobserved characteristics. The capital and labor per mining unit are relatively fixed, so productivity per section corresponds unambiguously to the cost of mining.

To examine how productivity varies as observed and unobserved characteristics vary, an equation of the following form can be specified:

$$q_i = f(G_1, \epsilon_i) \; , \tag{18-4}$$

where q = production per unit; G = set of observable characteristics; and ϵ_i = random disturbance term. The observed characteristics are seam thickness, size of mine, and number of openings to the seam for underground mining. The unobserved characteristics are represented by ϵ_i. These are left out of the equation, but their collective influence is measured as the "unexplained variance" in the statistical analyses. The data are observations in 1975 of all underground mines that produced more than 100,000 tons per year.[5] The estimation is described in detail elsewhere; here the results are presented and some important issues underscored.

The actual equations and results are as follows:

$$\sqrt{s} \; \log q = .7568 \; \sqrt{s} + 1.1071 \, (\log Th) \, \sqrt{s} - .2185 \, (\log s) \, \sqrt{s}$$

$$\tag{18-5}$$

(Std. error)	(.4842)	(.1205)	(.0594)
(t-statistic)	(1.5630)	(9.1906)	(-3.6762)

$$+ .0283 \, (\log Op) \, \sqrt{s}$$

(.0655)

(.4314)

Standard error of the regression (SER) = .9799

Chi-square   = 232.4285

Number of observations   = 244

where q = production per mining section; s = number of producing sections; Th = seam thickness in inches; and Op = number of openings (shift, slope, drift) to the seam. The equation is multiplied by $\sqrt{s}$ for econometric reasons explained in Zimmerman 1978.[6] The inclusion of log s as a variable reflects economies or diseconomies of scale. Even though the units are producing separately, they share common equipment for haulage, ventilation, and other purposes. As the size of the mine increases, logistical problems might be expected to lower productivity per unit. This decline would proceed only to the point at which it became profitable to sink another shaft and reverse declining productivity. The addition of Op as a variable captures the latter effect.

Several important facts emerge from the statistical analysis. Seam thickness, the variable for which information exists, has an important effect on deep mining productivity and hence on costs. In deep mining a 1 percent increase in seam thickness results in a 1.1 percent increase in productivity. A second subtle but equally salient point emerges from the statistical analysis. Although the observed factors are important, they do not give the complete picture. Seam thickness, together with size variables, only partially capture the variability in deep mining. That is, all the other factors for which no information is available collectively account for a substantial variance in productivity. This result is reproducible with widely different samples. Similar results were obtained with a sample of mines from 1954.[7] Seam thickness is important, but other factors taken together also have a large influence on costs, an influence that can be measured by the standard error of the regression equation. The standard error of log ϵ_i implies an estimated variance of ϵ_i equal to 4.21.[b] This is a very substantial dispersion. If only the seam thickness for a given mine is known, a 90 percent confidence interval includes productivity levels almost seven times greater than those predicted by Eq. (18-5). Think of ϵ_i as an index of the favorability of all other factors. High values mean productivity levels higher than average, and low values mean levels lower than average.

Long-Run Cost Function. Using the productivity equation (18-5) as a base, one can estimate an average cost function for a mine as a function of seam thickness and ϵ_i. As I have already done this, the method need only be outlined here.[8]

Equation (18-5) provides a relation between productivity per unit, the number of units, and geology. In Zimmerman 1978 it is shown that expenditures for capital, material, and labor can be approximated as a function of the number of units. For any given Th and ϵ_i it is known how many units are necessary to produce a given amount of output. From the expenditure equations the cost is known for that many units. Combining equations yields an equation for the average cost as a function of geology and mine output:

$$AC = g(Th), \epsilon_i, Q \ . \tag{18-6}$$

What output rate is correct for measuring long-run marginal cost? In Chapter 17 it was maintained that it is the rate that minimizes AC. This rate can be obtained by differentiating Eq. (18-6) with respect to Q, setting the result equal

---

b. The relation between the variance of ϵ_i, the lognormal variate, and log ϵ_i, the normal variate, is the following: Let $\sigma^2 = \mathrm{var}\,(\log \epsilon_i)$, s = number of sections; $\mu = E\,(\log \epsilon_i) = 0$ by assumption; then var $\epsilon_i = (e^{2\mu + s\sigma^2})\,(e^{s\sigma^2} - 1)$. If $s = 1$, Var $(\epsilon_i) = e^{\sigma^2}\,(e^{\sigma^2} - 1) = 4.2$.

to zero, and solving for Q^*, the output rate that minimizes AC. Given Q^*, Eq. (18-6) can be used to estimate AC^*, the minimum average cost. The resulting equation is

$$AC^* = \frac{2,567}{Th_i^{1.1071} \epsilon_i} . \qquad (18-7)$$

The locus AC^* is drawn in Figure 18-1. An average cost equation exists for each value of thickness and ϵ_i, the proxy used for all other mining conditions. The importance of the locus AC^* is that it gives the minimum average cost for mining a given seam, that is, a seam for which thickness and ϵ_i are known. The question of how to measure the long-run marginal cost can now be answered. Only the correct output level is needed to measure cost, and that is given by AC^*.

Figure 18-1 summarizes the results of my previous study. As the seam thickness declines (and as ϵ_i declines) so does the output rate that minimizes average cost. To see this, assume that cost were to be measured by holding the output

Figure 18-1. Locus of Minimum Average Cost Points AC^*.

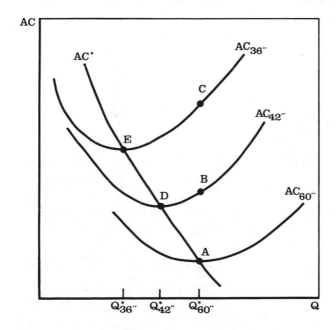

Source: Martin B. Zimmerman, "Estimating a Policy Model of U.S. Coal Supply," in *Materials and Society*, vol. 2 (London: Pergamon Press, 1978), pp. 67–83.

rate at Q^*_{60}. The increase in cost due to poorer mining conditions would be measured as the increase in cost from A to B and then to C. The actual cost increase would only be A to D to E. Depending upon the shape of the average cost, this could be a serious error. As mining conditions worsen, cutting back the output rate mitigates some of the worsening conditions. In summary, the rate of output at which we choose to measure cost is not arbitrary, and it interacts strongly with geology.

Again, previous studies indicate how wrong assessing the economic attributes of coal reserves can be if no attention is paid to the phenomenon just described. Table 18-2, column (2) gives the average cost of producing underground in a seam of specified thickness, assuming an output rate held constant at 3 million tons per year and setting ϵ equal to its median value of 1. Note also the estimated AC^* for that seam thickness when the output rates are adjusted to minimize cost. The last column yields the ratio of average cost at 3 million tons per year versus AC^*. As can be seen from the table, the error can be large, and it gets bigger as seam thickness declines.

Cost of mining cannot be measured unless the interaction of geology and economics is taken into account. As geological conditions deteriorate—that is, as seam thickness declines or as ϵ_i gets smaller—the cost increase is moderated by reducing the rate of output. When cost is measured, that must be taken into account. Mine size just reflects the underlying economies of scale. Therefore, an understanding of the technology must be part and parcel of reserve interpretation.

IS THE RESERVE BASE COAL AVAILABLE AT CURRENT COST?

The Incremental Mine

Bringing together the analyses of the previous sections, we can now compare the incremental cost of mining today to the costs implied by the cutoff criteria in

Table 18-2. Costs of Mining at 3 Million Tons per Year Rate versus the Optimal Output Rate.

(1) Thickness inches	(2) AC, $/ton	(3) AC*, $/ton	(4) AC/AC*
28	98.67	64.16	1.54
42	56.13	49.96	1.37
60	34.13	27.60	1.25
72	26.18	22.55	1.19

Source: Zimmerman, "Estimating a Policy Model of U.S. Coal Supply."

the Reserve Base. Since measuring cost requires first looking at minimum average cost, which is a function of seam thickness and ϵ_i, the long-run marginal cost of coal today can be measured if the seam thickness and ϵ_i of new mines opening today are known. Simply substitute the values of Th and ϵ_i into Eq. (18-7) to estimate cost.

In theory the costs in the last mine that had to be opened in order to satisfy demand are to be estimated. The minimum average cost of that mine would be the long-run marginal cost. If depletion in any time period were small, there would be many new mines opening with approximately the same cost. These mines would all be on a contour such as in Figure 18-2. Each contour in that diagram represents combinations of ϵ and Th that yield a given cost. The cost represented by C_0 is lower than that represented by C_1, since for any thickness the value of ϵ_i is greater along C_0 than C_1. The search for the incremental mine must be done by region because transportation costs are high and mine-mouth costs therefore differ at the margin among regions. We must also control for coal quality. Low-sulfur coal sells at a premium; therefore, mine owners will accept less favorable mining conditions for low-sulfur coal than they would for high-sulfur coal.

Figure 18-2.   Relation between Th and ϵ for a Given Level of Cost.

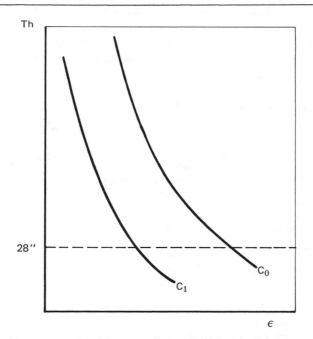

In practice ϵ_i cannot be observed. Only Th_i can be observed. The problem then is to estimate long-run marginal cost for a given region and for a given sulfur content using observations only on thickness for new deep mines. Because only observations on thickness are available, "typical" new seam thickness must be chosen for a set of new mines.

If it is assumed that in any region in any given sulfur category depletion has been small over the past three-year period, new mines opened during this period can be used to establish the "typical" seam thickness. Depletion being small, these new mines represent approximately equal cost. Equation (18–7) implies that new mines obey the following rule:

$$\frac{2,567}{Th_i^{1.1071} \epsilon_i} = \bar{C} , \qquad (18-8)$$

where $\bar{C}$ is the current incremental cost of coal. The best estimate of the incremental thickness depends upon the way ϵ_i is distributed in nature. Intuition suggests that some central value is correct, since the extreme low values of Th_i must represent the most favorable value of ϵ_i. If all the observed mines are of equal cost, then the thinner seams are compensated by more favorable sets of other mining conditions. In Zimmerman 1978 the assumption was made that the disturbance—that is, ϵ_i—is lognormally distributed. Adopting this assumption means that the best measure of the incremental mine is obtained by choosing the geometric mean of new seam thicknesses. Table 18-3 presents information on the incremental deep mine. The bulk of the observations are for mines opened in the 1963-1976 period.[9] In some areas these mines were supplemented with observations on planned mine development.[10]

Table 18-3.   Incremental Deep Mines.

Area/ Sulfur content/Type of mine	(1) Number of Observations	(2) Range	(3) Geometric Mean of Th	(4) Ratio of Cost of Cutoff Mine to "Incremental" Mine
Northern Appalachia				
High sulfur (1.5 percent), shaft	7	42–85	56.8	2.19
High sulfur (1.5 percent), drift	10	36–78	53.3	2.04
Southern Appalachia, drift				
Low sulfur (0.8 percent)	29	35–72	44.1	1.65
Medium sulfur (0.9 to 1.5 percent)	16	31–131	49.6	1.89
High sulfur (1.5 percent)	17	39–108	53.2	2.04
Midwest				
High sulfur, shaft	5	49–87	74.1	2.94

Sources: Keystone Coal Industry Manual, various years and U.S. Bureau of Mines, Projects to Expand Fuel Sources in Western States, Information Circular 8719 (Washington, D.C.: U.S. Government Printing Office, 1976).

The data in Table 18-3 suggest several interesting points. The importance of "other" characteristics is confirmed by the range in the seam thicknesses of new mines. If only seam thickness mattered, the thicknesses of new mines in any given region and sulfur category would be highly concentrated. The fact that the range is wide reflects the influence of other mining conditions.

The next fact of importance is that the central value of thickness is, in some cases, far away in value from the cutoff criteria for the Reserve Base. Column (4) shows what the percentage increase in cost would be if the central tendency in the observed distribution in new mines were equal to the cutoff criterion of 28 inches. A central tendency equal to the cutoff of 28 inches would imply almost a threefold increase in cost in the Midwest and more than doubling in northern and southern Appalachian high-sulfur coal costs. Low-sulfur southern Appalachian coal is closest to the cutoff criterion, but even there costs would rise by 65 to 90 percent if the central tendency in new mines were equal to the cutoff criterion.

Even the comparison of the central tendency in new mines to the cutoff thickness is not quite correct. The Reserve Base includes all coal in 28-inch or thicker seams. When the incremental mine is 28 inches, a lot of coal at 28 inches in the Reserve Base still has a higher cost than the expected cost for 28-inch seams. The influence of other factors ensures this. The conclusion then is inescapable: The Reserve Base includes coal that is far above current levels of cost; it does not represent coal available at today's level of costs.

This misunderstanding has arisen because the influence of unobserved characteristics of deposits has been ignored. In some areas of southern Appalachia, 28-inch seams are mined, which led the Bureau of Mines to conclude that thickness is the economic limit. In fact, it is the economic limit when other conditions compensate for the thin seam, as shown by the dotted line in Figure 18-2. In summary, the Reserve Base is simply coal that meets the physical and certainty criteria established by the Bureau of Mines. It is not coal available at today's level of cost.

STRIPPABLE RESERVE BASE

The total Reserve Base is calculated for all coal meeting the requirements of thickness, depth, and the limitation on certainty category. The total is then subdivided into coal that is exploitable by strip mining methods and coal that is available by deep mining methods. This subdivision does not affect the total Reserve Base, only the strip/deep breakdown. What is not strippable is left in the underground Reserve Base. There are two criteria for a strip reserve. The first (with local exceptions) is that the coal lies within 120 feet of the surface.[c] The

c. In the West the limit is 200 feet. In Illinois the limit is 150 feet. These local exceptions occur when mining has already gone to these depths. See Ref. 3 of Chapter 17.

second is that the overburden ratio, the ratio of overburden to coal removed, usually expressed as feet of overburden per foot of coal seam, be no greater than a specified value. Clearly the more material that must be removed, the more expensive the coal. The specified overburden ratios are presented in Table 18-4.

The rationale for the first criterion is that strip mining machinery cannot go below that depth. This fact is changing as larger machines are built; in some areas stripping is proceeding to 200 feet. Thus, more coal is potentially strippable and will become available as larger machines are built. The distinction between strip and deep coal has meaning only in light of present technology. As technology changes so will the depth at which deep-mining methods are considered needed. Nevertheless, for a given technology the cutoff depth does have meaning. A discontinuous jump in costs of strip mining beyond a given depth is captured by the cutoff. The physical limit on depth is only one of the stated criteria for strippable coal. In addition to this technological distinction, there is an apparent eco-

Table 18-4.   U.S. Bureau of Mines Criterion for Maximum Depths and Stripping Ratios for the Coal Reserve Base in Different States.

State	Maximum Depth, Feet	Stripping Ratio
Maryland, Ohio, Pennsylvania, West Virginia, Virginia, Kansas, Missouri, and Oklahoma	120	15/1
Alabama	120	24/1
Illinois	150	18/1
Indiana	90	20/1
Eastern Kentucky	120	14/1
Western Kentucky	150	18/1
Michigan	100	20/1
Tennessee	120	19/1
Alaska	120	n.a.
Arizona	130	8/1
Arkansas bituminous	60	30/1
Arkansas lignite	100	30/1
California	100	10/1
Colorado	50-120[a]	4/1-10/1
Iowa	120	18/1
Montana	60-125	2/1-8/1
New Mexico	60-90	8/1-12/1
North Dakota	50-75	3/1-12/1
Oregon	40	4.75/1
South Dakota	100	12/1
Texas	90	15/1
Utah	39-150	3/1-8/1
Washington	100-250	10/1
Wyoming	60-200	1.5/1-10/1

a. Where a range is shown for states, different cutoffs are applied among coal fields.

Source: U.S. Bureau of Mines, The Reserve Base of U.S. Coals by Sulfur Content, Information Circulars 8680 and 8693 (Washington, D.C.: U.S. Government Printing Office, 1975).

nomic distinction, a condition that, to be considered strippable by the Bureau of Mines, the coal must meet the overburden ratio criterion.

Three aspects should be considered in relation to the overburden ratio criterion: Should an overburden ratio criterion be used? If it were used, would the actual value chosen be a meaningful one? And was any overburden criterion used at all in defining the strippable Reserve Base?

Should the Overburden Ratio Be Used?

Adding an overburden ratio in the Reserve Base confuses the definition of strippable reserves. The cutoff for the overburden ratio is an economic limit that changes as coal prices change. All that should be considered is whether the coal is to be mined by strip techniques or deep techniques. To answer that technological question the depth limitation is adequate. As prices increase, the overburden ratio will increase past the present criterion of profitability.

Deep reserves are calculated as a residual by subtracting strip reserves from the total Reserve Base. Coal with overburden ratios greater than the cutoff ratio will be mined by strip methods as prices increase, although they are part of the residual called deep reserves.

Is the Current Cutoff the Economic Cutoff?

Just as the cutoff criteria for the Reserve Base can be analyzed as a whole, the analyst can ask whether the overburden ratio cutoff represents the marginal cost today. Again, it is a question of whether overburden ratios alone are sufficient to explain cost, and, again, statistical analysis provide the answer.

The Costs of Strip Mining. The bulk of the cost of strip mining is the cost of removing the overburden. Almost 60 per cent of the capital cost is for overburden removal equipment.[11] The more productive that equipment is, the less costly it is to produce coal. Let us now examine how productivity in strip mining is affected by the observed and unobserved characteristics of the deposit. The reserve classification scheme relies upon overburden ratios as the key economic parameter. Again, my earlier study is used to examine how much of the cost is explained by overburden ratios and how much is left unexplained.

Overburden removal is increasingly being accomplished with large draglines or shovels. The largest of these machines has buckets capable of moving up to 220 cubic yards of overburden in a single machine cycle. The size of the overburden equipment is related to the overburden ratio in the following way.

Mining engineers measure the size of dragline in terms of the *maximum usefulness factor* (MUF).[12] This measure of now much work the dragline is capable of performing is the product of the bucket size and the reach of the dragline, as

Figure 18-3. Walking Dragline.

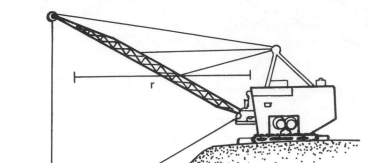

Source: G. Boulter, "Cyclical Methods – Draglines and Clamshells," in *Surface Mining*, ed. E.P. Pfleider (New York: American Institute of Mining, Metallurgical and Petroleum Engineers, 1968), p. 448.

shown in Figure 18-3. The size of the dragline is related to the amount of over-burden that must be removed:

$$MUF = A\,(Ov)^{\alpha}\,N^{\beta}\;,\tag{18-9}$$

where MUF = the total capacity of overburden shovels and draglines; A, a, β = constants; N = the number of machines; η = a disturbance term; and Ov = cubic yards of overburden removed per year.

The justification for this functional form is presented in detail in Zimmerman 1978. Here, a brief explanation of the role of each of these variables should suffice. Variable N is the number of machines being used. Two machines of equal size have the same theoretical capacity as one machine twice as large, but actual capacity could be different. For example, draglines that are larger and have longer reach need to move less often. To capture these effects N is included as an explanatory variable. The disturbance term η reflects factors not observed. Among these are the soil stability and the swell factor, which is a measure of the increase in the volume of the overburden when it is removed. In essence, the procedure is analogous to that for deep mining. Equation (18-9) explains the production of a unit, where a unit is defined as an MUF-unit. The productivity of the unit depends on observed and unobserved characteristics.

Where does the overburden ratio enter? It enters in the variable Ov, the cubic yards of overburden removed per year. The amount of overburden moved can be written as follows:

$$Ov = Z \cdot R \cdot Q , \qquad (18-10)$$

where $Z = \text{constant}^d = 0.89$; R = overburden ratio in feet of overburden to feet of coal; and Q = annual output rate. Upon substitution of the expression for Ov, the equation for MUF becomes

$$MUF = (AZ^{\alpha})(RQ)^{\alpha} N^{\beta} \eta . \qquad (18-11)$$

The resulting estimates are

$$\log MUF = -.446684 \quad + \quad .612306(\log RQ) + .506967 (\log N) \qquad (18-12)$$

(Std. error)	(2.31895)	(.145663)	(.229134)
(t-statistic)	(-.192623)	(4.20357)	(2.21254)

(F-statistic (2/7)   = 21.4089

        SER   = .287762.

The important result in Eq. (18-12) is the standard error of the regression (SER). When the result was tested with a widely different data set, a similar result was obtained.[13] Again there is dispersion, as measured by the standard error, although much less than in the case of deep mining. It would appear that in surface mining the overburden ratio captures more of the variation in cost than does seam thickness for deep mining.

Equation (18-12) also forms the basis for estimating a cost function. Combining the equation with estimates of how expenditures vary with MUF produces an average cost equation.[14] Equation (18-12) implies continuous economies of scale. The coefficient of $\log(RQ)$ indicates that doubling the amount of overburden removed only increases the necessary dragline size by 60 percent. The coefficient of $\log N$ indicates that one machine is more efficient than two.

---

d. The value 0.89 results from the following calculation:
(1) Total cubic footage of overburden = (feet of overburden × acres mined × 43,560)/27, where 43,560 = square feet per acres; and 27 = cubic feet per cubic yard. (2) Acres mined = Q/Th × 1,880 where Q = annual output; Th = thickness of seam in feet; 1,800 = tons per acre-foot. Substituting (2) into (1) yields:

$$\text{Total cubic footage of overburden} = \frac{\text{feet of overburden} \times (Q/Th \times 1,800) \times 43}{27}$$

$$= 0.89 \; QR.$$

These results are consistent with the observed increasing size of machines. Theoretically the limit to these economies of scale has not yet been reached. As a practical matter, at any time there are barriers to the exploitation of these economies of scale. These barriers can be technical (such as limits on the ability to transport larger draglines) or they can be topographic. Large draglines cannot be moved easily and are unstable in hilly terrain. These limits on dragline size were estimated using observations on new mines.[15] Values of RQ were calculated for new mines in various regions. Setting $N = 1$, the geometric mean of RQ was substituted into Eq. (18-12), to solve for the binding MUF value in each region. The result MUF value was combined with the average cost equation to yield a set of points corresponding to minimum average cost AC^*. The results differ by region. Costs are lower in the West than in the Midwest and East because there it is possible to take advantage of greater economies of scale in the large, thick-seam mines. The equations for AC^* are as follows.

West (Powder River Basin):
$$AC^* = .52R\, \eta^{1.63317} + .96 \quad ; \tag{18-13}$$

Appalachia:
$$AC^* = .82R^{1.66317} + .96 \tag{18-14}$$

Midwest:
$$AC^* = .67R^{1.63317} + .96 \tag{18-15}$$

The Incremental Strip Mine. Estimating the cost of an incremental strip mine is similar to estimating for incremental deep mines. But because data are scarcer, only approximations can be obtained. (The problem is that only a range of overburden and a range of seam thickness are given. Here the ratio of the midpoint of overburden thickness to the midpoint of seam thickness is used in each case.) Again, the goal is to find for the central tendency. Assuming η is distributed log-normally, the geometric mean is again what we want. Table 18-5 presents the

Table 18-5. Incremental Strip Mines.

	Strip Mines	Range of Overburden Ratio (R)	Geometric Mean of R	Ratio of Cost of Cutoff to Incremental R
Midwest (high sulfur)	4	7.5-22	17.0	1.06
Montana-Wyoming (low sulfur)	23	.07-11.5	5.2	.96[a]

a. Variable ratios were used as cutoff criteria, depending on area. Five was the average cutoff for the region.

Source: U.S. Bureau of Mines, *Projects to Expand Fuel Sources in Western States.*

geometric means of overburden ratios of new strip mine by area as well as the cutoff ratio used in delineating the Reserve Base.

It appears that the cutoff ratio is close to the incremental value. The cutoff criterion seems to capture much more closely the economic cutoff for strip mining than for deep mining. There is a caveat, of course. Even if the cutoff criterion is close to the current incremental mine, the criterion alone does not determine cost. Since η accounts for a significant portion of cost, all reserves meeting the cutoff criterion are not of equal cost. Reserves with more favorable conditions but higher overburden ratios have been excluded. Conversely, reserves with low overburden ratios but very unfavorable other conditions have been included. The net direction of the bias is not clear. The dispersion as measured by the standard deviation of η is small, however, and the problem introduced by ignoring η consequently is smaller than for deep mining.

Was the Ratio Actually Applied?

A question remains as to whether the ratio was actually applied in calculating the Reserve Base. Previous researchers suggest that the ratio was not used.[16] Evidence discussed in Appendix D also suggests that it was not. As the appendix details, it appears that the 28-inch thickness criterion and the 100-foot depth criterion were applied independently, and the ratio of overburden to coal thickness was not considered in determining strippable coal.

Summary on Strip Reserves

In summary, the division into strippable coal and deep coal has three dimensions. Should a ratio be applied at all? Does the ratio reflect current economics? And, finally, was the ratio criterion actually applied? The answer to the first question is that such an application has meaning only for current technology and prices. As prices or technology change, the relevant ratio must be changed. The answer to the second question is that the ratio criterion more closely captures th the present economic cutoff than the thickness criterion in deep mining. And finally, as detailed in Appendix D, there is some doubt as to whether the ratio was applied at all.

REFERENCES

1. William C. Peters, *Exploration and Mining Geology* (New York: Wiley and Sons, 1978), p. 540.
2. Compare early estimates of strippable coal resources in Montana (U.S. Bureau of Mines, *Strippable Reserves of Bituminous Coal and Lignite in*

the United States, Information Circular 8531 [Washington, D.C.: U.S. Government Printing Office, 1971] to current estimates (Robert E. Matson and John W. Blume, *Quality and Reserves of Strippable Coal, Selected Deposits, Southeastern Montana*, Bulletin 91 [Butte: Montana Bureau of Mines and Geology, 1974].

3.   American Institute of Mining Engineers, *Mining Engineering Handbook* (Summer 1973).

4.   Martin B. Zimmerman, "Estimating a Policy Model of U.S. Coal Supply," *Materials and Society* 2 (1978): 67–83.

5.   The data were collected by the Mining Enforcement Safety Administration in connection with regulatory activities.

6.   Zimmerman, "Estimating a Policy Model of U.S. Coal Supply."

7.   Ibid., p. 70.

8.   U.S. Bureau of Mines, *Basic Estimated Capital Investment and Operating Costs for Underground Bituminous Coal Mines* (Information Circulars 8632, 8641 (Washington, D.C.: U.S. Government Printing Office, 1974).

9.   *Keystone Coal Industry Manual* (New York, McGraw-Hill, various years).

10.   U.S. Bureau of Mines, *Projects to Expand Fuel Sources in Western States*, Information Circular 8719 (Washington, D.C.: U.S. Government Printing Office, 1976).

11.   U.S. Bureau of Mines, *Cost Analysis of Model Mines for Strip Mining of Coal in the United States*, Information Circular 8535 (Washington, D.C.: U.S. Government Printing Office, 1976).

12.   H. Rumfelt, "Computer Method for Estimating Proper Machinery Mass for Stripping Overburden," *Mininh Engineering* 13 (1961): 480–87.

13.   Seth Woodruff, *Methods of Working Coal Mines*, vol. 3 (Elmsford, N.Y.: Pergamon Press, 1966).

14.   Zimmerman, "Estimating a Policy Model of U.S. Coal Supply," p. 35.

15.   Ibid.

16.   ICF Incorporated, The National Coal Model: Description and Documentation, submitted to Federal Energy Administration, Contract No. C0–05–50198–00, Washington, D.C. 1976. This point was also confirmed by discussion with J. Eyster, now of the U.S. Department of Energy and formerly with ICF Incorporated.

19 WHAT WE NEED TO KNOW

Ignoring the collection of important cost-determining characteristics has led to an incorrect interpretation of reserve data. In effect, it has been assumed by the U.S. Bureau of Mines that seam thickness and depth are all that matters, but the statistical analysis presented in the preceding chapter shows that this is not true. The question now becomes: How can resource data be used to determine future cost increments? What do we need to know?

Turning first to observed characteristics, it is clear that we need more information about the distribution of coal according to seam thickness. The statistical analysis suggests that the seam thickness categories are too broad. The difference in cost between 28 inches and 42 inches, all other things constant, is approximately 60 percent. Seams greater than 42 inches include coal seams with mineable thickness up to 108 inches. The difference in cost between these extremes, all other things constant, is almost 300 percent. We need to know more about the distribution between cutoff points.

The data necessary for a more complete description of the distribution according to seam thickness are not publicly available. The individual observations used to categorize the current Reserve Base must contain the information necessary to do a more complete job. To use those observations to present much finer detail for the United States would be a huge undertaking.[1]

A possible alternative would be to use well-known statistical distributions to approximate the distribution for test areas. This approach has been attempted.[2] Some disaggregated data are available for important coal-producing areas.[3] These data appear to represent skewed distributions and are approximated fairly well by lognormal or displaced lognormal distributions. In some areas similar results have been obtained for distribution of strippable reserves according to overburden ratio.[4] These indications are clearly preliminary, and more work is needed to establish the distribution of tons of coal in the ground according to seam thickness. Questions as to the proper unit for analysis—state, basin, or some other area—remain to be answered.

The distribution according to seam thickness or overburden ratio is not enough. We need to know how "other factors" are distributed. These cannot be ignored. Ultimately, we want to know how coal is distributed in the ground according to the cost of mining. And cost depends on other factors besides seam thickness.

A typical distribution might resemble the solid line in Figure 19-1. The vertical axis in Figure 19-1 measures tons of coal available. The horizontal axis measures the cost at which the coal is available. Thus, point B means that there are T_1 tons at cost C_1. The shape of the distribution depends on the distribution of coal in the ground according to the characteristics of the seams. If all the coal were in a single seam with a constant thickness and ϵ_i, the distribution would be a spike at a given level of cost. All the coal would be available at that given cost. The more variable the distribution of seam thickness and of ϵ_i, the more dispersed will be the distribution; that is, the more spread out it will be. The dotted line is an example of such a dispersed distribution.

The categorization of reserves according to the observable variables ignores other factors. Any attempt to use that data without allowing for the omitted variables would ignore the dispersion in the cost distribution of Figure 19-1 accounted for by the dispersion in other factors. Thus, the distribution might appear as the solid line, whereas the true distribution is more accurately represented by the dotted line. Each distribution implies different rates of cost in-

Figure 19-1.   Two Hypothetical Distributions of Coal According to Mining Costs.

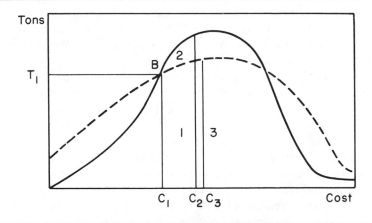

crease. In answer to the question of the cost of mining x million tons, each distribution gives a different answer. In Figure 19-1, assume areas $1 + 2 = x$ million tons. Starting at C_1, the solid distribution says that after x million tons are mined the cost will be C_2. The dotted distribution yields a higher cost for the same tonnage.

One way to rectify the lack of information on other factors is to collect information on mining conditions other than seam thickness. This has been attempted for some small areas.[5] A nationwide inventory is again likely to prove a massive undertaking. Another approach is to treat these factors as done here — as a collective variable — and explicitly incorporate it as a probability distribution. This approach has been attempted, but the results are preliminary.[6] We need to know more about how these other factors are distributed, at least on a regional basis, in order to improve our forecasting of future costs.

INFERRED RESERVES

Inferred reserves are not based on actual observations but are extrapolated from known deposits. The importance of inferred reserves is suggested by the information on strippable coal reserves discussed earlier. Appendix D shows that the strippable Reserve Base differs from strippable resources in that the latter includes inferred deposits whereas the Reserve Base does not. Table 19-1 shows by state the ratio of the strippable Reserve Base to the strippable resource. It indicates that inferred resources constitute a significant amount of coal even in the Eastern United States, where previous mining activity has been heaviest.

Inferred reserves, by definition, are farther away from current mining areas than the Reserve Base and are likely to require more infrastructure investment

Table 19-1. Ratio of Strippable Reserve Base to Strippable Resources for Selected States.

Pennsylvania	.96
Maryland	.97
West Virginia	.46
East Kentucky	.75
Tennessee	.66
Alabama	.24
Ohio	.66
Illinois	.65
Indiana	.61
West Kentucky	.82
Total	.61

Sources: U.S. Bureau of Mines, *The Reserve Base of U.S. Coals by Sulfur Content*, Information Circulars 8680 and 8693 (Washington, D.C.: U.S. Government Printing Office, 1975); and *Strippable Reserves of Bituminous Coal and Lignite in the United States* (Washington, D.C.: U.S. Government Printing Office, 1971).

before they can be exploited. Since mining has been conducted extensively in the Eastern United States, we would expect that inferred reserves have remained inferred because they are, on average, of much higher cost. The distribution in Figure 17-1 provides some evidence to support this, showing that on average inferred reserves are in thinner seams. As the cost of coal rises, these resources would become profitable. The extent to which inferred resources will slow down the rise in cost in the Eastern United States is unknown.

The inferred category is much more important in the West, where large-scale mining is just beginning. The comparison of Table 19-1 does not include the Western states because of a dramatic change of data as a result of drilling between the completion of Ref. 7 and Ref. 8. The already massive amount of Western coal in the Reserve Base at relatively low overburden values makes the inferred category less interesting. Addition of inferred reserves will not dramatically change expectations about future cost increases in the Western states.

TECHNOLOGICAL CHANGE

Everything said about estimating coal reserves and costs thus far has been postulated upon a given technology. We have seen how geology and economics interact. Over time this interaction will be subject to another factor—technological change.

Technological change has an impact upon the interpretation of reserve data in two ways. First, it changes the mean of the distribution of coal according to cost shown in Figure 19-1. Technological progress makes it cheaper to mine under any circumstances, and therefore it shifts the mean down. But technological change can affect the relative penalty associated with thinner seams, for example, thus changing the shape of the distribution. If it becomes possible to produce gas from coal underground, for example, the importance of seam thickness will be greatly reduced. Any seam will be as good as any other, so depletion will be insignificant. In sum, because technology affects forecasts based on reserve data, the economic analyst must be sensitive to in technology changes. The use of reserve data in an economic sense is predicated upon a technology base.

REFERENCES

1.   The government is updating U.S. Geological Survey Bulletin 1450 to provide guidelines for more detailed subdivision of thickness and depth categories. However, recalculation or in some cases initial calculation of Reserve Base according to those new categories is indeed a massive undertaking, likely to take twenty to thirty years to accomplish thoroughly (according to a personal communication of J. Platt with Gordon Wood in 1980).

2.   Martin B. Zimmerman, "Estimating a Policy Model of U.S. Coal Supply," *Materials and Society* 2 (1978): 67–83.

3.   Paul Weir Company, *Economic Study of Coal Reserves in Pike County, Kentucky and Belleville, Illinois*, Job No. 1555, January 1972. J. A. Simon and W. H. Smith, "An Evaluation of Illinois Coal Reserve Estimates," in Proceedings of the Illinois Mining Institute, 1968.

4.   Zimmerman, "Estimating a Policy Model of U.S. Coal."

5.   Paul Weir Company, *Economic Study of Coal Reserves in Pike County Kentucky and Belleville, Illinois.*

6.   Zimmerman, "Estimating a Policy Model of U.S. Coal."

7.   U.S. Bureaa of Mines, *Strippable Reserves of Bituminous Coal and Lignite in the United States*, Information Circular 8531 (Washington, D.C.: U.S. Government Printing Office, 1971).

8.   Robert E. Matson and John W. Blume, *Quality and Reserves of Strippable Coal, Selected Deposits, Southeastern Montana*, Bulletin 91 (Butte: Montana Bureau of Mines and Geology, 1974).

20 RECOVERABILITY

The chapters in this part have treated coal in the ground as if it were all potentially available at some price. In fact, for many reasons a large proportion of coal is not available. Let us consider some of the factors that limit recoverability.

Recoverability refers to the percentage of coal in a deposit that is actually extracted. The percentage varies from mine to mine due to geologic, economic, and legal restrictions. Recoverability is also used in an additional sense. Land-use patterns create limitations on mining. In certain places, as under railroads, towns, and other sites, mining cannot take place. In addition to making some areas totally inaccessible, land-use patterns render some areas uneconomical. The most often cited example is that because of economies of scale in mining, a minimum parcel of reserves is necessary. By interrupting the continuity of reserves land-use patterns thus render much coal economically unavailable. It would take very large increases in the cost of coal to make very small-scale mining attractive.

The U.S. Bureau of Mines has assumed in its studies of reserves that average recovery rates, reflecting all of these influences, are 50 percent for underground mining and 80 percent for strip mining. The lower recovery for deep mining reflects the necessity of leaving pillars of coal underground to support the roof of an underground mine.

RECOVERABILITY OF DEEP RESERVES

There have been a few studies of recoverability in deep mines. But these typically have focused on the mine itself and therefore exclude issues of land use, parcel

Table 20-1.   Unrecovered Coal Recorded in Survey of 200
Underground Mines.

Type	Range	Average Losses for the 200 Mines
Unavailable coal	0.1–26.4	1.8
Oil and gas wells	0.1– 8.0	0.3
Property boundaries	0.2– 6.9	0.7
Surface features	0.1–15.4	0.6
Other reservations	0.2–11.5	0.2
Economic-technologic losses	0.1–24.0	10.5
Bad top, thin, wants, geologic	0.4–23.2	4.4
Haulageway and miscellaneous	0.4–23.2	4.4
Top coal	3.7–15.9	1.5
Bottom coal	2.0–10.4	0.2
Unmeasured losses	1.6–60.7	30.7
Total losses	8.6–70.8	43.0

Source: Raymond Lowrie, *Recovery Percentage of Bituminous Coal Deposits in the United States*, Part 1, Underground Mines, U.S. Bureau of Mines Report of Investigation 7109, Washington, D.C. 1968.

size, and so on. The most comprehensive study involved 200 large deep mines.[1] The results are presented in Table 20-1. The average recovery rate was 57 percent, with the 95 percent confidence interval being 55.3–58.7 percent. The table is interesting for what it does not say.

Unmeasured losses, for which it could not be determined why the coal was not removed, were 75 percent of total lost coal. Economic and technologic reasons accounted for roughly 25 percent of the total losses; such losses would be reduced as prices rise.

The last category—unavailable because of land use, legal restrictions, and so on—was only a small percentage. It surely has a downward bias, however. The study examined existing mines, but legal, institutional, and land-use prohibitions against mining would undoubtedly prohibit opening mines in many areas. A loss figure on an areal basis rather than on a mine basis therefore would undoubtedly be much larger.

Examining existing mines to estimate recoverability has other biases. Seams lying below exhausted seams often are not recoverable. Existing mines obviously do not provide a sample of these. In sum, a recovery figure below the average of 57 percent is justified. The figure of 50 percent is probably not a bad estimate of overall recoverability, but there is no way of judging. Furthermore, the economic-technologic losses would diminish as price rise. Evidence for this is provided by a more recent survey of recoverability in underground coal mines.[2] That study shows average recovery factors higher for metallurgical than for steam coal mines, indicating that when coal sells at a premium, economic–technological losses can be overcome.

Finally, technological developments change recovery ratios. The more recent U.S. Bureau of Mines study, while not entirely comparable in coverage and in classification to the earlier study, indicates an average recovery figure above 60 percent, depending upon mining technique.

RECOVERABILITY OF STRIP RESERVES

The situation with strip reserves is even worse. Surface interference with mining is apt to be great. Areas under railroads, towns, and highways are not minable. Furthermore, in many areas land-use patterns make it difficult to assemble parcels large enough to take advantage of the large economies of scale in strip mining. The evidence on the latter point is limited, but many participants in the coal market appear to share this attitude. The problem is primarily a Midwestern one. In the Eastern fields because of the hilly topography, strip mining is done on a smaller scale. In the West, mining takes place in vast, unpopulated areas.

REFERENCES

1. Raymond Lowrie, *Recovery Percentage of Bituminous Coal Deposits in the United States*, Part 1, Underground Mines. U.S. Bureau of Mines Report of Investigations 7109, Washington, D.C.: U.S. Government Printing Office, 1968.
2. Robert G. Reese, Binod B. Dash, and Patrick Hamilton, *Coal Recovery from Underground Bituminous Coal Mines in the United States, by Mining Method.* U.S. Bureau of Mines Information Circular 8785, Washington, D.C.: U.S. Government Printing Office, 1978.

SUMMARY

Reserve and resource data reveal that there is a great deal of coal in the ground. Even the narrowest reserve classification is broader than anyone intends it to be. Needed information is lacking. Attempts to discover the missing information to fill the gaps have yielded some tentative conclusions, but more information is needed before more confident predictions can be made. More data are needed on unobserved characteristics and on the distribution of seam thicknesses.

If a particular form for the distribution of ϵ, is assumed the data at hand can be used to establish approximate upper limits on cost increases. Present estimates are for coal in seams at least 28 inches thick. Because the Reserve Base excludes inferred reserves, it is a conservative estimate of all coal in 28-inch or thicker seams. The estimated cost functions reported here must be supplemented by information on other factors, represented by ϵ.

Assuming c is distributed lognormally (the large number of factors that enter ϵ and the multiplicative fashion in which they interact suggest this distribution) and assuming further that ϵ is independent of thickness, we can bound the likely cost increase as output cumulates, as done in Table IV-1.

The incremental mine seam thickness is reproduced in column (1) of Table IV-1. Column (2) presents the ratio of cost of a mine in a 28-inch seam with $\epsilon = 1$ to the cost of the incremental mine. Recall that 28 inches is the cutoff seam thickness for the Reserve Base. Given that the mean of log ϵ is zero and that log ϵ is normally distributed, 50 percent of the total Reserve Base must have a value of $\epsilon > 1$. That means that at least 50 percent of the Reserve Base has a cost less than that of a mine in 28-inch seam with $\epsilon = 1$. In other words, at least 50 percent of the Reserve Base is available at less than the ratio of the cutoff cost to current cost shown in column (2). In the last column this reserve total is expressed as a multiple of 1975 rates of output.

Table IV-1.   Limit on Increase in Cost due to Depletion
in the Reserve Base.

Region and Type of Coal	(1) Incremental Seam Thickness	(2) Ratio of Cost of Incremental Mine to Cost of Mine in 28-inch Seam, $\epsilon = 1$	(3) Reserve Base as a Multiple of 1975 Output Rates
Northern Appalachia high sulfur (1.5 percent)	56.8	2.19	140
Southern Appalachia low sulfur (.84 percent)	44.1	1.65	40
Medium sulfur (0-1.5 percent)	49.6	1.89	85
High sulfur (0-1.5 percent)	53.2	2.04	150
Midwest (high sulfur)	74.1	2.94	220
Montana-Wyoming[a] low sulfur (.84 percent) strip reserves		1.0	853

a. For Montana and Wyoming, only strip reserves were considered because of the much higher cost of deep mining there.

Sources: Columns (1) and (2) from Table 18-3; Column (3) is the demonstrated reserve base of the U.S. Bureau of Mines, *The Reserve Base of U.S. Coals by Sulfur Content*, Information Circulars 8680 and 8693 (Washington, D.C.: U.S. Government Printing Office, 1975).

Clearly, it will take a long time for costs to double solely due to depletion at approximately current rates of output. In some segments of the supply function, however, the increase of costs will be rapid. As the rate of output is expanded beyond 1975 levels while low-sulfur coal is substituted for high-sulfur coal, the increase in low-sulfur coal costs could be quite large in southern Appalachia. Of course, Western low-sulfur coal is in very elastic supply and that will set a limit to the cost increase for low-sulfur coals. As Eastern prices rise, consumers will turn to Western coal.

In summary, an upper limit on cost increases by category of coal has been established. But it is a high upper limit and it depends on key assumptions. To know more about the trajectory of costs as that high level of cost is approached, we need to know more about coal resources, both the distribution according to seam thickness and the distribution according to other mining conditions. One last caveat is in order. Reserve data tell us only about depletion. Cost patterns over time will be affected and most likely dominated by factor price behavior and technological change.

V URANIUM Estimates of U.S. Reserves and Resources

John C. Houghton

INTRODUCTION

Although estimates of uranium resources have not changed as drastically as those for oil, interest in domestic uranium resources increased sharply during the 1970s. The National Uranium Resource Evaluation (NURE) program was established in 1974 to develop estimates of uranium resources in the United States. The program, under the direction of the Grand Junction Office (GJO) of the U.S. Department of Energy, included a great expansion of information for resource evaluation and a concerted effort to improve resource estimation methodologies.[a] And outside GJO the Committee on Mineral Resources and Environment (COMRATE), Battelle, the Committee on Nuclear and Alternative Energy Systems (CONAES), the Nuclear Energy Policy Study Group (the MITRE Corporation for the Ford Foundation), Stoller, and others have all added their interpretations to resource estimates.[1] Interest has increased because resolution of scheduling of breeders, recycling of plutonium, and other crucial issues hinge on estimates of supply. Other issues, such as the opening of native American and public lands, tax policies for extractive industries, and environmental requirements, can be examined in the light of their effect on supply.

The Grand Junction Office's estimates of domestic reserves and resources influence all other estimates. The GJO publishes the only aggregate reserve estimates that are widely accepted as accurate approximations of the true reserve

---

a. At the time of publication, the NURE program was being sharply curtailed due to the relatively low price and demand for uranium and overall stringent fiscal goals of the Reagan administration.

333

base. Furthermore, other experts usually discuss domestic uranium supply by interpreting the GJO's estimates. These reserve and potential resource estimates also are used as a standard of comparison for other estimates. The GJO's estimates are respected because they represent careful analysis based on a large volume of disaggregated data, much of it proprietary, and on results of drilling and mining, as well as detailed regional information.

In light of the influence of their estimates, it is critical to examine the GJO's estimation procedure closely. Data generated by GJO are often misinterpreted by others, and in some ways the estimation procedures could be improved. Chapter 21 discusses the estimates, and Chapter 22 is devoted to how others interpret the GJO statistics. The GJO's method of estimating resources may be categorized as geological analogy, although it is becoming an amalgam of various procedures.

The remaining chapters describe various other methods. For each method, one application or more is presented as a case study. Chapter 23 describes what can be called trend projection models. Such models estimate uranium supply by extrapolating trends; included are econometric models as well as discovery rate and life-cycle models. Trend projection models are best applied to short-run forecasts.

A very different approach is used in crustal abundance models (Chapter 24). Whereas historical records of discoveries, production, and drilling rates are included directly in trend projection models, crustal abundance models are more concerned with grades, tonnages, and, to some extent, geological processes. The analyses depend heavily on successfully modeling properties of the earth's crust, placing less importance on economic parameters such as price. Abundance models tend to be applied to long-run forecasts.

Chapter 25 presents subjective probability assessment as a particular type of estimation methodology. To be sure, subjective assessment is an integral part of other approaches, such as geologic analogy. But an almost entirely subjective approach, such as that presented in Harris,[2] is quite different from geologic analogy and can overcome difficulties such as exact model formulation and explicit data requirements.

URANIUM GEOLOGY

The readers of this book are probably familiar with the habitats of oil, gas, and coal, but may be less familiar with that of uranium. Hence the following brief discussion: The average crustal abundance of uranium is around three parts per million (3 ppm), which means it is less abundant than nickel or copper but more abundant than tungsten or tin. Geologic processes have concentrated uranium into deposits with average grades several orders of magnitude higher than the crustal average; deposits in the United States typically contain between 500 ppm (0.05 percent) and 2,000 ppm (0.2 percent) uranium.

About 90 percent of the uranium resources in the United States occur in sandstone environments; in contrast, about 85 percent of the rest of the world's uranium resources occur in nonsandstone environments. This fact is the basis of controversy. Some scientists argue that the discrepancy between United States' uranium environments and those in the rest of the world is due largely to differing geology, whereas others argue it is due to the tendency to explore in areas similar to those where the best success has been achieved. This issue will be resolved as the results of domestic exploration in atypical areas are gathered. Understanding the genesis of the deposits is still in an early stage. The state of the art in uranium exploration has been likened to that in oil when oil explorationists were drilling surface expressions of anticlines. The previous stage in uranium exploration, prospecting with a hand-held Geiger counter, is compared to drilling at sites of oil seeps.

Sandstone uranium deposits occur in a variety of shapes and presumably are produced by various means, but most deposition processes have certain main features in common. Uranium is dissolved by water from a source rock with a relatively low concentration or uranium. In the sandstone beds, often interbedded with mudstone, groundwater containing uranium in solution moves down-dip by gravity. The ore forms slowly at shallow depths a few miles from the source rock, where adequate reducing conditions exist. The uranium generally is precipitated in the interstices around each of the sand grains in the sandstone. Two types of deposits are common. Tabular deposits, found in the Colorado Plateau, vary in size but average a few feet thick and as much as hundreds of feet across. They have been described as discrete masses, "like raisins in a loaf of raisin bread."[3] The second type, roll-front ore bodies, found for example in the Powder River Basin of Wyoming, are crescent-shaped in cross section and have a front (facing down-dip) and a back. The convex front side usually has a relatively narrow but gradational boundary extending from ore to protore (protore is mineralized rock with a grade less than the cutoff grade) to the background levels of the unaltered host sandstone. The concave side has a sharp boundary between the altered and barren rock, up-dip, and the ore. This geometry may be complicated by the presence of long, thin, mineralized limbs, themselves sharply bounded, stretching up-dip from the ore. Each roll ore body is a part of a larger frontal tongue of altered sandstone; the deposits can be quite irregular but are described as "widely spaced elongate beads on a string."[4]

Processing uranium for utilization occurs in several steps. Reserve and resource estimates usually incorporate assumptions about the steps preceding the packaging of uranium as yellowcake (U_3O_8), including exploration, mining, and milling. Uranium mining is generally organized in districts, partly because of the geological proximity of the deposits and partly because of the economies of scale in milling and the high cost of transporting ore. A rule of thumb is that 50 miles is the maximum distance ore can be hauled economically. Once the uranium is available as U_3O_8, it is converted to UF_6, then enriched. Enrichment is a nonchemical, energy-intensive process that separates uranium into its different isotopes. This is a particularly sophisticated stage, because the isotopes have identical chemical properties and can be separated only by a minute weight difference in each atom. The fissionable isotope U-235 is enriched from a concentration of 0.7 percent in nature to approximately 3 percent. Finally, enriched uranium is fabricated into fuel elements for a nuclear reactor, and the spent fuel may or may not be reprocessed.

Occasionally some of the process steps are excluded. Some deposits with grade too low to be mined in a conventional manner are leached in situ and the uranium extracted directly from the uranium-rich water. Uranium is also produced as a by-product from other operations, such as phosphate or copper mining. These unconventional methods are included in GJO's estimates, although projections of by-product uranium are tabulated in a class separate from the reserve and resource estimates. Readers who would like more background on the geological and physical aspects of uranium exploration, mining, and milling should consult the works listed in Ref. 5.

REFERENCES

1. Committee on Mineral Resources and Environments (COMRATE), *Reserves and Resources of Uranium in the United States* (Washington, D.C.: National Academy of Sciences, 1975). Battelle–Northwest, *Assessment of Uranium and Thorium Resources in the United States and the Effect of Policy Alternatives* (Richland, Wash.: December 1974). Committee on Nuclear and Alternative Energy Systems (CONAES), *Problems of U.S. Uranium Resources and Supply to the Year 2010* Report of the Uranium Resource Group (Washington, D.C.: The National Research Council, 1978). Nuclear Energy Policy Study Group, *Nuclear Issues and Choices*, MITRE Corporation for the Ford Foundation (Cambridge, Mass.: Ballinger 1977). S.M. Stoller Corporation, *Uranium Data*, EPRI EA-400, and *Uranium Exploration Activities in the United States*, EPRI EA-401, (Palo Alto, California: Electric Power Research Institute, June 1977).

2. D. P. Harris, "A Subjective Probability Appraisal of Metal Endowment of Northern Sonora, Mexico," *Economic Geology* 68 (1973): 222–42.

3. R. P. Fisher, "Exploration Guides to New Uranium Districts and Belts," *Economic Geology* 69 (1974): 364.

4. Ibid.

5. J. D. Hansink, "Supply Factors and Their Implication on the Domestic Uranium Market," M. S. Thesis, Massachusetts Institute of Technology, Cambridge, Mass., June 1977. International Atomic Energy Agency, *Formation of Uranium Ore Deposits* (Vienna: 1974), p. 744. W. I. Finch et al., in *United States Mineral Resources*, ed., D. A. Brobst and W. P. Pratt, U.S. Geological Survey Professional Paper 820 (Washington, D.C.: U.S. Government Printing Office), pp. 456–68.

21 GRAND JUNCTION OFFICE METHODOLOGY

HISTORICAL PERSPECTIVE

Much of the way the Grand Junction Office views uranium supply can be explained by examining the procurement activities of its predecessor, the Grand Junction Operations Office of the U.S. Atomic Energy Commission (AEC), since World War II.[1] The AEC was a direct offspring of the wartime effort to develop the atom bomb. While most of the uranium for the first two atom bombs was mined in what was then the Belgian Congo, the United States government was secretly evaluating uranium resources in the Southwest, under the name Union Mines Development Corporation, a subsidiary of the Manhattan Project. The long-standing close relationship between the AEC and the uranium industry began at this time. From 1948 through the late 1950s the U.S. government provided numerous incentives to encourage prospectors and mining companies.

In 1958 the AEC withdrew its guaranteed price schedule, replacing it with negotiated price agreements based on forward costs (a rough measure of costs, to be defined shortly). Only ore reserves defined by that date were eligible for government purchase. This began in earnest the AEC's practice of asking that mining companies supply complete information on production statistics and industry reserves.[a] On the basis of the industry's raw data, the AEC then made independent estimates of tonnage and forward cost. The acquisition program

a. Production and reserves data acquisition had begun in 1947, however, when the AEC started purchasing uranium ore and concentrate. Upon completion of the AEC's purchasing program, the uranium companies continued to provide data voluntarily.

was highly successful, so by the late 1950s the uranium industry was booming and supplies of uranium were adequate for defense needs. The AEC extended incentives from 1962 through 1968, allowing deferred incentives during the so-called stretch-out plan in order to fill the gap until the creation of the non-military nuclear power market for uranium. The cutoff date for the establishment of reserves eligible for AEC purchase after 1962 was November 1958. But incentives were kept minimal. The AEC expressly did not want to pay for exploration leading to discoveries, and it wanted to compensate mining companies only for costs needed to produce the delineated ore. It was clearly in the interest of the uranium industry to provide as much positive information as possible to increase the size of the ore bodies.

The GJO's current concept of forward cost and distinction between reserves and potential resources stem from this period. Uranium reserves were defined by GJO as ore that had been delineated by development drilling. Forward cost was the expense of producing a given uranium deposit, not including investments made before the analysis. The AEC also began using geologic analogy as a method of estimating potential resources at this time: Geologists would study unexplored regions that were geologically similar to producing regions. An empirical comparison of the geological attributes of the unexplored and of the producing areas was the basis for judgments of the likelihood of finding uranium. Production capability analysis also had its beginnings in the procurement program.

ORE RESERVE ESTIMATION

Reserves are made up of known deposits that can be recovered at costs equal to or less than the selected forward cost category. This means that a deposit is identified with some certainty; its grade and physical shape are usually delineated by developmental drilling. Tonnages and grades are calculated assuming mining dilution and recovery levels, but they are not adjusted for mill recovery. Uranium produced as a by-product is listed as a separate category from reserves and resources. A resource, on the other hand, is inferred by some process that gives an expectation of ore occurrences. The conversion of potential resources into reserves constitutes the exploration and development phase. Two additional terms are used to quantify uranium in deposits with grade greater than 0.01 percent $U_3 O_8$, irrespective of economic cost. *Endowment* refers to undiscovered deposits, of which potential resources are a subset, and *inventory* refers to discovered deposits and includes reserves.

Once a deposit has been delineated by development drilling or mine workings, all previous information, including exploration drilling, local geology, and so on, is used to estimate the ore reserves. The GJO acquires gamma-ray logs, chemical assays, maps, geologic information, and engineering and geophysical data from uranium companies. The bulk of these usable data is gamma-ray logs from development and exploration drill holes. These logs are usually an analog readout of a small detection probe (similar to a Geiger counter) lowered down the hole. The continuous records are digitized and converted by a computer program to an equivalent grade of U_3O_8 for each 6-inch or 1-foot interval for each of the holes that have been drilled in the uranium deposit. Research is being directed toward improving the accuracy of both the measurement and the conversion, especially for low-grade material.

Reserves are estimated for each forward cost by developing a sample tonnage/grade relation. Because the grade is sampled only in drill holes or mine exposures, the average grade for the entire reserve is estimated from a statistical analysis of the drill hole grade distribution data. A mining method is predicated and related costs are estimated, thereby transforming the geologic reserves into economic reserves. The choice of the optimal mining method and shape of the mine requires evaluation by personnel with much experience.

The GJO analyzes the tonnage/grade relation in basically the same way a mining company would analyze its own mine, except that private industry spends greater effort to make the analysis more precise. Typically, GJO estimates reserves for an ore body as follows: A minimum thickness (cutoff thickness), which is based on the projected mining method, and a minimum grade (cutoff grade), based on projected costs are determined at each cost category. For each drill hole, the 6-inch or 1-foot interpreted thicknesses and grades are combined into intervals that meet the cutoff thickness and grade for each cost category. If the average grade of the cutoff thickness interval is below the minimum grade, the hole is excluded from the estimate. If the cutoff thickness and grade criteria are met, other 6-inch or 1-foot increments are added to the vertical interval to include as much material as possible, so long as the grade of the marginal increments equal or exceed the cutoff grade.

An example of this analysis is shown in Figure 21-1. Solid circles denote holes with intercepts exceeding cutoffs for grade and thickness, while holes with less than these cutoffs are shown as open circles. The GJO staff contours the solid circles to determine the area of the deposit. Total tonnage and average grade for the deposit are then estimated using one or more methods. One such method is to average the grades of all the vertical intervals, weighted by their thicknesses. Another method, commonly used for underground mines, is to fit the distribution of grades, within a minimum thickness that approximates the average mining height (usually 6 to 8 feet), with a three-parameter lognormal distribution. The fitted distribution allows reserves to be determined at all cutoff grades, and a table showing the grade-tonnage distribution is provided. Several methods may be tried and compared for the best fit for any individual deposit. The process is labor-intensive, for although computer programs are used, many adjustments must be made by hand, incorporating geologic and data other than drill hole information provided by the industry. New techniques using geostatistics will be used increasingly in the future.

Figure 21-1.   Plan View of Drill Holes in Typical Uranium Mine.

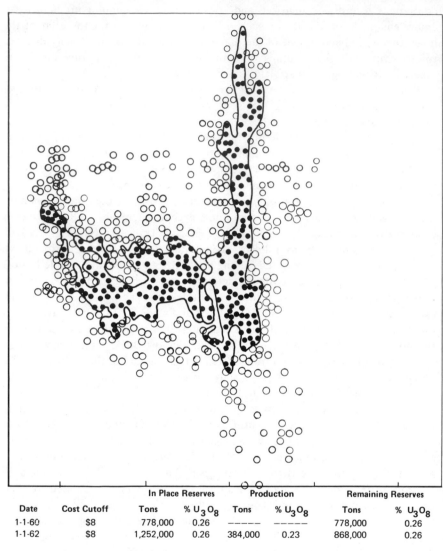

		In Place Reserves		Production		Remaining Reserves	
Date	Cost Cutoff	Tons	% U_3O_8	Tons	% U_3O_8	Tons	% U_3O_8
1-1-60	$8	778,000	0.26	-----	-----	778,000	0.26
1-1-62	$8	1,252,000	0.26	384,000	0.23	868,000	0.26

● Ore-Grade Uranium Drill Hole @ $8 Cost Cutoff
○ Low Grade Uranium Drill Hole

Source: R.J. Meehan, "Ore Reserves," in *Uranium Industry Seminar*, GJO–108(77), Grand Junction Office, U.S. Department of Energy, October 1977.

For any single deposit, the tonnage, average grade, and cutoff grade are calculated for each forward-cost category. Private companies often compute average grade and tonnage for only one cutoff grade, which is determined by the price of their contracts for $U_3 O_8$. The GJO compares the single-point industry estimate with one of its own estimates based on comparable cutoff grades or average grades. If the two numbers do not agree, then differences are reconciled during periodic visits to company offices.

POTENTIAL RESOURCE ESTIMATION

The GJO estimates potential uranium resources on the basis of geologic analogy. Estimates of potential resources are made for an area under evaluation by comparing its attributes with a geologically similar control area. These individual estimates are combined to provide estimates of the total domestic potential uranium resource.

It is here that some misinterpretation of GJO estimates occurs. The potential resource estimates include the amount of uranium available in areas that contain enough information to be evaluated given present knowledge and technology. Resource estimates are introduced as "domestic potential resources."[2] According to the GJO, "The results [of the estimation methodology] are believed to reasonably express the magnitude of potential resources based on the existing data base."[3] The GJO fully expects resource estimates to change, however, as data are collected and knowledge of new uranium provinces grows. At this point it is important to understand what these domestic resource estimates are *not*. Estimates are not attempted for areas with some indications of favorability, but where GJO believes it has insufficient data for estimation of resources. Thus the resources that such areas may contain will be part of a future resource base when adequate additional data are available.

Categorizing Resources

In Part I of this book one of the most basic notions in categorizing resources for any mineral was presented. There are at least two orthogonal axes for subdivisions.[b] The vertical axis measures the cost of extraction; the horizontal measures relative assurance or confidence in the estimate. For example, ore occurrences may be well defined but expensive to extract, or they may be highly speculative but expected to be of very high grade. The horizontal axis divides the estimates along a scale of decreasing levels of knowledge of total endowment. The GJO

---

b. Resource terminology is discussed in much greater detail in Schanz,[4] and the reader might refer to Figure 2-1 in Part I for another look at the McKelvey diagram.

uses forward cost categories to measure the cost of extraction and reserve and potential resource definitions for the assurance or confidence dimension.[c] Potential resources are further divided into three subcategories in order of decreasing knowledge: probable, possible, and speculative. The standard GJO definitions of these three categories are as follows:

> *"Probable" potential resources* are those estimated to occur in known productive[d] uranium districts: (1) in extensions of known deposits, or (2) in discovered deposits within known geologic trends or areas of mineralization. *"Possible" potential resources* are those estimated to occur in undiscovered or partly defined deposits in formations or geologic settings productive elsewhere within the same geologic province. *"Speculative" potential resources* are those estimated to occur in undiscovered or partly defined deposits: (1) in formations or geologic settings not previously productive within a productive geologic province, or (2) within a geologic province not previously productive.[5]

Efforts to define more precisely the levels of reliability associated with these categories are discussed in Chapter 22, on interpretations of GJO estimates.

Estimation of Potential Resources

Estimates of potential resources[e] are calculated with a "conditional estimating equation"; subjective assessment is used to compare attributes between a control area and an evaluation area. This relatively new method is expected to be an improvement over previous methods for three major reasons. First, the measurement of resources in both the control area and the evaluation area is based on geological and physical attributes, such as number, size, and grade of deposits. Second, the variables in the conditional estimating equation are expected to be less correlated. And third, the resultant estimates are presented as distributions rather than as point estimates, which allows calculation of confidence limits for each of the evaluation areas.

The first part of the assessment is an estimate of the probability that any deposits exist in the evaluation area. The second part estimates potential as a product of four factors:

c. As will be explained later, the GJO system of subdividing potential resources is only a rough surrogate for assurance or confidence. The categories reflect divisions of *where* resources will be discovered, with the more frontier regions contributing to the more speculative categories.

d. *Productive* districts are those where past production plus known reserves exceeds 10 tons U_3O_8.

e. Property potential, a subset of probable potential, which consists basically of inferred extensions of deposits beyond the limits of reserves, is calculated in conjunction with reserve estimates.

A = projected surface area of favorable area in square miles,

F = fraction of A underlain by uranium-bearing rock, i.e., endowment,

T = tons of uranium-bearing rock per unit area within the fractional area of $A \times F$, and

G = average grade of endowment in decimal fraction form.

A deposit is defined as "a discrete lens or a cluster of smaller lenses of uranium-bearing rock containing at least 10 tons $U_3 O_8$ at a cutoff grade of 0.01 percent $U_3 O_8$."[6] The field geologist assigns modal and other fractile values for F, T, and G, conditional on there being at least one deposit. The fractiles and mode are used to define truncated, unimodal distributions of F, T, and G. (The distributions are defined between zero and a large, finite upper bound.) The product distribution for endowment, $A \times F \times T \times G$, is computed and then adjusted by the probability of at least one deposit. Correlation among variables and among regions is included in the following way. The GJO assumes that the estimates of A, F, T, and G are independent. Estimates of potential for different evaluation areas that use the same control area are assumed perfectly correlated, and areas based on different control areas are assumed to be independent. This results in a "moderately correlated" effect that is closer to being independent than to being perfectly correlated. The resulting tons $U_3 O_8$ are assigned to forward cost categories based on statistical relations between average grade, cutoff grade, and forward cost developed for the control region. Results of this process are shown in Figure 21-2. Distributions are presented for each resource category (probable, possible, etc.) and for each cost ($30 forward cost, etc.).

In current research,[7] Harris investigates alternative methods that allow experts to manipulate not only the subjective probability, but also the structure of compound guesses. The PROSPECTOR model, by SRI International,[8] propagates its subjective scores in a multiplicative manner using Bayes' Rule (discussion of which can be found in any elementary statistics text).

In common with most other methods, there is an underlying bias in the GJO methodology that tends to underestimate the true resource base. Projections can only be made from what is known. As technology increases and knowledge of new geological regimes grows, so do the estimates. The GJO is quite aware of this phenomenon. In the Southern Powder River Basin case study, the total estimated low-cost resource base "increased about twenty-fold as a result of acquisition of new knowledge of the region between the late 1950s and 1976."[9] A large increase of this kind is generally not the result of poor estimation, but rather should be interpreted as a caveat: Expect similar upward and downward readjustment of resource estimates in various other regions in the future. Most improvements in exploration or development techniques, or the discovery of new uranium host environments, can be expected to increase the resource base; however, other factors, particularly higher costs, lower it, as reflected by recent downward trends in estimates for the lower cost categories.

Figure 21-2. Cumulative Probability Distributions of Potential
Resources of the United States by Cost Category and Resource Class.

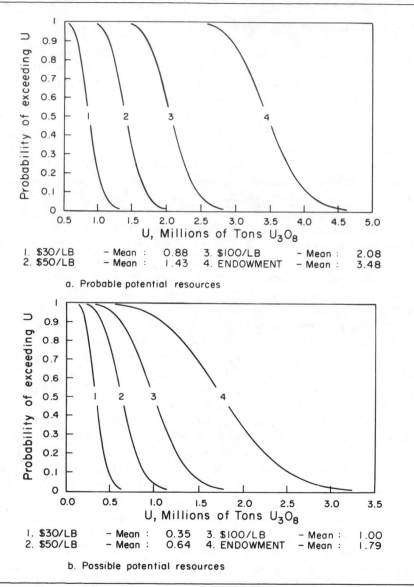

1. $30/LB   – Mean :   0.88   3. $100/LB   – Mean :   2.08
2. $50/LB   – Mean :   1.43   4. ENDOWMENT   – Mean :   3.48

a. Probable potential resources

1. $30/LB   – Mean :   0.35   3. $100/LB   – Mean :   1.00
2. $50/LB   – Mean :   0.64   4. ENDOWMENT   – Mean :   1.79

b. Possible potential resources

Source: An Assessment Report on Uranium in the United States of America, GJO–111(80), Grand Junction Office, U.S. Department of Energy, 1980, p. 18, 19.

Figure 21-2. continued

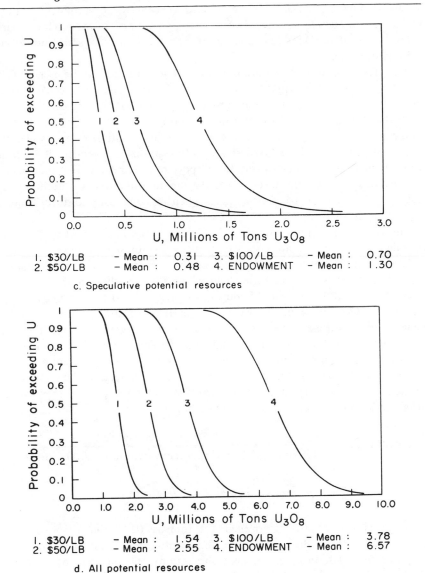

1. $30/LB	– Mean :	0.31	3. $100/LB	– Mean :	0.70
2. $50/LB	– Mean :	0.48	4. ENDOWMENT	– Mean :	1.30

c. Speculative potential resources

1. $30/LB	– Mean :	1.54	3. $100/LB	– Mean :	3.78
2. $50/LB	– Mean :	2.55	4. ENDOWMENT	– Mean :	6.57

d. All potential resources

FORWARD COST METHOD

Uranium is unique among minerals in that the reserve and resource estimates are presented by a particular measure of cost (except for by-product uranium, which is not presented by cost). This measure, *forward cost*, was developed early in the period of the AEC's purchasing arrangements with the private producers, and it is still being used.

The presentation of forward cost causes some confusion to the users of GJO data. Some users misinterpret forward cost as an approximation of the economic cost to a producer to exploit uranium, although the GJO is very careful not to claim that forward cost measures economic cost. A transformation by the GJO of the physical properties of the reserve and resource estimates into a measure of cost of production is highly desirable, although this is very difficult given the form of the data, as will be shown. It is important to remember that the concept of forward cost was introduced when the uranium industry was declining and had excess capacity. It has shortcomings when applied mostly to steady-state or expanding industry.

The GJO explanation of forward cost is as follows: In the short run, a period considerably less than the life of a particular mine, most decisions are based on variable costs. These short-run variable costs are labeled operation and maintenance (O&M) costs. For each truckload of ore taken to the mill, the benefits of the yellowcake produced should equal or outweigh the costs of mining and milling the ore. Ore with a grade lower than a particular cutoff grade should not be mined. The lowest grade ore that is mined, the cutoff grade, is set at the value where the marginal benefits equal marginal costs. These O&M costs include costs of labor, supplies, power, royalties, and taxes for mining, hauling, and processing the ore. The cutoff grade is calculated from a formula relating the variable cost to the forward cost. If, for example, the O&M cost is $38 per ton, then the cutoff grade for $30 (forward cost) ore would be:

$$\text{Cutoff grade} = \frac{\text{O\&M cost \$/ton} \times 100 \ (\%)}{\text{forward cost \$/lb} \times \text{mill recovery rate} \times 2{,}000 \ \text{lb/ton}}$$

$$= \frac{38 \times 100}{30 \times .95 \times 2{,}000} = 0.07 \text{ percent} \ . \tag{21-2}$$

Once the cutoff grade is determined, the tonnage (total tons of ore available with a grade greater than or equal to the cutoff grade) is calculated according to procedures described in the next subsection. The quantity of pounds of U_3O_8 in that tonnage of ore, however, is not necessarily economic to mine at the expressed forward cost. A determination must be made as to whether or not the mine will be profitable. The mine is profitable only if the average grade is sufficiently greater than the cutoff grade. The mining company must make capital expenditures to produce the uranium. If the mine is already functioning and the reserves are an estimate of ore to be mined in the future, then the costs might include sinking more shafts or driving more entries. If the mine is relatively new, however, it may be necessary to build a mill. The term *forward cost* is therefore used in uranium supply analysis to denote all expected future costs associated with production from the time of analysis.

Future capital costs are divided by tonnage to give average capital costs per ton, which are then added to the O&M costs per ton. The average grade, not the cutoff grade, is used to judge whether the deposit is economic at the given forward cost, in which case the following condition must be fulfilled:

$$\text{Forward Cost \$/lb} > \frac{\text{O\&M cost \$/ton + average capital cost \$/ton}}{\text{average grade lb/ton} \times \text{recovery rate}} . \quad (21\text{-}3)$$

Continuing the example, the average grade for a cutoff grade of 0.07 percent might be 0.12 percent (2.4 lb/ton). The average capital costs per ton might be $10/ton. Since

$$30 \text{ \$/lb} > \frac{38 \text{ \$/ton + 10 \$/ton}}{2.4 \text{ lb/ton} \times .95 \text{ recovery rate}} , \quad (21\text{-}4)$$

the tonnage associated with a cutoff grade of 0.07 percent would be included in the $30 category. If the average capital costs per ton had been $40 per ton, however, no reserves or resources would be included from this deposit in the $30 category.

Note that two identical ore bodies, one 10 miles and the other 100 miles from a mill, would certainly have different forward costs, and that two identical ore bodies might or might not have different forward costs if only one had an entry in it. In the first case, the variable costs would be different, and in the second case the future capital costs would be different.

Forward cost applied to potential resources is handled in the same fashion as reserves. The only difference with potential resources is that more capital costs need to be spent to produce the uranium; the capital investments are included in the forward cost calculation in the same way. That is, all future capital costs (for exploration, land acquisition, and so forth) are apportioned, undiscounted, over the total production of U_3O_8 to check whether the average grade is high enough to warrant the investment.

The GJO defines forward cost and then specifies its estimates of reserves and resources in those terms. But if one were estimating a supply curve from data GJO provides, one class of adjustments would need to be made, and a second might help the process. The first class of adjustments is relatively straightforward. Although it is difficult to agree on any specific discount rate, it is necessary in any cost calculation to discount cash flows because of significant time preference for money. Not applying discount rates to capital expenditures implies a zero rate of return on capital: Future revenues are not discounted either. Moreover, certain federal taxes and depletion allowances are standard enough to be included in calculations of cash flow. Because federal regulations are always changing, the ideal solution would be to have enough information available that the modeler could apply whatever adjustments were appropriate.

The second class of adjustments has to do with separating exploration costs from other capital costs. Forward cost includes an estimate of exploration costs based on effort/yield relations: So many pounds per foot are found and exploration costs so many dollars per foot. This is quite a simple exploration model. In creating a supply function, one might want to replace simple effort/yield relations with more realistic formulations. An exploration model, separate from resource estimates and geologic analogy, might add credibility to GJO estimates of possible and speculative categories. The resource estimates and the economic measure of cost would indicate only the likelihood of the existence of potential resources in the various cost categories. Exploration models would predict timely development and probabilities of discovery. If exploration costs could be separated, then a cost category such as $30/lb ore would actually be ($30 + E), where E is the exploration cost. The GJO is now addressing this problem through the funding of research by DeVerle Harris, who is seeking to develop a potential supply model for improved definition of costs required to discover and convert resources to reserves.

Forward cost is an idea whose time is past. In the "stretch-out period" it almost evaluated uranium reserves correctly. Discounting did not matter very much because the interest rates were low, and variable costs were incurred at almost the same time the ore was sold. Issues regarding the uncertainty and cost of exploration were not important. Forward cost was an appropriate measure for a declining industry with excess capacity. It is not applicable for a steady-state or expanding industry. In recognition of this, the GJO production capability studies are now being oriented more toward answering questions related to industry viability, relationship of domestic to foreign long-term uranium prices, and other specific problems. Also, estimates are now being developed for full economic recovery costs.

PRODUCTION CAPABILITY ANALYSIS

The GJO production capability projections are based on modifications of two analytical models: "Could Do" and "Need." Neither approach alone approximates the economic notion of a supply curve, but current analysis, including a hybrid approach, is moving in the direction of a truer economic analysis.

The "Could Do" case primarily refers to an upper bound on the amount of U_3O_8 that could reasonably be produced. Misunderstandings continually arise from the incorrect interpretation of the "Could Do" scenarios by outsiders.

They tend to assume that the "Could Do" case represents a realistic prediction of the future supply of uranium, whereas in fact the "Could Do" case presents what might happen if the marketplace and other factors were neglected.

The primary assumption underlying the "Could Do" analysis is that full production is achieved from a certain set of production centers as rapidly as possible, subject to lead time and capacity constraints, but independent of demand and from a limited subset of the resource base. This subset might be those $30 reserves and probable resources estimated within the production center areas. The "Need," on the other hand, assumes a demand schedule and assigns production to meet expected demand in a least-cost way. Current modifications of the "Could Do" projections largely respond to current conditions of low prices. These modifications include narrowing the resource base to exclude relatively lower grade ore and adjusting output for announced curtailments. Market price, in addition to forward cost, is being calculated on an experimental basis.

The results of production capability studies indicate long-term adequacy of the resource base. Figure 21-3 shows a modified "Could Do" projection, applied to (forward cost) $30 reserve plus probable resources and limited to pro-

Figure 21-3. Estimated Annual Near-Term Production Capability from Resources Available at \$30/lb U_3O_8 or Less, Scheduled to Meet Near-Term Nuclear Power Growth Demand.

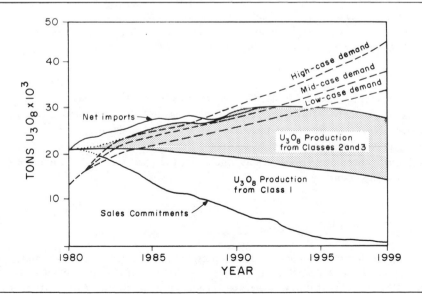

Source: An Assessment Report on Uranium in the United States of America, GJO–111(80), Grand Junction Office, U.S. Department of Energy, 1980, p. 129.

Figure 21-4.   Annual Production Capability from Resources Available at $50/lb U_3O_8 or Less, Projected to Meet Nuclear Power Growth Demand.

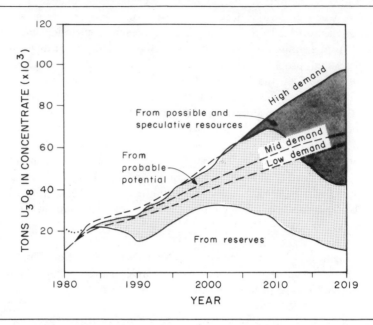

Source: An Assessment Report on Uranium in the United States of America, GJO–111(80), Grand Junction Office, U.S. Department of Energy, 1980, p. 138.

duction centers already in place or, where already-identified reserves justify construction of new facilities but commitment has not yet been made. This graph shows more than adequate production for at least a decade, even though the demand schedules are probably too high. Figure 21-4 shows that as higher cost and more speculative resources are included, the resource base is adequate well into the next century.

REFERENCES

1.   A much more detailed account is presented in Charles River Associates, *Uranium Price Formation*, EPRI–EA–498 (Palo Alto, Calif.: Electric Power Research Institute, October 1977) and in S.M. Stoller Corporation, *Uranium Data*, EPRI EA–400 (Palo Alto, Calif.: Electric Power Research Institute, June 1977).

2. D.L. Hetland and W.D. Grundy, "Potential Uranium Resources," in *Uranium Industry Seminar*, GJO-108(78), Grand Junction Office, U.S. Department of Energy, October 1978, p. 149.

3. D.L. Hetland and W.D. Grundy, "Potential Uranium Resources," in *Uranium Industry Seminar*, GJO-108(78), Grand Junction Office, U.S. Department of Energy, October 1978, p. 156.

4. J.J. Schantz, Jr., *Resources Terminology: An Examination of Concepts and Terms and Recommendations for Improvements*, EPRI-336 (Palo Alto, Calif.: Electric Power Research Institute, August 1975), p. 13.

5. D.L. Hetland and W.D. Grundy, "Potential Uranium Resources," in *Uranium Industry Seminar*, GJO-108(77), Grand Junction Office, U.S. Energy Research and Development Administration Contracts AT-05-1-16344 and E(05-1)-1665, January 1, 1977, p. 157.

6. D.L. Hetland and W.D. Grundy, "Potential Uranium Resources," in *Uranium Industry Seminar*, GJO-108(78), Grand Junction Office, U.S. Energy Research and Development Administration Contracts AT-05-1-16344 and E(05-1)-1665, October 1978, p. 157.

7. D.P. Harris, *A Critique of the NURE Appraisal Procedure as a Basis for a Probabilistic Description of Potential Resources, and Guides to Preferred Practice*, unpublished manuscript prepared at the request of U.S. Energy Research and Development Administration, 1976, p. 33.

8. SRI International, *Development of the Prospector Consultation System for Mineral Exploration* (Palo Alto, Calif.: SRI International, October 1978).

9. D.L. Curry, "Estimation of Potential Uranium Resources," in *Concepts of Uranium Resources and Producibility* (Washington, D.C.: National Academy of Sciences, September 1977).

22 INTERPRETATIONS OF GJO ESTIMATES

Several studies were conducted in the 1970s to assess the adequacy of the domestic uranium resource base and estimation methods. In 1974 Battelle-Northwest accumulated information on domestic uranium supply, especially as it relates to possible federal policy options.[1] In 1975 the Committee on Mineral Resources and Environment (COMRATE), under the auspices of the National Academy of Sciences, reported on problems relating to resource estimation and the uranium industry.[2] A unique asset of this study is the quantity and variety of participants; divergent views are expressed in a collection of papers appended after the consensus report. The S. M. Stoller Corporation produced two analyses of uranium supply. The first assesses the adequacy of resources to meet long-term needs, produced in 1976 for the Edison Electric Institute (EEI).[3] The second is a pair of reports that discuss the data and methodology of various resource estimates, especially those of GJO, produced in 1977 for the Electric Power Research Institute (EPRI).[4] Taylor, of Pan Heuristics, argues that because uranium resources are abundant and that uranium prices are likely to drop and remain low, plutonium recycling and breeder reactors are unjustified.[5] In 1977, the Nuclear Energy Policy Study Group, sponsored and administered by the Ford Foundation and the MITRE Corporation, published a book on nuclear power issues.[6] One chapter, written primarily by Landsberg, discusses the adequacy of uranium supply and some of the important issues dealing with uranium production and estimation. Finally a report released in 1978 by the subpanel of the Committee on Nuclear and Alternative Energy Strategies (CONAES) presents estimates of the domestic resource base and projections of discovery rates.[7]

These reports represent a wide variety of views and include many recommendations. The Ford/MITRE and CONAES studies will be discussed in greater detail. Not only are they the most recent, they represent polar views on the extent and adequacy of the uranium resource base. The Ford/MITRE study claims that the resource base is adequate, whereas the CONAES study concludes that GJO has dramatically overestimated the resource base and that expected time lags and potential lack of incentives make future production levels uncertain at best. An investigation into these two approaches provides perspective on the various interpretations of reserve and resource estimations.

Landsberg is a senior economist at Resources for the Future. His perspective on the current status of uranium is consistent with his background in resource economics, and he draws on such studies as Barnett and Morse.[8] This leads him to discard a common but misleading notion that scarce minerals are naturally growing scarcer (more expensive) and that reserves and resources are progressively being depleted without replacement. He says,

> The history of most mineral industries reflects a period of pessimism after the cream has been skimmed and the obvious locations investigated, and before new ideas have turned explorers to new structures and minerals. The search for petroleum was extended to ever-deeper strata, different traps, and from onshore to offshore fields; for many years reserves rose faster than needed to replace production. Copper ore mined one hundred years ago contained 3.5 percent copper; today it contains 0.5 percent and is produced at real costs not much higher. The immature uranium industry, which has barely gone commercial, is not likely to prove the exception.[9]

Landsberg presents a concise background of many of the issues relating to uranium supply. He defines reserves and resources, introduces the McKelvey diagram and describes some of the important characteristics of uranium production, such as exploitation process, the history of the market, and a definition of forward costs.

There is no modeling in Landsberg's chapter. To back his assertions about the sufficiency of the domestic resource base, he calculates cumulative demand up to a particular year according to three projected demand scenarios. He then matches these three totals of required tons of U_3O_8 with subsets of the GJO reserve and resource estimates. For example, he finds that 1.2 million tons of U_3O_8 will be required to meet the expectations of the U.S. Energy and Research Administration (ERDA) in 1976 for mid-range growth up to the year 2000. This figure is equivalent to $30/lb reserves (reserves available at a forward cost of $30/lb or less) plus half of the $30/lb probable resources, an amount Landsberg is confident the United States can produce. As explained in Chapter 21, it is often misleading to choose a particular subset of the McKelvey diagram and make specific assumptions about that subset's availability. Landsberg is careful not to give the impression that the McKelvey diagram represents a static situation, however. He states, "Reserve and resource estimates rise along with rising production."

Although Landsberg concludes that the resource base is sufficient to supply fuel for light water reactors beyond the twentieth century at no more than current spot prices ($40/lb), he admits that the actual production of the uranium presents a problem. Given likely projections of growth in demand, he anticipates increases on the order of 10 percent per year, admitting that these rates are not easily sustained. He notes, though, "Fortunately, society has better means of dealing with industrial constraints than with lack of natural resources."

The CONAES subpanel presents a radically different picture from Landsberg's. Their interpretations arise from what might be inferred as the philosophy "What uranium resources can we rely on?" They rationalize their low assessment of the potential resource base as follows:

> It is the general opinion of the subpanel that there is, as yet, no reasoned philosophical or geological analysis for providing a quantitative value for the total "Potential Resources" as defined by ERDA. We have considered the model basis for each of the three potential resource classes—"Probable," "Possible," and "Speculative." We recognize the qualitative basis for ERDA's description of each of these classes, but the "Possible" and "Speculative" classes, by definition, have no history of production or identification of reserve associated with them. Hence, there is no basis in experience for quantification.
>
> ERDA has offered a plausible method, based on accepted geological models and its large library of industry production and reserves information, for developing estimates in the class of "Probable" potential resources. It is our judgment that ERDA's figure of 1,060,000 tons in this class is the closest approach to a quantitative estimate available for all potential resources at this time.
>
> We have placed a reliability figure of no better than ±50 percent on this estimate. (This should be compared with ERDA's judgment of ±20 percent reliability for its $10/lb reserves.) We see a conservative lower limit of 500,000 tons in this class. At the same time, the imponderables of the "Possible" and "Speculative" potential resources add a much larger range in the higher-limit estimate. We accept a figure of 3 million tons (ERDA's estimate) as a possible maximum upper limit for all potential resources at $30/lb forward costs.
>
> For the three cases, we have summed the reserves, by-products, and potential resources. Our best quantitative estimate of domestic $30/lb uranium resources is 1.76 million tons. The lower-limit estimate is 1.0 million tons; the upper-limit estimate is 3.78 million tons. These estimates will not change through time unless additional exploration generates new information.[10]

These estimates are presented in Table 22-1. The CONAES subpanel believes that:

> ... ERDA's estimates are among the best available. However, ERDA is not committed to apply rigorous mining engineering practice in defining reserves.

Table 22-1.   CONAES Resource Estimates for Domestic Uranium
Resources, Tonnage as of January 1, 1976.[a]

	Reserves	By-products	Potential Resources	Sums
Best estimate	640,000 (ERDA)[b]	60,000	1,060,000	1,760,000
Lower limit estimate	480,000	20,000	500,000	1,000,000
Higher limit estimate	640,000	140,000 (ERDA)	3,000,000 (ERDA)	3,780,000 (ERDA)

a. Tons U_3O_8 at \$30/lb U_3O_8 cutoff cost.

b. (ERDA) indicates ERDA estimates, which are taken from D.L. Hetland. "Discussion of the Preliminary NURE Report and Potential Resources Uranium Industry Seminar." Grand Junction Office Report No. GJO-108(76). Washington, D.C.: U.S. Department of Energy, October 1976.

Source: Committee on Nuclear and Alternative Energy Systems (CONAES), Problems of U.S. Uranium Resources and Supply to the Year 2010, Report of the Uranium Resource Group (Washington, D.C.: National Academy of Sciences, 1978). Reprinted with permission.

In estimating potential resources, it has made extensive use of geological analogy and subjective favorability analysis. This is good exploration procedure. Our interpretation of their assessment practices is that they may tend to reflect more enthusiasm and optimism than more critical and rigorous analysis. Until a far more extensive factual base of domestic exploration data is developed, this will probably continue to be the case.[11]

In order to assess the CONAES results properly, it is useful to impute various assumptions CONAES might have used but did not make explicit, to establish their estimates. The first assumption is that economic incentives to encourage sufficient exploration may be absent. Second, uranium resources may not be brought into production rapidly enough to satisfy projected demand. And third, the uranium resources identified by GJO in the possible and speculative categories simply do not exist.

The first assumption (that either the future price of uranium will be too low or the uncertainty will be too high to encourage adequate exploration) is consistent with many of the recommendations proposed by CONAES. However, an economist would view problems of an adequate incentive in terms of a market. Unless federal regulations actively limit prices or production, a market in the long run will set a price for the product consistent with actual supply and demand.

The second assumption (even if incentives are adequate, the lead times are too long to prevent shortages) is supported by associated analysis and is raised as a potential problem by Landsberg, but it does conflict with the GJO production capability projections discussed in the last chapter. The actual GJO production capability curves, which are difficult to interpret for reasons outlined in Chapter 21, show no lack of projection capability in the short run. In the long run, lead times are not an issue.

The final assumption might be that the resources GJO allocates to possible and speculative categories are simply nonexistent. The CONAES group may have good evidence that in fact virtually all of the uranium in these speculative categories does not exist, but they do not present this evidence.

It is instructive to describe an often committed error in providing availability statements on the basis of a McKelvey box diagram. It is dangerous to use the McKelvey diagram to provide subsets for which availability statements can be made. Remember that the box represents only a snapshot of a flow; resources move from column to column and row to row as effort is made to learn more about the resource base. The following unspoken misconception arises from this column-by-column assessment:

> The probability is high that the domestic supply of uranium includes at least reserves plus probable resources. Given some luck (the probability is lower), we can include possible resources in that total. And given a great deal of luck (the probability is very low), the speculative resources will be found; however, we should not count on being that lucky.

But this has not been the historical pattern, nor are the GJO estimates, for instance, generated in such a way for this to be true. Not all development will be concentrated exclusively on one category of probable resources until those resources are depleted. Exploration is generally more successful in the class of probable resources than in the class of possible or speculative resources. Nevertheless, we can expect that some future development will occur simultaneously from ore that is contained now in each of three categories. This can be better understood with a simple diagram (Figure 22-1). The coefficients of variation as portrayed are fictitious. They serve only to represent a possible accurate assessment of the uncertainty surrounding any point estimate of the resource category. This distribution could have been obtained by subjective probability or by some other means.

If a model first arrived at such a set of distributions in order to describe the estimates of a particular category of resources, then a point estimate for any category could be obtained by choosing some central measure of tendency. It is true that low percentiles for each category—say, the 5 percent left-hand tail—are much lower for low-assurance categories than for high-assurance ones, but unless it is known that the error of overestimation is much more serious than the reverse, such low percentiles are not the appropriate base for inference. A more central measure of all four categories should be used. If it is necessary to combine categories, then the use of the more speculative categories should be con-

Figure 22-1. Hypothetical Density Functions as the Basis for Reserve and Resource Estimates.

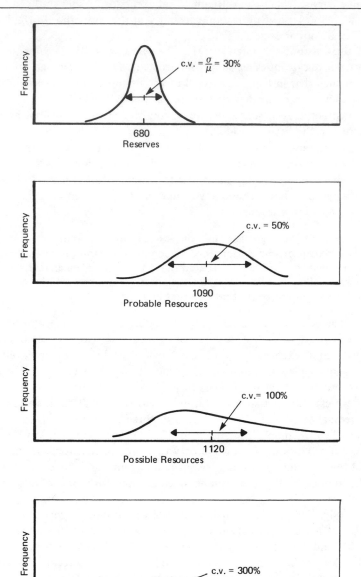

sidered along with the fact that it is likely to take a longer time for them to be exploited, and they have a higher variance. In the long run, however, even the central measures of speculative categories are likely to be underestimates.

REFERENCES

1. Battelle–Northwest, *Assessment of Uranium and Thorium Resources in the United States and the Effect of Policy Alternatives* (Richland, Wash.: December 1974).
2. Committee on Mineral Resources and Environment (COMRATE), *Reserves and Resources of Uranium in the United States* (Washington, D.C.: National Academy of Sciences, 1975).
3. S. M. Stoller Corporation, "Report on Uranium Supply: Task III of EEI Nuclear Fuels Supply Study Program," Appendix II in *Nuclear Fuels Supply* (New York: Edison Electric Institute, 1976).
4. S. M. Stoller Corporation, *Uranium Data*, EPRI EA–400, and *Uranium Exploration Activities in the United States*, EPRI EA–401 (Palo Alto, Calif.: Electric Power Research Institute, June 1977).
5. V. Taylor, "The Myth of Uranium Scarcity," unpublished manuscript by Pan Heuristics, Los Angeles, April 25, 1977.
6. Nuclear Energy Policy Study Group, *Nuclear Issues and Choices*, MITRE Corporation for the Ford Foundation (Cambridge, Mass.: Ballinger, 1977).
7. Committee on Nuclear and Alternative Energy Systems (CONAES), *Problems of U.S. Uranium Resources and Supply to the Year 2010*, Report of the Uranium Resource Group (Washington, D.C.: National Academy of Sciences, 1978).
8. Harold J. Barnett and Chandler Morse, *Scarcity and Growth* (Baltimore: The John Hopkins Press, 1963).
9. Nuclear Energy Policy Study Group, *Nuclear Issues and Choices*, p. 82.
10. CONAES, "Problems of U.S. Uranium Resources and Supply to the Year 2010," p. 17.
11. Ibid., p. 19.

23 TREND PROJECTION MODELS

Another common method of estimating uranium availability uses a class of models based on trend projection.[a] This class ignores the direct assessment of undiscovered resources; instead, future discoveries are projected from past trends in exploration, effort, and discovery. This category includes: (1) life-cycle models, as Hubbert made famous for oil; (2) discovery rate (rate-of-effort) models, as in Lieberman[1]; and (3) econometric models. Although all of these models are based on rates of discovery and production, an estimate of ultimately recoverable resources is sometimes calculated as a by-product of projecting future discovery rates.

The underlying concept of life-cycle models is that the level of discovery and production of any mineral, plotted against time, approximates one of a family of bell-shaped curves. This curve is postulated to rise slowly from zero as the mineral is first introduced, reach a single peak somewhere in the middle of its life history, and then approach zero again as the resource base is depleted and priced out of the market. It is assumed that this process can be approximated by a particular analytical distribution, and the parameters of that distribution can be estimated from past history. Discovery rate models (also called rate-of-effort models) replace the time dimension with a measure of effort, such as cumulative drilling. Thus, as the rate of discovery declines with cumulative output, the resource base is exhausted when the discovery rate equals zero. Econometric models are more complicated versions of the discovery rate models and include more market information.

a. Much of the theory regarding these approaches is presented in Part III of this book, on undiscovered oil and gas. This class of models is not as important for uranium as it is for oil and gas, so some of the detail has been omitted here.

Trend projection models afford some advantages. They tend to be simple and can be applied with a very small amount of data. The equations are easily expressed, and the data can be displayed. If there is doubt about certain assumptions, the model or data can be manipulated.

On the other hand, trend projection models have several drawbacks. Part III of this book discusses in detail the major faults of trend projection models: (1) The choice of the analytical form of the curve used to make projections is ambiguous and can significantly affect the result. (2) The procedure for estimating parameters is often ad hoc and not robust. (3) Economic factors such as price are not adequately portrayed, especially for life-cycle and discovery rate models. And (4) critics argue that it is impossible to represent such a complex process with a single function.

As might be expected, many of these weaknesses stem from the very simplicity of the model. It is difficult to aggregate a heterogeneous process across a region as large as the United States and still portray reality. The application of a simple equation to the complicated process of resource exploitation requires a host of implicit assumptions, many not adequately stated or known. And the data that are available to the person doing the estimating often are not a good measure of model requirements. All methods of resource estimation (trend projection, as well as others) suffer from the necessity of using the present to provide perspective. None can adequately predict the future or unscramble the situation in unexplored regions. But trend projection models have a closer link to past trends in economics and productivity than do other types of models. This ties them to a history that may bear little relation to the future. One example of this is the typical reliance of trend projection models on the discovery rate. If a year's drilling is divided by the resulting reserves discovered, the result is some measure of the ratio of effort to yield, usually expressed in terms of pounds U_3O_8 per foot drilled. This discovery rate is the core of most of these models, yet it is fraught with difficulties.

First, the discovery rate is not a robust measure of the depletion of a nation's mineral stock. Stoller[2] presents plots of discovery rates for several areas, which are shown in Figure 23-1. They show the wide variation over time for any one area and a lack of correlation from one area to the next. Aggregating all the dissonant exploration histories in the disparate regions is not an adequate method of measuring the ratio of effort to yield for the entire nation.

Second, the GJO quantifies uranium by forward cost in current dollars. Adjusting for inflation is not completely straightforward, due to the nature of reporting by the GJO. But an attempt can be made, as in Klemenic and in Harris, but Lieberman does not do so.[3]

Third, irrespective of whether uranium is measured in current dollars or in constant dollars, productivity changes and changes in factor prices are an inherent part of any trend applied to the historical pattern. These changes are necessarily extrapolated into the future. For example, if the average dilution encoun-

Figure 23-1.   Tentative Discovery Rate Plots. (The vertical axes are
pounds per foot drilled; the horizontal axes are years as on the graph
for the United States.)

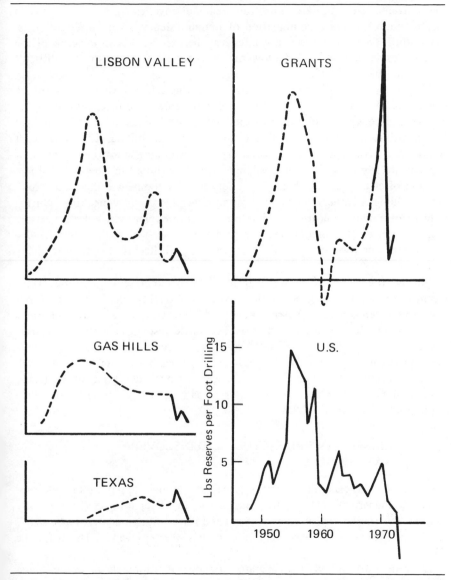

Source: S.M. Stoller Corporation, *Uranium Exploration Activities in the United States.*
EPRI-401. (Palo Alto, Calif.: Electric Power Research Institute, June 1977), Reprinted
with permission.

tered in mining all deposits in the United States has increased by 2 percent per year (it probably has not), then this pattern would be indistinguishable in the past record and extrapolated as an invisible part of future trends.

Fourth, the different grades, sizes, and depths of uranium have correspondingly different costs of extraction. Any year's discoveries are not all of a single cost uranium but are a combination of uranium deposits with different costs. Depletion, or declining discovery rates, can be expressed both in terms of the mixture inside each set and the lowering of the yield effort ratio. It is difficult to separate these two effects.

Fifth, there are serious problems in representing effort. Either feet of drilling or number of holes drilled is commonly used to represent effort, but drilling has not been in constant proportion to total effort. Additional difficulty is entailed in separating development drilling from exploration drilling. The measure of effort should correspond to the discovery of new deposits, not the extension of deposits already identified, so development drilling should not be included in a measure of effort. Yet the breakdown between development drilling and exploration drilling is ambiguous. Indeed, certain federal taxes influence the reporting in favor of exploration drilling. Moreover, because exploration holes are getting deeper, equivalent uranium deposits located deeper would decrease the discovery rate per foot. So discovery rate per foot drilled is also an inadequate measure.

Finally, there are also problems in the data that represent discoveries. Until recently, the figures reported by the GJO did not distinguish between additions due to new deposits and enlargements of old deposits. In other words, additions to reserves in any particular year may bear little resemblance to the deposits found in that year. Further, the GJO classifies a uranium occurrence as a reserve, as distinct from a potential resource, if its shape has been determined by developmental drilling. The choice and sequence of deposit development are highly influenced by an individual firm's strategies, as well as by external factors (such as the price of uranium).

LIEBERMAN'S LIFE-CYCLE AND DISCOVERY RATE MODELS

The best-known example of trend projection models appeared in a *Science* article by Lieberman.[4] His result prompted a pair of replies in *Science*, one by Gaskins and Haring and another by Searl and Platt.[5] Subsequently, Lieberman's article on life-cycle and discovery rate models was reviewed by Harris for the GJO.[6]

Lieberman attempted two separate estimates of ultimately recoverable uranium. The first is based on rate of effort and the second on life-cycle. The first method assigns an exponential:

$$R(h) = R_0 e^{-bh} \quad , \tag{23-1}$$

where h = cumulative drilling in feet; R = discovery rate in pounds per foot; and R_0, b = constants to be determined, to the declining discovery rate. Fitting the curve to past records and constraining it to pass through the actual figures for 1974 determines R_0, b, and, by integration, ultimately recoverable resources for $8 uranium of 630,000 tons. Substracting cumulative production and identified reserves leaves only 87,000 tons of undiscovered resource, about one-third of that estimated by GJO at the time.

Serious problems arise in Lieberman's application of the rate-of-effort model, including all the shortcomings in measures of discovery rate discussed previously. His very low estimates of undiscovered $8 resources had already been surpassed by the time Harris wrote his rejoinder. Basically, there is little reason to believe that the past record fits a negative exponential and even less reason to believe it will follow that trend in the future.

In his second method Lieberman applied a life-cycle model to uranium. He admits that his fit of a logistic curve to cumulative uranium production and cumulative discoveries is not good. The most potent criticism of his model is the one that can be properly applied to all trend projection models: The dynamics of the process of exploitation in changing market conditions are not adequately represented by past trends.

Figure 23-2. Production of Primary Lead in the World and in the United States.

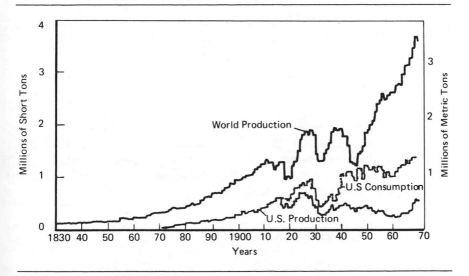

Source: Adapted from D.P. Harris, *Estimation of Uranium Resources by Life-Cycle or Discovery-Rate Models: A Critique*, GJO–112(76), Grand Junction Office, U.S. Energy Research and Development Administration, 19.

Figure 23-3.   United States Production of Coal and Lignite.

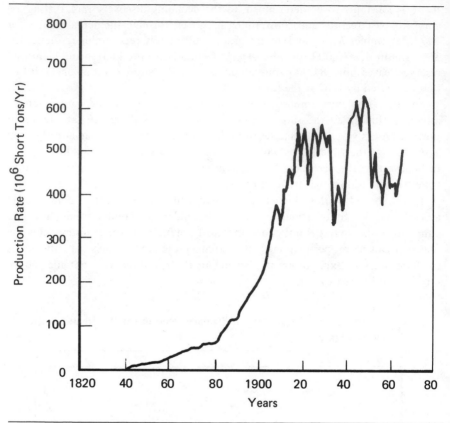

Source: Adapted from Harris, *Estimation of Uranium Resources*, p. 8.

Consider the application of a life cycle to lead or to coal, as did Harris.[7] As can be seen from Figure 23-2, in the 1920s production of lead in the United States might have been thought to have peaked and to be tailing off. But the pattern did not follow that cycle. Similarly one might expect from Figure 23-3 that coal production is currently peaking, but that is almost certainly not true.

ECONOMETRIC MODEL BY PERL

The econometric study of domestic uranium supply prepared by Perl in preparation for testimony at General Environmental Statement Metal Oxide (GESMO) hearings[8] does not have a wide distribution. Nonetheless, it is a careful study and one of the few econometric analyses of uranium supply available. Despite its sophistication relative to other trend projection models, it still illuminates some of the difficulties in applying trend projection models to supply of depletable resources.

Perl's purpose is to calculate the savings in fuel cost due to the lower fuel requirements of uranium and plutonium recycling. This requires several forecasts of price given several demand scenarios, a difficult task. Perl accomplishes this by calculating a true economic cost as a function of forward cost, then forecasting future additions to reserves by projecting past trends in discoveries due to exploration. He is careful to introduce his models with sections on the assumptions that are involved; these are conventional economic topics that tie in with the modeling effort. He also divides the forecasting period into short-, medium-, and long-run intervals. Each of these intervals has a set of assumptions about its resource base and hence its supply function.

The analysis can be divided into three parts. First, using mining cost data presented in Klemenic,[9] federal taxes, depletion allowance, and discounting are added to forward cost to estimate economic cost. Second, supply curves for the short and medium run (1975-1985) are calculated from the distribution of known reserves. And third, long-run supply curves (1985-2000) are calculated on the basis of projection effort/yield relation.

Perl calculates that the ratio of economic cost to forward cost for reserves in production is approximately 1.3 to 1; for reserves not in production, 1.5-1.8 to 1; and for resources, exclusive of exploration costs, approximately 2 to 1. Short- and medium-run supply curves are distinguished by production of the reserve. In the short run, only reserves in production are counted, and in the medium run, reserves that are not in production are counted. Perl uses a GJO table of reserves, classified by forward cost and production. Assuming that each mine is exploited evenly over twelve years, annual production from the table of reserves can be predicted by dividing each estimate by 12.

Trend projection methods are used in the estimation of supply in the long run. Perl is careful to use added reserves that have been adjusted for inflation and in which, for the earliest years, the higher cost categories are extrapolated from lower cost categories, as suggested by Klemenic.[11] For drilling, Perl uses total exploratory and development drilling.

Both additions to reserves and drilling data used in this way are subject to most of the problems mentioned earlier. However, the reserves added are adjusted for inflation and each price category is considered separately. Perl is careful also to assess expenditures in drilling as a proportion of overall expense of exploration.

Perl's formulation of the effort/yield relation is interesting. He postulates that

$$R_{p,t} = e^{a_0 + a_1 T} \; R_{p,t-1}^{a_2} \; D^{a_3} \; P^{a_4} \quad , \qquad (23-2)$$

where

R = reserves $U_3 O_8$ discovered in tons with extraction cost less than P,

D = drilling thousands of feet,

P = marginal costs of extraction,

T = time in years,

p, t = indices for price and year, and

a_0, a_1, a_2, a_3, a_4 = parameters to be determined by regression.

In other words, additions to reserves for a particular price category decline exponentially over time at a rate of a_1 per year, and the elasticity of drilling is a_3 ($a_3 < 1$ means diminishing returns for extra drilling). One can interpret a_2 as the parameter in a Koyck transformation[12] for lagged reserves; $(1 - a_2)$ is the fraction of discoveries associated within a year of the effort, and a_2 is the fraction associated with the past few years. Fewer reserves are found in the low-cost category than in the high-cost category for any particular year; a_4 measures the relation between high- and low-cost discoveries. Because P represents only the extraction costs, exploration and development costs, which increase over time, are added to extraction costs to derive a selling price. The optimization problem, which determines the proper trade-off between higher extraction costs and more drilling, is solved analytically. Presumably this correctly includes reserves that were discovered but that are too expensive to extract and put into inventory for future price rises.

It should be noted here that Perl assumes that exports will balance imports. In view of recent foreign discoveries, especially Australia and Canada, this may not be true. Also, according to current projections, the demand, with 570 gigawatts of electricity installed in the year 2000, is too high. The resulting supply functions are shown in Table 23-1. One striking result is that the predicted prices at the endpoints of each interval are not close to each other. For instance, the predicted price with recycle for 1978 (part of the short run) is $33/lb—a price significantly greater than the $15/lb predicted for 1979 in the medium run. The same discrepancy is true of the transition from medium to long run. This would lead one to question the accuracy of the results; the effect of depletion over time is almost certainly too great for each separate supply function. Interestingly, Perl had already arbitrarily adjusted the a_1 parameter (yearly decline in discoveries) from a calculated 4.3 percent to an assumed 3 percent to reflect his personal opinion.

The short- and medium-run analyses suffer from a static rather than a dynamic interpretation of the resource base; this problem is apparent many times in Perl's report. Almost certainly some production in the medium run will come from the resource base that is now identified as reserves. For example, "property resources" is a subcategory of "probable resources," which is used internally in the GJO, but is not distinguished from other probable resources in their

Table 23-1.   Equilibrium Price.

Year	Required Production, Tons	Equilibrium Price, 1975 $/lb
Short-Run Supply Curve		
1975	12,500	22.03
1976	15,900	31.75
1977	17,100	35.18
1978	16,500	33.46
Medium-Run Supply Curve		
1979	22,200	15.49
1980	22,900	16.19
1981	25,000	18.32
1982	33,000	26.40
1983	32,000	25.39
1984	40,900	34.39
Long-Run Supply Curve		
1985	43,900	20.09
1986	44,600	21.12
1987	52,400	25.72
1988	53,700	26.00
1989	59,000	29.56
1990	62,900	31.37
1991	64,200	33.33
1992	69,600	36.94
1993	72,900	39.72
1994	74,700	41.31
1995	74,300	43.12
1996	78,500	47.51
1997	79,400	49.24
1998	81,600	52.65
1999	83,700	55.88
2000	85,500	59.10

Source: Perl, "Testimony on U_3O_8 Prices," Table 8.

publications. Property resources refers to extensions of known deposits in producing properties; the area has not yet been drilled extensively so they are not counted as reserves. It is much more likely that U_3O_8 from this part of the resource base will be produced in the medium run than from reserves in currently nonproducing areas. By confining the production to the reserve base, Perl has significantly overestimated the effects of depletion.

The long-run analysis also has problems. High-cost reserves found in early years when low-cost uranium was being mined are very much underestimated because reserves measure ore that is delineated, and high-cost ore simply would not be included in those reserve statistics. The lack of distinction between the expansion of working deposits and the discovery of new ones biases the data substantially.

Even assuming the data measured what they were supposed to measure and fit the discovery relation well (high R^2), they would still be a weak base from which to project forward twenty-five years. It might be called "looking forward through a rear-view mirror."[13] Because the United States is composed of many different kinds of environments, and technology changes rapidly, it is difficult to assume Eq. (23-2) will hold true for twenty-five years.

REFERENCES

1.   M. A. Lieberman, "United States Uranium Resources – An Analysis of Historical Data," *Science* 192, no. 4238 (April 30, 1976): 431–36.

2.   S. M. Stoller Corporation, *Uranium Exploration Activities in the United States*, EPRI–401 (Palo Alto, California: Electric Power Research Institute, June 1977).

3.   J. Klemenic, "Analysis and Trends in Uranium Supply," in *Uranium Industry Seminar*, GJO–108 (76), Grand Junction Office, U.S. Energy Research and Development Administration, October 1976. D. P. Harris, *The Estimation of Uranium Resources by Life-Cycle or Discovery-Rate Models: A Critique*. GJO–112(76), Grand Junction Office, U.S. Energy Research and Development Administration, October 1976.

4.   Lieberman, "United States Uranium Resources."

5.   D. W. Gaskins, Jr., and J. R. Haring, letter to *Science* in response to "United States Uranium Resources – An Analysis of Historical Data" by M. A. Lieberman, *Science* 196, May 6, 1977. M. F. Searl and J. Platt, letter to *Science* in response to "United States Uranium Resources – An Analysis of Historical Data," by M. A. Lieberman, *Science* 196, May 6, 1977.

6.   Harris, *Estimation of Uranium Resources by Life-Cycle or Discovery-Rate Models*. The Harris article and the *Science* replies in Ref. 5 constitute a comprehensive review of Lieberman's work.

7.   Ibid.

8.   Lewis J. Perl, "Testimony of Dr. Lewis J. Perl on Behalf of the GESMO Utility Group on U_3O_8 Prices," National Economic Research Association, New York, 1977.

9.   J. Klemenic. *Examples of Overall Economies in a Future Cycle of Uranium Concentrate Production for Assumed Open Pit and Underground Mining Operations.* Grand Junction Office, U.S. Energy Research and Development Administration, October 19, 1976.

10. Perl, "Testimony on U_3O_8 Prices," p. 13.

11. Klemenic, 1976, *Examples of Overall Economies in a Future Cycle of Uranium Concentrate Production.*

12. See, for example, H. Thiel, *Principles of Econometrics* (New York: Wiley & Sons, 1971) or some other advanced econometrics text for a description of Koyck transformations.

13. Marshall McLuhan, *The Medium Is the Message* (New York: Random House, 1967).

24 CRUSTAL ABUNDANCE MODELS

Unlike trend projection models, crustal abundance models did not originate with oil and gas supply estimation. Crustal abundance models are commonly applied to hard-rock minerals, and much of the model effort has been applied to non-fuel minerals such as copper. Rather than measuring and projecting effort/yield indices, as in trend projection, crustal abundance models emphasize statistical properties of physical attributes, such as distribution of tonnage and grade. This provides a grander view of the world. Modelers are not so concerned with a description of recent exploitation processes as they are with what they consider to be more salient attributes: the original mineral endowment and how much of it has been already discovered. The models are typically applied to a large region for a long term. As with trend projection models, however, the data requirements of crustal abundance models are low, and predictions are highly dependent on the underlying structure of the model. Crustal abundance models are often praised for their ability to estimate uranium occurrence in unfamiliar modes, a characteristic other models, such as subjective probability and trend projection, seem to lack. This ability is highlighted by the disparate results often obtained by crustal abundance and other models. Table 24-1 shows an example; although both crustal abundance and subjective probability models agree on the prediction of tonnage above the approximate current cutoff grade, they differ by a factor of about 300 for estimates of low-grade ore.

The common characteristic in crustal abundance models is that the relative frequency of occurrence of the grade of uranium is lognormal. That is, if one samples the grade of a number of volumes of rock independently, then the resulting histogram would approximate the lognormal distribution. Proponents

Table 24-1.   Comparison of Crustal Abundance with Subjective
Probability for Average and Low Cutoff Grades.

	Endowment, Million Tons U_3O_8	
Method	.10%	.01%
Brinck	1.10	440
Subjective probability	1.26	1.4

Source: D.P.Harris, "Undiscovered Uranium Resources and Potential Supply: A Non-technical Description of Methods for Estimation and Comment on Estimates Made by U.S. ERDA, Lieberman, and the European School (Brinck and PAU)," Part VI of *Mineral Endowment, Resources, and Potential Supply: Theory, Methods for Appraisal, and Case Studies*, MINRESCO, Tucson, Ariz., 1977, p. 4-9.

believe that this proposition is true for a wide variety of scales, from geochemical samples on the order of inches in a particular deposit to mineral provinces throughout the earth's crust. The attributes being measured—tonnage of ore, pounds of metal, size of deposits, and so on—also seems to be lognormal.

There are three basic explanations for these findings. The first is theoretical. Just as the normal (Gaussian) distribution approximates the sum of a large number of independent uncertain quantities,[a] so the lognormal distribution approximates the product of a large number of independent positive-valued uncertain quantities. If a geological process separates an originally homogeneous block into two pieces, the first piece with a relatively higher concentration of uranium than the second, then a series of these separations would result in blocks that have a frequency approximating the lognormal.

The second reason for lognormality of the distributions is empirical. Many reports of crustal abundance models cite work that supports the lognormality assumption. Later in this section I will discuss B. Skinner's contention that the assumption is not true at least for other minerals; however, it is difficult to find empirical evidence that refutes lognormality and supports Skinner.

The third explanation is offered by the authors of this book. Many events in nature have frequencies that are positively skewed and contain only nonnegative values. Goodness-of-fit tests, tests that determine the likelihood of a particular sample being drawn from the particular distribution, are not very powerful for asymmetrical (positively skewed) distributions. This is exacerbated by the fact that the parameters of the distribution usually need to be estimated from the sample itself, thus negating the effectiveness of many of these tests. In short, it is very difficult to test whether a sample is lognormal. Even if good data were available, which is not often the case, the actual tests of significance are not sensitive enough to properly reject samples that should be rejected.

A relatively simple example of the crustal abundance approach is shown in Figure 24-1. This plots the relative frequency of grade x for some region, say

a. See the central limit theorem in an elementary statistics test.

Figure 24-1. Frequency Density of Grade for Uranium.

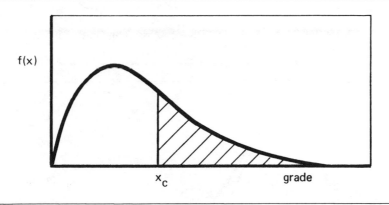

the United States. The two parameters of the underlying distribution are estimated from two pieces of information: the average crustal abundance (the *clarke*) and the total reserves plus past production for some specified cutoff grade. The clarke is equated to the mean of the lognormal distribution. The second parameter is determined by setting the amount of reserves plus production equal to the area above some cutoff grade x_c. Hence, estimates of resources may be inferred by determining the area under the curve between some lower cutoff grade and the original cutoff grade. Statements about potential resources are made, such as: "A decrease of a factor of ten in grade implies an increase of about a factor of 3000 in the tonnage of uranium ore and a factor of about 300 in the tonnage of available uranium."[1] This kind of statement combines processes that actually take place simultaneously. The five margins mentioned in Part I (the rate of mining, the ultimate recovery, the production of marginal deposits, more exploration, and more research) all vary with changes in price.

Skinner dissents from the hypothesis of lognormality of the distribution of grade, at least for certain geochemically scarce metals such as lead, copper, and tin.[2] He suggests that instead of the lognormal distribution (as shown in Figure 24-1), the true distribution is bimodal (as shown in Figure 24-2). This has intuitive appeal if one assumes the right-hand mode represents occurrences of oxides and sulfides, and the rest of the distribution is represented by silicates. The usual counterargument is that there may be several different distributions, which when added together approximate a lognormal distribution (as shown in Figure 24-3). Singer has stated, "The greater the variety of environments in which a metal occurs, however, the more likely that its grade distribution will be unimodal."[3]

A primary example of the application of crustal abundance methods is the work provided by Drew,[4] who is a member of the Programmes Analysis Unit (PAU) of the United Kingdom Atomic Energy Authority. His underlying

Figure 24-2.   Hypothetical Density Assuming Grade Bimodal Distribution.

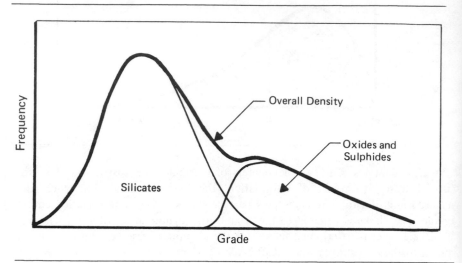

Figure 24-3.   Overall Lognormal Grade Density as the Sum of Several, Separate Environments.

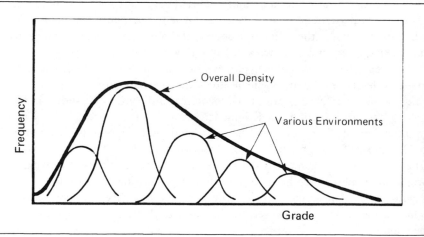

assumption is that any geological environment can be completely partitioned into a set of discrete uranium deposits, each of which has a grade g and a size s, measured in tons of ore. He assumes the joint density function is bivariate lognormal, and that g and s are independent.

Production costs as a function of size and grade are calculated for deposits that have already been identified. Drew chooses a Cobb-Douglass form for the cost equation, a common assumption that results in linear isoquants in log grade-log size space. This is sketched in Figure 24-4.

The formula for cost, using 1970 dollars is

$$C = .066s^{-.159} g^{-1} \qquad (24-1)$$

Figure 24-4. Schematic Representation of the PAU Model.

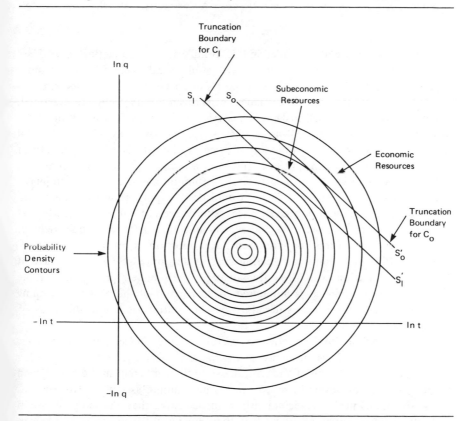

Source: Harris, in *Mineral Endowment, Resources, and Potential Supply: Theory, Methods for Appraisal, and Case Studies*, © MINRESCO, Tucson, Ariz., 1977, titled "Schematic Representation of the PAU Model," p. 4-15. Reprinted with permission.

Table 24-2.   Input Requirements for the Deposit Distribution Model (U.S. Data).

Mass of the environment under consideration	2.12×10^{16} tons (U.S. to a depth of 1 km)
Overall mean grade (clarke)	2 parts per million (U_3O_8 equivalent)
Cost function	$C = 0.66 s^{-0.159} g^{-1}$
Average size, grade and metal content of the known deposits	100,000 tons 2.6 parts per thousand 190 tons U_3O_8
Average exploration cost	$0.8/lb U_3O_8 discovered
Base price	$8/lb $\dot{U}_3O_8$ (January 1, 1970)
Total reserve + past production	3.06×10 tons U_3O_8

Source: M.W. Drew, "U.S. Uranium Deposits: A Geostatistical Model," *Resources Policy* (March 1977), p. 68.

Exploration costs are not included in this formulation of production costs. They are determined by a combination of average historical exploration costs and a function relating the proportion of deposits discovered to the average exploration costs per pound; the PAU model needs seven inputs. They are described, along with the values used in Drew's analysis, in Table 24-2. The results of the analysis, shown in Figure 24-4, can be interpreted two ways. First, potential discoverable resources as a function of price can be calculated for a variety of arbitrary exploration expenditures. This is in no sense optimal; the assumption is that if spending were confined to $8/lb for uranium in the exploration phase, then the dashed line labeled "$8/lb" would be the appropriate curve determining quantity available at any given price. Alternatively, total profits can be calculated by integrating over the entire economic region (to the upper right of the economic truncation line) and then the proportion of profits, where "profits" are $8/lb price minus production cost, is accounted for by an average of $0.80/lb exploration cost. Drew finds that for the numbers presented in Table 24-2, the exploration costs amounted to 25 percent of the profit; continuing that level results in resources described by the continuous line in Figure 24-5 marked "25 percent."

A model such as PAU's is not very useful in addressing specific policy questions. Its main value is probably in its grand view and its presentation of a unique perspective. Some of the critical assumptions inherent in the model need to be evaluated however. First, cost in uranium mining is sensitive to depth, a factor the model neglects. Second, little evidence exists that size and grade are in fact independent, even assuming that they are jointly lognormal. Third, estimates of $8/lb reserves plus production generally increase each year; there is little reason to think that any year's estimate can be equated to the integral over the "economic region" in Figure 24-5. In addition, the deflated price of ura-

Figure 24-5. Potential Discoverable Reserves of Uranium in the United States with Different Exploration Expenditures. (Percentages indicate the proportion of average profit devoted to exploration; dollar values indicate actual exploration expenditure per pound of U_3O_8 discovered.)

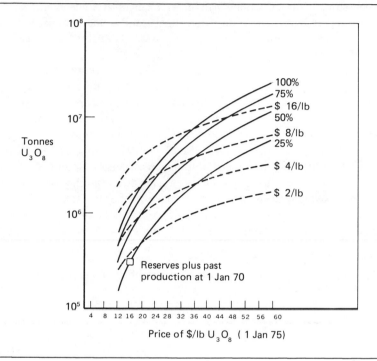

Source: Drew, "U.S. Uranium Models," p. 69.

nium has not been $8 throughout the history of uranium production; this affects actual reserve and production levels. Fourth, the manner of presenting the results is unusual. Of what significance is an isoquant reflecting an artificial constraint on the amount that is spent on exploration? Exploration is margin 4 in the list of margins in Part I, and it is a function of price. As price rises, so does the amount that can be spent on exploration. Moreover, the concept that profits are not assigned to exploration costs is troublesome. An economist would argue that long-run profit (rent associated with greater than return on investment) is out of equilibrium and would need to be explained in some way.

Many other crustal abundance models have been presented. Some of these have been reviewed in some detail by Harris.[5] Brinck's work is well known.[6] He uses assay samples as a basic unit, rather than the average of large blocks, which results in a different estimate of the variance of the lognormal distribution, one that can be adjusted by a function of the volume and shape.

Agterberg and Divi, of the Canadian Geological Survey, refine the general Brinck approach.[7] Rather than estimating the second lognormal parameter on the basis of a single cutoff grade, they make several estimates based on different cutoff grades. As shown in Figure 24-6, the highest few cutoff grades measure reserves plus production, a statistic that captures the true underlying resource base. The estimate of the second parameter changes at some point, however, because adding reserves and production significantly underestimates the real resource base. The best estimate, then, is based on that subset of estimates that are approximately equal.

McKelvey's hypothesis[8] is somewhat different from those used by Brinck, PAU, and Agterberg. McKelvey empirically noted that for well-explored minerals, the total amount of reserves plus production was approximately 10^9 or 10^{10} times the clarke. Hence minerals with reserves plus production substantially less than that number are implicitly expected to have great resource potential.

A method proposed by Searl and Platt[9] combines some attributes of the trend projection models with crustal abundance aspects.[10] Resource estimation is done in two steps: the first applies to producing areas and the second to favorable but nonproducing areas. An effort is made in the first step to remove the bias from estimates of production plus reserves as a basis for extrapolation. The portion of the GJO's estimates that refers to the top 400 feet is assumed to be

Figure 24-6. Estimating the Variance of the Lognormal Distribution Using Several Cutoff Grades.

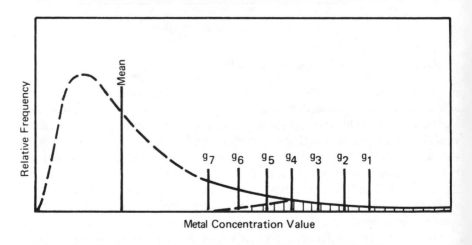

Source: F.P. Agderberg and S.R. Divi, "A Statistical Model for the Distribution of Copper, Lead, and Zinc in the Canadian Appalachian Region." *Economic Geology* 73, no. 2 (March–April 1978): 230–45. Reprinted with permission.

an adequate representation for that part of the resource base. The occurrence of uranium is postulated to be uniform down to 4,000 feet; therefore, the estimates for the top 400 feet are expanded by a factor of 10. The second step extrapolates from producing areas to favorable, but as yet nonproducing areas. A size distribution is postulated for all regions, with the producing regions constituting the largest amounts. The potential that is assigned to each of the rest of the nonproducing areas tapers off according to an exponential decline.

REFERENCES

1. H. D. Holland and E. Tulcanaza, "Concentrations and Ores of Uranium," abstract, GAAPBC 10, no. 7, Geological Society of America (Fall 1978): 423.

2. B. F. Skinner, "A Second Iron Age Ahead?" *American Scientist* 64, no. 3 (1976): 258-69.

3. D. A. Singer, "Long-Term Adequacy of Metal Resources," *Resources Policy* (March 1977): 60-70.

4. M. W. Drew, "U.S. Uranium Deposits: A Geostatistical Model," *Resources Policy* (March 1977).

5. D. P. Harris, "Undiscovered Uranium Resources and Potential Supply: A Nontechnical Description of Methods for Estimation and Comment on Estimates Made by U.S. ERDA, Lieberman, and the European School (Brinck and PAU)," Part VI of *Mineral Endowment, Resources, and Potential Supply: Theory, Methods for Appraisal, and Case Studies.* MINRESCO, Tucson, Ariz., January 1, 1977.

6. J. W. Brinck, *The Prediction of Mineral Resources and Long-Term Price Trends in the Non-Ferrous Metal Mining Industry*, Section 4, 14th IGC, 1972.

7. F. P. Agterberg and S. R. Divi, "A Statistical Model for the Distribution of Copper, Lead, and Zinc in the Canadian Appalachian Region," *Economic Geology* 73, no. 2 (March–April 1978): 230-45.

8. R. L. Erickson, *Crustal Abundance of Elements, and Mineral Reserves and Resources*, U.S. Geological Survey Professional Paper 820, 1973, pp. 21-25.

9. M. F. Searl et al., *Uranium Resources to Meet Long-Term Uranium Requirements*, EPRI SR-5 (Palo Alto, Calif.: Electric Power Research Institute, 1974). M. F. Searl and J. Platt, "Views on Uranium and Thorium Resources," *Annals of Nuclear Energy* 2 (1975): 751-62.

10. An extensive critique of Searl's approach, as well as Searl's rebuttals, appears in a report by the S. M. Stoller Corporation (*Uranium Data*, EPRI EA-400 [Palo Alto, Calif.: Electric Power Research Institute, June 1977]).

25 SUBJECTIVE PROBABILITY APPRAISAL

There is no distinct dividing line between subjective appraisal and many other methods.[a] On the one hand, the CONAES study presents a poll of panels' opinions about total domestic uranium resources. On the other, GJO's geologic analogy might be considered a subjective assessment, because many crucial calculations—the estimation of the fraction underlain by uranium bearing rock—are essentially subjective. In some sense, any method of resource estimation includes a certain amount of subjective probability; one would hope an analyst would guide the models to an answer that the analyst thought was reasonable. However, the archetypal subjective probability appraisal involves a selection of a group of experts, an anonymous polling of their best guess concerning uranium of specified characteristics in a certain area, guesses as to the uncertainty of the best estimate, and sometimes another iteration after they learn the estimates of their peers.

Subjective probability has certain advantages. It allows an informed scientist to accumulate and digest a wide variety of data and experience, and it has the potential for representing a much more complicated model than could be expressed explicitly. It can be done quickly, without many of the complications of analysis.

The major disadvantages of subjective probability methods are the biases inherent in the process; the methods generally underestimate the resource base because it is difficult to imagine resources in unknown environments. In addition, it is hard to both attack and defend results based on subjective probability.

---

a. Subjective probability techniques are discussed in more detail in Part III.

One's best guess is one's best guess, and there is little possibility for arguing about measures such as discovery rates or the time effect of inflation. It is difficult even to choose experts. How does one determine whether or not some of the observers slant their answers one way or another for strong personal reasons? And who determines whether one person is more expert than the next?

One recurring issue in subjective assessments is how much disaggregating should be done in the appraisal. A curious result happens when the appraisal is disaggregated. The smaller the area, in general, the lower the estimate relative to the area. That is, an estimate covering an entire region is likely to be larger than the sum of the estimates for all the smaller districts that make up a region. Also, the analyst might want to disaggregate by physical characteristics—for example, to simultaneously predict such attributes as grain size and dip, as well as the effect of these attributes on the resource base. By subjectively appraising several characteristics and constructing a procedure for combining the separate estimates, a whole is estimated from its parts. But at what point does the estimation of yet another characteristic or the building of an even more complicated structure add more noise than signal?

Harris, long a proponent of subjective probability techniques, is currently researching a hybrid between traditional subjective probability techniques and geologic analogy. Although detailed descriptions of the research are not available, the procedure allows the appraiser to assess scores for various characteristics, as well as to structure the convolution of the scores in whichever way the assessor thinks is appropriate. Through computer data retrieval, any appraiser can obtain certain anonymous parts of other appraisers' evaluation schemes.

A Delphi procedure is sometimes added after the first round of anonymous appraising. Each appraiser is able to match his own estimate with the groups'. Then each assessor is asked to reassess the original guess. This tends to bunch the answers, tightening the consensus. However, there is no certainty that the distribution is contracting around the true number. The original spread probably is significant, and to reduce it via a Delphi procedure will gain little other than a false sense of security.

A CASE STUDY OF SUBJECTIVE PROBABILITY

It is useful to examine the application of subjective probability appraisal by Ellis et al. in 1975.[1] Thirty-six geologists were questioned anonymously about sixty-two cells,[b] which together constitute the state of New Mexico. They also answered the same questions about a single group of eight cells, which represents the San Juan Basin. Each geologist responded for each study area with a sub-

---

b. Each cell is approximately 0.5 degree latitude by 1 degree longitude and averages about 2,000 square miles.

jective probability assessment of the number of deposits, tonnage of ore per deposit, and grade for the given tonnage class. In addition, depth and grade were estimated so cost equations could be used. Along with these estimates of physical characteristics, each geologist gave a subjective weight that represented a self-appraisal of familiarity with that particular cell. This method is designed to allow geologists to make estimates for all sixty-two cells and then, for the cells with large uncertainty due to a lack of knowledge, weight those cells relatively less. After the first round of answers a second-round Delphi technique was applied.

Aggregating the geologists' answers over all of the cells gives a total of 226 million tons of ore containing 455,000 tons of U_3O_8 for the state of New Mexico. The overall subjective probability distribution for the state is calculated as well; the 5 percent tails are 380,000 and 675,000 tons. The spread around the predicted number of deposits is even smaller, from 500 deposits to 565 deposits. Depth and thickness information can be combined with assumptions about distance to the mill to make inferences about cost distribution. For example, 80 percent of the U_3O_8 is estimated to be available at a cost of $15 (1974 dollars) or less.

In the first round, the geologists were asked to consider implicitly multiple modes of uranium occurrence. But the results showed that the characteristics of the predicted resource base represent economic ore almost exclusively. Less than 1 percent of the estimated resources have an average grade between 0.01 percent and 0.04 percent. And only 14 percent have an average grade in the 0.05–0.14 percent range. The second-round answers were directed to specifically consider subeconomic grades. After this Delphi reassessment, grade versus log cumulative ore was plotted for the predicted (plus past production) ore grades, then extrapolated to subeconomic grades. The estimates are still rather low, only about 1.3 million tons of U_3O_8 for all grades as low as 0.01 percent.

The results of this careful study are enlightening. The estimates of economic resources are probably conservative; Ellis and the other authors agree with this assertion. The estimates of subeconomic resources are even more conservative, so one must conclude that this technique did not assess low-grade resources very well. Given the bias inherent in the subjective assessments, extrapolation of grade versus log cumulative tonnage is probably unjustified. The spread of uncertainty around each of the point estimates also is suspiciously low. One relatively minor point is bothersome. The authors compared a histogram of the subjective assessment of tons of U_3O_8 for the entire area (which appeared to be approximately lognormal) to the classic lognormal assumptions used in crustal abundance models. In fact, however, there should be little connection between the shape of a probability assessment of a single value, such as undiscovered resources, and the shape of the distribution of available metal as a function of grade.

Despite these shortcomings and aided by the authors' candid critique of the study, this is a useful document of an important procedure. Subjective probability is failsafe. If models or data are unbelievable or unavailable, about the best that can be done is to obtain the consensus of a panel of experts.

REFERENCE

1. J.R. Ellis et al., "A Subjective Probability Appraisal of Uranium Resources in the State of New Mexico," Grand Junction Office, U.S. Energy Research and Development Administration, GJO–110 (76), 1976.

VI COMPARISONS, RECOMMENDATIONS, AND CONCLUSIONS

Martin B. Zimmerman

26 RESOURCES AND RESERVES

A CONTRAST IN TERMINOLOGY

In this part the analyses of Parts III–V are brought together. The goal is to put resource and reserve estimates into the context of a cumulative cost curve. The focus is on concepts, not on estimation methodology. What inferences can be made about the future evolution of costs for each of the fuels considered? It is clear that reserves and resources mean different things for each of the minerals. The terms have been used loosely, and as we have seen, this has led to erroneous interpretations of the data. The misinterpretation arises from the different role exploration plays in the supply of each of these fuels. For coal, exploration is not a significant factor in U.S. supply. Hence attention focuses on the cost of developing already discovered deposits, which in turn focuses interest on the reserve base. The reserve concept for oil and gas comparable to the reserve base for coal is proved plus probable reserves. The comparable concept for uranium is simply reserves, since they are limited to already discovered deposits, developed and undeveloped.

The question arises as to whether a comparison of tonnage or barrel estimates means anything. Does it mean anything, for example, to divide reserves by production rates to get a life index? Would it be meaningful to then say we have twenty-five years of oil left and 300 years of coal? The answer to both these questions is no, even if we make "reserve" numbers comparable. This is so not because of incomparability of concept but, rather, because of the large differences in the supply process.

For coal, exploration is unlikely to make a large difference in future cost evolution. In oil or in uranium supply, it will. Depletion in known deposits leads to higher cost as mining proceeds to lower grade ore, or as reservoir pressure falls, or as coal seams thin. Exploration is costly and is worthwhile only when discovery of a new deposit forestalls the increased cost that comes with more intensive exploitation of known deposits. The exploration cost is equated at the margin to the increased development cost. Thus at any time it pays to discover only a fraction of potential deposits. Exactly how much it pays to discover depends on prices and on both development and exploration costs. For coal, to take one extreme, it does not pay to explore. The increase in development costs in known deposits is sufficiently mild so the expectation of finding new lower cost deposits is sufficiently small to discourage exploration. For oil this is not true. Part III showed how oil development costs are increasing, as are exploration costs and prices. For uranium, declining grade in known deposits provides ample incentive for exploration.

Where exploration is important, a comparison of known reserves is not meaningful. We are looking at only one segment of the supply process and we cannot conclude anything about future cost evolution.

Even if exploration were not a factor, comparison of reserve totals would still have no meaning. The reason is that, as was shown in Part I, the ultimate arbiter of the use and life of the resource stock is cost. A large amount of known high-cost reserves is not comparable to a small amount of known low-cost reserves. Reserve data are useful only to the extent that they shed light on the evolution of cost.

The foregoing objection to comparing reserve data carries over, a fortiori, to comparisons of resource data. Only for uranium is there a cost dimension to the resource estimate. Ultimately recoverable oil (or gas) reserves do not have an estimated cost attached to them. Estimates of ultimate reserves for specific areas can convey information about costs, but aggregate total estimates tell us only about a point we are unlikely ever to reach. Discovery process models offer some hope of explaining the evolution of cost, but a great deal of work remains to be done incorporating information on the size distribution of deposits to estimate costs of oil supply. Furthermore, the probability of finding a deposit is never made explicit. A comparison of uncertain quantities with different probabilities of success is difficult to interpret.

In sum, resource and reserve data are useful only to the degree that cost numbers can be attached to them. In the following sections the main conclusions of each part are summarized and or knowledge of future cost developments for each of the minerals is compared.

FUTURE COST INCREASES

The dichotomy between the known and the unknown portion of the resource stock is an oversimplification. The extent and characteristics of a known deposit are never completely known. Conversely, something *is* known about what potential deposits will look like. Nevertheless, the distinction is meaningful. Once a deposit is known to exist, concern turns to development cost. Before discovery of mineral in place, existence is the prime concern.

The distinction between the discovered and the undiscovered was maintained in each part. The discussion of the discovered portion of mineral resources in the United States consists of the chapters on proved reserves of oil, coal, and what the Grand Junction Office of the Department of Energy calls uranium reserves. The undiscovered portions were discussed in the chapters on undiscovered oil and in the chapter on uranium resources.

In the present discussion two questions are asked: Do the reserve/resource data reveal how fast costs will rise with cumulative output? If not, what do the reserve data reveal about the evolution of costs? For example, can upper limits be set on cost increases for likely levels of cumulative output?

COAL

Attempts at providing economically meaningful reserve concepts leave much to be desired. The reserve base, as seen in Part IV, includes coal of widely differing cost characteristics. There it was shown that the percentage increase in cost between the current average seam thickness and the cut-off criteria was over 200 percent. When mines with 28-inch seams represent the incremental mine, costs will have risen by 200 percent. There will still be coal in the reserve base available only at higher cost. However, at current rates of output, the 200 percent level would take many years to reach.

A great deal of coal is available in the interval between current costs and double current costs. But the trajectory at which these higher costs will be approached is not known. To know it we need to know the distribution of coal in the ground according to physical characteristics. Part IV covered some attempts to fill this gap by statistical estimation of the distribution of coal according to physical cost-determining characteristics. The supporting data are scarce and the conclusions tentative, however.

The behavior of costs in the future will be affected by interregional substitution. As depletion raises costs in the Eastern United States, Western output will expand and moderate the cost increase. The trajectory of cost increases actually experienced will depend on the extent of this interregional substitution.

Coal reserves indicate that depletion will not raise coal costs by more than 200 percent, for a long time. Long-range perspective may be comforting. For intermediate policymaking it begs all the interesting questions. The rate at which different regions develop will depend on the rate of cost increases in each segment of the cumulative cost curve in each area. The effect of antipollution regulations will depend upon the trajectory of cost increases in the low-sulfur portion of the stock. The effect of higher wages and/or capital costs will depend on depletion in the strip-mining sector versus depletion in the deep-mining sector. And ultimately the impact of depletion will be affected by technological change and factor price changes. Nevertheless, reserve data and the information they convey about depletion can help understand how technological change and factor price change will alter the industry.

OIL

The information conveyed by oil reserves is at the same time more precise and more ambiguous than that conveyed by coal reserves. Coal reserves represent mineral in place. They are known to exist, but the costs of development are unknown and only hinted at by the reserve classification.

Oil reserves can be also divided into those that are known to exist—the proved and probable—and the undiscovered. Part II showed that proved reserves are simply a shelf inventory. They are what will be recovered with existing facilities. The bulk of the cost has already been expended. Since the cost of lifting from the already installed capacity is small, proven reserves are a firm estimate of what will be produced at current or higher prices. The extraction cost at any time can also be calculated by using the decline rate as was done in Part II and Appendix B. That sets an upper limit on cost. The cost of bringing these reserves out of the ground faster can also be calculated. If we know the investment cost per annual barrel, again we can use the economics of decline to calculate how cost will behave.

But cost estimates of extracting proved reserves would refer only to a very limited portion of the stock. Proved reserves is such a small portion of the total stock that the upper limit calculation is like asking what would happen if steel mills shut down and all the shelf inventory were consumed. Prices would rise, but so would the incentive to open the mills.

The key issue is the cost of converting probable reserves into proved reserves. In the United States, the growth of proved reserves can be partitioned for analysis into discoveries, changes in recovery rates, and development of new oil in old fields. All that can be demonstrated with historical data is that the conversion of probable into proved reserves in the United States has experienced diminishing returns over the last thirty years. Future costs for this conversion are not known, nor can the reserve data be used to establish an upper limit. We need to know probable reserves, more about investment cost, and how cost varies by field before we can do that. Then reserve data could be combined with cost data to yield a cumulative cost curve.

Proved oil and gas reserves are a relatively precise estimate of the national inventory. The issue of supply in the intermediate run depends on the cost of developing proved reserves out of probable reserves. Reserve data alone indicate little about the process. Available information suggests only that it has been getting costlier to "prove" reserves. The historical information is useful in confirming that fact. More must be known about the distribution of reserves according to cost before predictions about the future can be made.

UNDISCOVERED OIL AND GAS

If reserve estimates are regarded as points on a cumulative cost function, proved reserves comprise the point closest to the origin. The last point is the ultimate amount recoverable. That, unfortunately, is all that estimates of ultimate reserves say. They are devoid of cost information. Part III detailed the methods that have been used to determine the ultimately recoverable resources.

Methods that relied on time-series techniques share the problem that information on the past is contaminated by special circumstances. Extrapolation from the past is based on previous prices, reserve characteristics, and so on. Thus, even as estimates of ultimately recoverable resources, these methods are not satisfactory. The discovery process models provide a method for estimating ultimates that is not mere extrapolation. The detail these models include on the size distribution of reservoirs can be used too to establish cost of development. Because these models require a great deal of data, they are most useful where much prior knowledge has been gathered about the play. For areas where little or no exploratory activity has been undertaken, analysts must rely on geologic analogy or judgment. A promising area of current research is the integration of statistical methods with the judgment that ultimately must be used in the exploration process.

The point that should be stressed is that we are less interested in ultimately recoverable reserves than in the cost trajectory to be followed in moving toward that total. Methods for estimating the total ultimate recovery are useful only insofar as the detail uncovered by the method informs us about cost. The size distribution of reservoirs is interesting because it translates into cost. The total content of the reservoirs, without the detail, is a number of much less importance.

So far as petroleum reserves are concerned, a good body of knowledge about proved oil reserves exists. But that is little more than knowing the inventory. Ultimately recoverable resources reveal little more than an absolute limit. In between there is the process of discovery and then the transformation of probable reserves into proved reserves, about which little is known.

URANIUM

Uranium reserve estimates are the most comprehensive reserve estimates. They include not only proved reserves but also the undiscovered resource. What is unique about uranium reserve estimates is that the analysts who devised the estimating scheme started out by attempting to attach a cost estimate to the reserve quantity. They began by trying to estimate the cumulative cost curve.

The Discovered Portion (Reserves)

The reserves estimated for uranium are in already defined deposits. In this respect they are akin to mineral in place. They correspond to the proved plus probable oil reserves, since they include discovered undeveloped as well as developed deposits. Besides quantities available, the forward cost at which they are available is also estimated. Pointed out in Part V were the technical questions about the correct cost calculation and the use of average cost not marginal cost.

Furthermore there are two central conceptual difficulties with the forward cost concept. In the first place expenditure and revenue streams are not properly discounted. Since revenue is earned as uranium is produced, the value of output must account for the time value of money. The further into the future output is postponed, the less its present value. Properly reckoned cost would express capital expended per "present value" pound of uranium. In fact the forward cost estimates count capital cost per pound of ultimately produced reserves with no consideration given to the rate at which the reserve is produced.

The second conceptual difficulty is the confusion of points on the short- and long-run cost function. Some $40/lb reserves are calculated as $40 reserves only because there is already capital in place. This is a correct estimate of cost in the short run when there might be excess capacity. It is not the cost of digging and equipping a new mine. Therefore, cost estimates attached to the reserve data do not indicate how costs will increase in the long run as new mines are opened. The huge jump in uranium prices in the early 1970s was due in part to a change from excess supply to excess demand. The price necessary to bring forth new mines was far above the forward cost estimate.

The Potential Resources

The nonreserve category consists of estimates of undiscovered resources. The Grand Junction Office also guesses about the ultimate cost of discovering and developing these deposits. Since no investment has taken place in these deposits, all costs are forward costs.

The difficulties discussed in Part III, on undiscovered oil deposits, carry over to estimates of potential uranium resources. The estimates are established by methods that amount to geological analogy combined with judgment. The estimation process involves biases, as Part V showed. Nevertheless, the exercise is useful. The resulting estimates do have economic meaning, although this meaning has been widely misinterpreted.

The resource estimates are presented as the amount of ore to be discovered at a given total cost. The resource estimates can be regarded as central values in a probability distribution. The GJO expects so many tons of ore in the $30/lb category to be discovered. The actual realized total may be more or it may be less. The difference in certainty reflected in the probable, possible, and speculative categories can be captured by the variance in the estimate. The 90 percent confidence interval would be much wider for speculative than for probable resources. In other words, the estimate is more likely to be wrong the more speculative the category. Even for the least speculative category there is a substantial risk of error. Unfortunately no quantitative estimate of the variance exists but only the qualitative notion that it increases the more speculative the estimate.

The misinterpretation of uranium reserves concepts is to assume that probable reserves or even possible plus probable constitute an upper limit, or even a planning estimate of the amount to be found. The actual amount will be more or less and what we need to know is the confidence interval about the estimate. A confidence interval is more useful for planning purposes because it reflects the uncertainty surrounding the estimates. The Grand Junction Office of the U.S. Department of Energy is clearly pointing in the right direction, however, by attempting to attach a cost estimate to resource categories.

THE NEED FOR FURTHER RESEARCH

It is clear from the previous chapters that we are far from a complete understanding of the future behavior of costs. The discovered portion of energy resources in most cases tells us little about the future cost trajectory. The exception is coal, for which the discovered reserves represent a large stock relative to consumption and where cost increases in the known deposits will be sufficiently small to discourage large-scale exploration. For oil, gas, and uranium, methods are being developed to better estimate what lies yet to be discovered beneath the surface. For such estimates to improve our knowledge about future costs, however, we need to integrate the geological estimation with an economic understanding of costs and supply. The preceding chapters have elucidated this link and pointed out the gaps. It now remains to fill those gaps.

APPENDIXES

APPENDIXS

APPENDIX A

THE RATE OF INTEREST AND THE OPTIMAL DEPLETION RATE

A much simplified calculation of the optimal depletion rate in a single mineral deposit, and its relation to the rate of interest, are presented here. Contrary to the impression of many mineral economists, a rise in the discount rate does not necessarily imply a speed-up in the exploitation of a resource. Considered here is the example of an oil or gas reservoir, because it is realistic to assume the decline of output over time as a constant exponential, approximately equal to the depletion rate—that is, the percent removed each year. This permits a very simple mathematical scheme.

All types of *user cost* are included by implication (see Chapter 1). Short-term operating costs over the life of the deposit are assimilated into investment cost by converting them to a single present value. Hence there is no equating of operating cost to price at any moment. The higher the depletion rate of the finite reserve, the higher will be the cost per unit. One may assume that the price rises (or falls) over time; hence the fourth type of user cost, the present discounted value of higher prices, is also allowed for.

Thus the formula embodies the effects of higher or lower output in the individual deposits, or all deposits taken together, as an endogenous variable. It accomodates any desired assumption about the whole productive system, whose long-run marginal cost, or supply price, is taken as an exogenous fact.

R_0 = initial reserve to be depleted, R_t = reserve in year t.

Q_0 = initial level of output in units per year, Q_t = output in year t.

a = constant decline rate of output in percent per year.

$$R_0 = \int_0^\infty Q_t\, dt = Q_0 \int_0^\infty e^{-at}\, dt = Q_0/a \ , \qquad (A\text{-}1)$$

$$R_t = R_0 - Q_0 \int_0^t e^{-at}\, dt = Q_t/a \ . \qquad (A\text{-}2)$$

Hence $a = Q_0/R_0 = Q_t/R_t$.

If we define

I = investment needed to obtain Q_0.

K = constant annual operating cost, in dollars per year.

r = discount rate, risk-adjusted, in percent per year.

P_0 = initial price.

g = expected rate of price increase, in percent per year.

$P_t = P_0\, e^{gt}$, and taking T as infinity

Then,

$$\text{Net present value (NPV)} = \int_0^\infty P_t\, Q_t\, e^{-rt}\, dt - I - \int_0^\infty K\, e^{-rt}\, dt$$

$$\text{NPV} = \int_0^\infty (P_0\, e^{gt}\, Q_0 e^{-at})\, e^{-rt}\, dt - I - K \int_0^\infty e^{-rt}\, dt$$

$$\text{NPV} = \int_0^\infty (P_0\, Q_0\, e^{-(a+r-g)t})\, dt - I - K \int_0^\infty e^{-rt}\, dt$$

$$\text{NPV} = ((P_0\, Q_0)/(a+r-g)) - I - (K/r) \ . \qquad (A\text{-}3)$$

If the reservoir is barely profitable, then NPV = 0 and

$$P_0^* = \frac{I + (K/r)}{Q} (a+r-g) \ . \qquad (A\text{-}4)$$

That is, the supply price of the incremental output is just equal to the total investment per unit of output, plus present value of the constant (per annum) operating cost of the new installation, multiplied by the combined discount rate. This supply price is the *average cost* of the incremental output.

To find the optimal rate of output, differentiate NPV with respect to Q_0:

$$\frac{\partial \text{NPV}}{\partial Q} = \frac{(a+r-g)\,P_0 - P_0 Q_0\,(\partial a/\partial Q_0)}{(a+r-g)^2} - \frac{\partial I}{\partial Q} - \frac{1}{r}\frac{\partial K}{\partial Q_0} \ . \qquad (A\text{-}5)$$

Recall that $a = Q/R$; hence $\partial a/\partial Q = 1/R = a/Q$; denoting the derivatives of I and K by primes, we have

$$\frac{\partial \text{NPV}}{\partial Q_0} = \frac{P_0\,(r-g)}{(a+r-g)^2} - I' - \frac{K'}{r} \ . \qquad (A\text{-}6)$$

To ensure that we have a maximum, we take the second derivative:

$$\frac{\partial^2 \, \text{NPV}}{\partial Q_0^2} = \frac{-P_0 \, (r-g) \, [2 \, (a+r-g) \, (a/Q)]}{(a+r-g)^3} - I'' - \frac{K''}{r} \, . \tag{A-7}$$

If we assume that I and K increase more than proportionately with Q_0, the first and second derivatives are positive. Therefore, their contribution to Eq. (A-7) is always negative. The first term is also negative unless g exceeds r, which would imply that the deposit should not be operated at all. Putting this case aside, Eq. (A-7) is negative and NPV is maximum where its derivative is zero. At that point marginal supply price is:

$$P^{**} = \frac{[I' + (K'/r)] \, [a+r-g]^2}{(r-g)} \, . \tag{A-8}$$

Either way, considering P^* as average cost and P^{**} as marginal cost,

$$P^{**} = P^* \, (a+r-g)^2 / (r-g) \, . \tag{A-9}$$

This marginal cost is the obstacle to increasing the rate of exploitation. Returning to Eq. (A-6), we set $\partial \text{NPV}/\partial Q$ to zero for maximum present value and rearrange to obtain the optimum depletion rate, which as shown before approximates the corresponding decline rate:

$$(a+r-g)^2 = \frac{P_0 \, (r-g)}{I + (K/r)} \, ,$$

$$a^* = \sqrt{\frac{P_0 \, (r-g)}{I' + K'/r}} - (r-g) = (P_0 \, (r-g)/I' + K'r^{-1})^{\frac{1}{2}} - (r-g) \, . \tag{A-10}$$

As a test of consistency: If $g = (a+r)$, NPV is infinite (Eq. [A-3]); also, the average cost is zero (Eq. A-4). Consider now the economic meaning of this scheme. With stable prices, the time value of money is r. If the price is expected to rise, it is $(r-g)$, except that the present-value-equivalent of the operating costs (K/r), or at the margin (K'/r), is not affected by expected price changes.

The choice of the depletion rate a^* is the key investment decision. Average cost is a result, equal to the combined discount rate $(a+r-g)$ multiplied by the unit investment $(I + K/r)/Q$. If $r = g$, or $r = 0 = g$, then $a^* = 0$. This is consistent, because with zero time value of money it makes no sense to invest in order to produce from a mineral deposit. Since the reserves cannot be increased, increasing the rate of output only accelerates ultimate recovery but cannot increase it, thus meaning additional expense. In other words, with no premium on the present compared with the future, there is no incentive to produce in the present, and the system is at rest with zero output. Production requires a disequilibrium.

If $a*$ is negative, this would imply that disinvestment would be optimal, putting mineral back into the ground—if that were possible.

Now let us examine the effect on the optimal depletion rate of changes in the underlying variables, expressed in Eq. (A-10). For any given discount rate and expected rate of price change, a higher price or lower investment or operating cost always raises the optimal rate of depletion. (Cf. p. 20 above.) The more profitable the operation, the faster it should proceed. But marginal investment and operating costs are increasing functions of a. We can think of the ascertainment of $a*$ as an iterative process; one tries higher and higher values of a, and profit increases; but at some point, rising marginal costs will halt the increase.

Thus the higher the current price, and the lower the cost, the faster the depletion; contrariwise, lower prices and higher costs will lower the depletion rate. This is consistent with what was said above (p. 20). But a higher discount rate r works both ways. It lowers the denominator of the fraction, and also increases the numerator, tending to increase the depletion rate $a*$. But it increases a negative term, tending to decrease depletion.[1]

The higher expected rate of price increase g raises the supply price, hence lowers the optimal depletion rate. But the addition to the supply price is the higher stream of expected future benefits, and its present value depends on the discount rate. This leaves its amount uncertain, hence the importance of the effect.

The mathematical formulation is of course greatly oversimplified, but it captures two basic equivocations. The discount rate r affects present value of both revenues and costs, both in the same direction, with a net effect upon the depletion rate which may be positive or negative. A higher rate of expected price increase lowers the depletion rate, but its importance depends on the discount rate.

Recent experience is suggestive. During the frenzied oil and gas investment boom of 1979-81 in the United States, real oil prices were universally expected to rise at about 3 percent per year, real, and gas prices with them.[2] Yet for oil, there is no indication of any tendency to reduce depletion rates, while for gas, rates of deliverability were considerably increased, with considerable embarassment in 1982 and later, when gas consumption stagnated or even declined.

The ownership of a mineral may be public or private. It is often said that difference of ownership makes a difference in user cost and therefore in production policy, because governments have typically longer time horizons, and discount at lower rates. The proposition seems very implausible. A ruling party, family, or junta will probably not have a long time horizon. Suppose there is a probability P that in any given year the ruling group will be overthrown and their control of the mineral revenues will be irretrievably lost. Then the present value of that year's net receipts is: $PV_t = (1 - p)^t / (1 + r)^t$.

In other words to receive the payments in year t they must also have received payments in every year previous to t, and the probability of still being in power is $(1 - p)^t$. In that case the discount rate adjusted for the loss of power is r', such that

$$1/(1 + r') = (1 - p)/(1 + r) \ , \tag{A-12}$$

$$r' = (r + p)/(1 - p) \ .$$

Suppose it appears that the X family has a 50–50 chance of being in power ten years from now. Then $(1 - p)^{10} = 0.5$, $(1 - p) = 0.933$, and $p = 0.067$. A risk-adjusted discount rate r' must be

$$\frac{r + 0.067}{0.933} = 1.072r + 0.072 \ . \tag{A-13}$$

Thus, if 10 percent per year were a discount rate allowing for normal commercial risk, the family ought to calculate with 17.9 percent. But no one can tell, without more data, whether this means faster or slower depletion.

REFERENCES

1. For a fuller discussion of the rate of discount, see Adelman, "OPEC as a Cartel," in Griffin and Teece, eds., *OPEC Behavior and World Oil Prices* (1982), pp. 40–42, 57–59.
2. For some light on this extraordinary episode, see *Oil & Gas Journal*, September 27, 1982, pp. 118, 210–216.

APPENDIX B

DECLINE RATES AND DEPLETION RATES

The exponential decline curve is

$$Q_t = Q_0 e^{-at} \quad , \tag{B-1}$$

where Q_t = production in year t; Q_0 = initial production; a is the decline rate, in this case constant.

The hyperbolic decline curve is

$$Q_t = Q_0 (1 + na_0 t)^{-1}/n \quad , \tag{B-2}$$

and

$$a_t = a_0 K Q_0^n \quad , \tag{B-3}$$

where K is a scale factor and n is an empirical constant between zero and unity, usually below 0.4.[1] Thus the decline rate itself declines.

Cumulative production, or reserves, for constant percentage (exponential) decline is

$$R_0 = Q_0 \left(\frac{1 - e^{-at}}{a} \right) = \frac{Q_0}{a} \left(1 - \frac{Q_b}{Q_0} \right), \quad \text{and} \quad a = \frac{Q_0}{R_0} \left(1 - \frac{Q_b}{Q_0} \right) \quad , \tag{B-4}$$

when Q_b is production at time of abandonment. For hyperbolic decline

$$R_0 = \frac{Q_0^n (Q_0^{1-n} - Q_b^{1-n})}{(1 - n)a_0} \quad , \quad \text{and} \quad a_0 = \frac{Q_0^n (Q_0^{1-b} - Q_b^{1-n})}{(1 - n)R_0} \tag{B-5}$$

403

It is convenient to neglect final output levels, which are typically very small percentages of initial output, in which case:

$$\text{Exp } a_0 = Q_0/R_0 \ . \tag{B-6}$$

$$\text{Hyp } a_0 = Q_0/R_0 \ (1 - n) \ . \tag{B-7}$$

Reserves under hyperbolic decline are initially somewhat larger than those under constant percentage decline, $\text{Hyp } R_0 = \text{Exp } R_0/(0.8 \pm 0.2)$.

Most actual decline curves are of the hyperbolic type. In practice, therefore, constant percentage decline is simply a conservative way of estimating reserves from known production rates. But as $\text{Hyp } a_t$ declines below $\text{Hyp } a_t$, the discrepancy decreases between $\text{Hyp } R_t$ and $\text{Exp } R_t$. In approximate work, or for groupings or reservoirs, exponential decline is assumed.

Consider Eq. (B-4). Where t is large, $(1 - [Q_t/Q_0])$ can be safely neglected; but as t decreases, the value of a must decrease, as the following table shows:

R_0/Q_0	t	a
7.69	∞	.1300
	50	.1299
	30	.1272
	25	.1242
	20	.1176

The life of a hydrocarbon reservoir is determined at the endpoint, where current operating costs exceed the net price. The higher the initial output the longer the life, although the larger well usually has higher operating expenses. The life of reservoirs is typically long; it is sometimes said that old oil fields never die. Probably the best rule of thumb would be twenty-five years.

Let us assume a twenty-five-year life and a true decline rate of 12.42 percent, to explore a second bias. Suppose we have a stable population of twenty-five reservoirs. Each year there is one new well to produce one barrel, and one being phased out after producing for twenty-five years. The lifetime production of each reservoir will be 7.69 barrels; this is also the annual production. The production of each reservoir in each year is equal to $Q_t = e^{-.1242t}$. In the first year (zero) production is one barrel; in the twenty-fifth year ($t = 24$) it is 0.051. The remaining reserves in each pool are equal to starting reserves less accumulated output; that is,

$$R_t = 7.69 \ \frac{1 - e^{-.1242t}}{.1242} = 7.69 - 8.05(e^{-.1242t}) \ . \tag{B-8}$$

In the first year ($t = 0$), reserves start at 7.69. In the twenty-fifth year ($t = 24$), reserves begin as equal to planned production of 0.05, then go to zero by the end of the year.

The Q/R ratio starts out at 0.13 (that is, $1/7.69$), a little over the decline rate. But it ends up as 100 percent at the start of the final year. Summing up for the whole group of reservoirs: Their remaining reserves are at any given moment:

$$R_t = 25(7.69 - 8.05) + 8.05 \int_0^{24} e^{-.1242(24)} \, dt$$

$$= -9 + 8.05 \; \frac{1 - e^{-.1242(24)}}{.1242} \; = \; -9 + 64.41 = 55.41 \; .$$

$$(B-9)$$

Therefore at any given moment the ratio of current production to current reserves for all reservoirs taken together is $7.69/55.41 = 0.1387$. This is the ratio that would be actually observed, not the .1242 decline rate postulated earlier. The error is 1.45 percentage points.

In general the longer lived the average well and the lower the ratio of final output to initial output, the less the error. Moreover, there is a factor that points in the opposite direction and makes the observed Q/R tend to understate the true decline rate. We assumed that production began at the peak and tailed off. But this would mean that complementary facilities (separators, tanks, flow lines, pipelines, and so on) work at capacity only at the start and are partly idle thereafter. This would be wasteful. There is in fact a quick buildup to initial capacity, then a plateau, following which production declines relatively rapidly. Output is kept in check at the start to lessen the investment above ground. Hence the R_t/Q_t we observe embodies increasing production in some reservoirs, stationary production in others, and decline in most—and the decline in many is correspondingly more rapid than the reserves/production ratio would indicate.

Two publications of the National Petroleum Council have touched on the decline rate: In 1961, there was an estimate somewhat *greater* than the observed Q_t/R_t, both for oil and gas.[2] The 1976 publication has already been referred to.[3] It therefore seems best to follow their example and use the unadjusted ratio to approximate the decline rate.

REFERENCES

1. John J. Arps, "Estimation of Primary Oil and Gas Reserves," in T. C. Frick and R. W. Taylor, eds., *Petroleum Production Handbook* (New York: McGraw-Hill, 1962), pp. 37–43 to 37–45.
2. National Petroleum Council, *Report of the National Petroleum Council on Proved Petroleum and Natural Gas Reserves and Availability*, Washington, D.C.: NPC, 1961, p. 27.
3. National Petroleum Council, *Enhanced Oil Recovery*, Washington, D.C.: NPC, 1976.

THE SOVIET RESERVE CLASSIFICATION SYSTEM

Although the Soviet Union considers oil reserve information an official secret, they have in the past published highly detailed gas reserve data. Their classification system was extensively revised in 1967,[1] and corresponds well to the flow concept advocated in this book.[2]

A. Completely developed reservoir; detailed information available on sands, drive, pressures, and properties of hydrocarbon. Parameters known within ± 10 percent.

B. Several producing wells; detailed reservoir information as in *A*; delineation incomplete but geological and geophysical indications of limits. Parameters known within ± 20 percent.

C-1. At least one producing well; reservoir properties must be closely similar to a nearby, well-explored pool. Parameters known within ± 50 percent.

C-2. Untested zones in producing wells; untested structures in an established province; untested fault blocks; or other closures in a proved field.

D. Predictions of a new area, largely by analogy and statistical estimation, within a larger area already productive (*D-1*) or unproductive (*D-2*).

Proved reserves would include category *A* and the immediate vicinity of the producers in *B* and *C*. Smaller pools classified as *B* would fall entirely within proved reserves, whereas larger ones would be mostly proved, partly probable; *C-1* would be mostly probable, partly proved; *C-2* would be entirely probable.

REFERENCES

1. R. W. Campbell, *The Economics of Soviet Oil and Gas*, Resources for the Future (Baltimore: The Johns Hopkins University Press, 1968).
2. V. V. Semenovich et al., "Methods Used in the U.S.S.R. for Estimating Potential Petroleum Reserves," in *The Future Supply of Nature-Made Petroleum and Gas*, ed. R. F. Meyer (New York: Pergamon Press, 1977), pp. 139–52.

APPENDIX D

THE RESOURCES AND RESERVES OF STRIPPABLE COAL

The first effort to classify strippable reserves and resources defined strippable resources as all coal lying in seams 28 inches or thicker and less than 120 feet below the surface in the measured, indicated, and inferred categories.[1] A portion of that resource was deemed "economically recoverable reserves." The latter portion was determined by applying an 80 percent recovery rate to the resource and eliminating all coal rendered inaccessible due to surface obstructions such as railroads and towns. Further, all coal under overburden ratios greater than a specified maximum is excluded from strippable reserves.

It is clear that the overburden ratio criterion was used for reserves. The question is whether strippable resources also meet the ratio restrictions. The U.S. Bureau of Mines' definition of the resource category[2] mentions that the overburden ratio criterion was used in defining strippable coal resources, the most general of the categories. However, the description of the methodology claims that to arrive at reserves, all coal in the resource category not meeting the overburden ratio criterion was excluded.[3] It is implied therefore that the resource category includes coal not meeting the overburden ratio criterion.

Estimates of strippable reserves and resources for major producing states are presented in Table D-1. Column (1) is the resource. Column (2) represents reserves; there is quite a reduction in tonnage between *resources* and *reserves*. It would be useful to know exactly what separates the two categories. How much of the reduction in the reserve category is due to inaccessible coal, and how much is due to the overburden ratio exclusion criterion? Information about these factors would illuminate the issue of availability discussed in Chapter 20.

Table D-1.   Strippable Resources and Reserves, 1971, in Selected States, Million Tons.

State	(1) Resources	(2) Reserves	(3) Reserve Base
Pennsylvania	2,272	752	1,191
Maryland	151	21	146.3
West Virginia	11,231	2,118	5,212
Virginia	1,556	258	679
East Kentucky	4,619	781	3,451.16
Tennessee	483	74	317.59
Alabama	667	134	15,724
Ohio	5,566	1,133	3,653.9
Illinois	18,845	3,247	12,222.9
Indiana	2,741	1,196	1,674.1
West Kentucky	4,746	977	3,914
Montana[a]	14,871	6,897	45,562
Wyoming	22,128	13,971	23,845

Sources: Columns (1) and (2) from U.S. Bureau of Mines, *Strippable Reserves of Bituminous Coal and Lignite in the United States*, Information Circular 8531 (Washington, D.C.: U.S. Government Printing Office, 1971). Column (3) from U.S. Bureau of Mines, *The Reserve Base of U.S. Coals by Sulfur Content*, Information Circulars 8680, 8693 (Washington, D.C.: U.S. Government Printing Office, 1974, 1975).

The next estimate of strippable coal reserves was in connection with the estimation of the reserve base.[4] The strippable reserve base disaggregated by state is presented in column (3) of Table D-1. The reserve base numerical estimates lie between estimated resources and the earlier estimate of reserves. The seam thickness of 28 inches and an overburden depth of 120 feet were used to delineate the reserve base as it was with strippable resources. The reserve base excludes inferred resources. There is an ambiguity about overburden ratios, however. Were overburden ratios actually used in defining the reserve base? The report says yes; other data say no. Some indirect evidence suggests that the reserve base is the same as the strippable resource minus the inferred category and neither considered the overburden ratio criterion.

The evidence consists of tracing through the various estimates for the state of Illinois. The Illinois Geological Survey is one of the best state level geological offices, and certain of its studies provide evidence corroborating the hypothesis that overburden ratios were not considered in the reserve base.

In a recent study,[5] the Illinois Geological Survey presents estimates of strippable coal resources in the two major Illinois coal seams—the Herrin (No. 6) and Harrisburgh-Springfield (No. 5). The study defines strippable coal as all coal 18 inches or thicker under less than 150 feet of overburden. Strippable coal is divided further into two classes, one corresponding to the total of measured and indicated and the other to inferred. No overburden ratio is considered. The totals by county are presented in column (1) of Table D-2. Column (2) reproduces the reserve base estimates for the same seams by county. The two columns

Table D-2.   Comparison of Different Reserve Estimates for Illinois #5 and #6 Seams, Million Tons.

County	Coal Seam	(1) Total Reserves	(2) Reserve Base
Bureau	6	163.337	161.1
Cass	6	19.700	19.7
Fulton	6	249.285	235.79
	5	802.386	439.38
Gallatin	6	114.506	114.5
	5	115.489	115.85
Greene	6	75.359	73.94
Henry	6	192.527	192.53
Jackson	6	149.318	142.02
	5	99.843	95.89
Jersey	6	50.521	41.56
Knox	6	257.066	244.5
	5	468.222	269.08
LaSalle	6	48.951	95.23
Livingston	6	25.053	25.05
Macoupin	6	191.602	163.1
Madison	6	392.025	392.02
Monroe	6	6.726	6.73
Peoria	6	881.821	858.98
	5	468.135	468.14
Perry	6	896.767	807
	5	181.430	166
Randolph	6	279.139	260.57
	5	160.477	156.62
St. Clair	6	1,249.123	1,162.75
Saline	6	284.572	260.29
	5	93.422	91.42
Schuyler	5	105.054	99.11
Scott	6	6.120	0
Stark	6	243.061	234
Tazewell	6	57.637	57.64
	5	28.147	28.15
Warren	5	.807	0
Williamson	6	290.718	266.99
	5	200.268	192.95

Sources: Column (1) from William H. Smith and John B. Stall, *Coal and Water Resources for Coal Conversion in Illinois*, Cooperative Resources Report 4 (Urbana: Illinois Geological Survey, 1975) and Column (2) U.S. Bureau of Mines, *The Reserve Base of U.S. Coals by Sulfur Content*, Information Circulars 8680 and 8693 (Washington, D.C.: U.S. Government Printing Office, 1975).

are extremely close, corresponding exactly in many instances. The Illinois estimates are generally slightly higher, which is to be expected, since they considered coal 18 inches or thicker. It would appear that the reserve base is simply coal available in seams at least 28 inches thick and at less than 150 feet.[a] The ratio appears not to have been considered.

a. Recall that in Illinois the depth criterion is 150 feet, not the usual 120 feet.

This evidence is suggestive, not conclusive. It is possible that all coal in the two seams greater than 18 inches also meets the overburden ratio, so that explicit consideration of the ratio was not necessary in Illinois. To confirm the hypothesis for the entire United States would require more data on individual states.

REFERENCES

1. U.S. Bureau of Mines, *Strippable Reserves of Bituminous Coal and Lignite in the United States*, Information Circular 8531 (Washington, D.C.: U.S. Government Printing Office, 1971).
2. Ibid.
3. Ibid., p. 4–3.
4. U.S. Bureau of Mines, *The Reserve Base of U.S. Coals by Sulfur Content* Information Circulars 8680 and 8693 (Washington, D.C.: U.S. Government Printing Office, 1975).
5. William H. Smith and John B. Stall, *Coal and Water Resources for Coal Conversion in Illinois*, Cooperative Resources Report 4 (Urbana: Illinois Geological Survey, 1975).

BIBLIOGRAPHY

PART I: INTRODUCTION TO ESTIMATION
OF PRIMARY ENERGY SUPPLIES

Adelman, M. A. *The World Petroleum Market.* Baltimore, Md.: The Johns Hopkins University Press, 1972.

Barnett, H. J., and Morse, C. *Scarcity and Growth.* Baltimore, Md.: The Johns Hopkins University Press, 1963.

Brinck, J. "Calculating the World's Uranium Resources" *Euratom Bulletin* 6, no. 4 (December 1976): 109–14.

Carlisle, D. "The Economics of a Fund Resource with Particular Reference to Mining." *American Economic Review* 44 (1956): 595–616.

Cook, E. "Limits to Exploitation of Nonrenewable Resources." *Science* 191 (February 20, 1976): 667–82.

Erickson, R. L. "Crustal Abundance of Elements, and Mineral Reserves and Resources." In *United States Mineral Resources*, U. S. Geological Survey Professional Paper 820. Washington, D.C.: U. S. Government Printing Office, pp. 455–468.

Fisher, A. C. "On Measures of Natural Resource Scarcity." Stockholm School of Economics and International Institute for Applied Systems Analysis, Stockholm, Sweden, 1977.

Gaffney, M., ed. "Editor's Introduction and Editor's Conclusion." In *Extractive Resources and Taxation.* Madison: University of Wisconsin Press, 1967.

Gordon, R. L. "A Reinterpretation of the Pure Theory of Exhaustion." *Journal of Political Economy* (1967): 274–86.

Gray, L. C. "Rent under the Assumption of Exhaustibility." In *Extractive Resources and Taxation.* Edited by Mason Gaffney. Madison: University of Wisconsin Press, 1967.

Harris, D. P. "Information and Conceptual Issues of Uranium Resources and Potential Supply." Paper presented to Workshop on Energy Information for the U. S. Energy Information Administration, sponsored by Institute of Energy Studies, Stamford University, Stanford, Calif. 1977.

Herfindahl, O. C., and Kneese, A. V. *Economic Theory of Natural Resources.* Columbus, Ohio: Bobbs-Merrill, 1974.

Hotelling, H. "The Economics of Exhaustible Resources." *Journal of Political Economy* 39, no. 2 (April 1931): 137-75.

Kneese, A. V. "Natural Resources Policy 1975-1985." *Journal of Environmental Economics and Management* 3 (1976): 253-88.

Koopmans, T. C. "Some Observations on 'Optimal' Economic Growth and Exhaustible Resources." In *Economic Structure and Development: Essays in Honour of Jan Tinbergen.* Edited by H. C. Bos. Amsterdam: North-Holland, 1973, pp. 239-56.

Lane, K. F. "Choosing the Optimum Cut-Off Grade." *Quarterly of the Colorado School of Mines* 59 (October 1964): 811-29.

Marshall, P. "Ford Foundation Report 'Too Optimistic' on U. S. Uranium Supplies." *Nuclear Engineering International* (August 1977): 36-38.

Mason, E. S. "Political Economy of Resource Use." In *Extractive Resources and Taxation.* Edited by Mason Gaffney. Madison: University of Wisconsin Press, 1967.

McKelvey, V. E. "Approaches to the Mineral Supply Problem." *Technology Review* (March/April 1974): 13-23.

Nordhaus, W. D. "The Allocation of Energy Resources." *Brookings Papers on Economic Activity* 3 (1973): 529-76.

Nordhaus, W. D., and van der Heyden, L. "Modeling Technological Change: Use of Mathematical Programming Models in the Energy Sector." *Cowles Foundation Discussion Paper No. 457.* Yale University, New Haven, Conn., April 18, 1977.

Peterson, F. M., and Fisher, A. C. "The Exploitation of Natural Resources, A Survey." *Economic Journal* 87 (December 1977): 681-721.

Schanz, J. J., Jr. "Mineral Economics-Perspectives of the Past, Present, and Future." In *Economics of the Mineral Industries,* 3rd ed., Edited by William A. Vogely. Washington, D. C.: American Institute of Mining, Metallurgical and Petroleum Engineers, 1976, pp. 807-20.

Scott, A. T. "The Theory of the Mine under Condition of Certainty." In *Extractive Resources and Taxation.* Edited by Mason Gaffney, Madison: University of Wisconsin Press, 1967.

Searl, M. F., and Platt, J. "Views on Uranium and Thorium Resources." *Annals of Nuclear Energy* 2: 751-62.

Singer, D. A. "Long-Term Adequacy of Metal Resources." *Resource Policy* (June 1977): 127-33.

Singer, D. A. "Mineral Resource Models and the Alaskan Mineral Resource Assessment Program." In *Mineral Materials Modeling, A State-of-the-Art Review.* Edited by William A. Vogely. Washington, D. C.: Resources for the Future, 1975, pp. 370-82.

Singer, D. A. "Properties of Mineral Resources and Information Requirements for Assessment." U. S. Geological Survey, Menlo Park, Calif., 1978.

Skinner, B. F. *Earth Resources.* 2nd ed. Englewood Cliffs, N. J.: 1976.

Smith, V.K., ed. *Scarcity and Growth Reconsidered.* Baltimore, Md.: The Johns Hopkins University Press, 1979.

Smith, V. K., and Krutilla, J. V. "Resource and Environmental Constraints to Growth." Discussion Paper D-17. Resources for the Future, Washington, D. C. 1977.

Solow, R. M. "The Economics of Resources or the Resource of Economics," Richard T. Ely Lecture. *Proceedings of the American Economic Association* 64, no. 2 (May 1974).

Solow, R. M., and Wan, F. Y. "Extraction Costs in the Theory of Exhaustible Resources." *Bell Journal of Economics and Management Science* 7, no. 2 (Autumn 1976): 359-70.

Uhler, R. S. "Economic Concepts of Petroleum Energy Supply." In *Oil in the Seventies* (Vancouver, B.C.: The Fraser Institute, 1977), pp. 3-42.

PART II: OIL AND GAS: ESTIMATION
OF DISCOVERED RESERVES

Adelman, M. A. *The World Petroleum Market.* Baltimore, Md.: The Johns Hopkins University Press, 1972.

Alberta Energy Resources Conservation Board. *Application to Alberta Energy Resources Conservation Board.* November 11, 1977.

American Association of Petroleum Geologists. "North American Drilling Activity." *AAPG Bulletin,* various issues.

American Petroleum Institute. *Joint Association Survey,* Section II. Washington, D. C.: API, March 1977.

American Petroleum Institute. *Organization and Definitions for the Estimation of Reserves and Productive Capacity of Crude Oil.* Technical Report No. 2. Washington, D.C.: API, 1970.

American Petroleum Institute. *Quarterly Review of Drilling Statistics.* Section II. Washington, D. C.: API, various years.

American Petroleum Institute–American Gas Association–Canadian Petroleum Association (API-AGA-CPA). *Reserves of Crude Oil, Natural Gas Liquids, and Natural Gas in the United States and Canada as of December 31, 19xx.* Washington, D. C.: API-AGA, annually, with slight title changes, since 1946.

Arps, J. J. "Estimation of Primary Oil and Gas Reserves." In *Petroleum Production Handbook.* Edited by T. C. Frick and R. W. Taylor. New York: McGraw-Hill, 1962.

Beydoun, Z. R., and Dunnington, H. V. *The Petroleum Geology and Resources of the Middle East.* London: Scientific Press, 1975.

Crandall, M. S. "The Economics of Iranian Oil." MIT Energy Laboratory Working Paper No. MIT-EL 75-003WP, Massachusetts Institute of Technology, Cambridge, Mass., March 1975.

Craze, R. C. "Development Plants for Oil Reservoirs." In *Petroleum Production Handbook.* Edited by Thomas C. Frick, New York: McGraw-Hill, 1962.

Federal Power Commission National Gas Survey, *National Gas Reserves Study.* Washington, D.C.: FPC, revised September 1973.

Garcia, M. N. "Correlacion de Registros Geofisicos de Pozos Perforados en el Area Tabasco–Chiapas y la Plataforma Continental de Campeche y Yucatan." In *Energeticos.* Comission de Energeticos, Secretaria del Patrimonio, Mexico, January 1978.

Independent Petroleum Association of America. *Report of the Cost Study Committee.* Washington, D.C.: IPAA, May 1975 and May 1980.

IGT Highlights (July 31, 1978), Institute of Gas Technology.

Lovejoy, W. F., and Homan, P. T. *Methods of Estimating Reserves of Crude Oil, Natural Gas, and Natural Gas Liquids.* Washington, D.C.: Resources for the Future, 1965.

Mansfield, D. "Saudi Arabia: Uncertainty Hits Forward Planning." *Petroleum Economist* (August 1978).

Ministerio de Energia y Minas, Republica de Venezuela. *Petroleo y Otros Datos Estadisticos.* Caracas, Venezuela: Ministry of Energy and Mines, various years.

Muskat, M. "The Proved Crude Oil Reserves of the United States." *Journal of Petroleum Geology* (September 1963): 915–21.

National Petroleum Council. *Enhanced Oil Recovery.* Washington, D.C.: NPC, December 1976.

National Petroleum Council. *National Petroleum Council Committee on U.S. Energy Outlook: An Interim Report. An Interim Appraisal by the Oil Supply Task Group.* Washington, D.C.: NPC, 1972.

National Petroleum Council *Report of the National Petroleum Council Committee on Proved Petroleum and Natural Gas Reserves and Availability.* Washington, D.C.: NPC, May 15, 1961.

National Petroleum Council. *U.S. Energy Outlook – Oil and Gas Availability.* Washington, D.C.: NPC, 1973.

Oil and Gas Journal (various issues).

Petroleos Mexicanos. *Comparacencia del Sr. Ing. Jorge Diaz Serrano . . . ante el H. Congress de la Union,* Pemex, October 1977.

Petroleos Mexicanos. *Generalidades Del Proyecto de Construccion del Gasoducto Cactus-Reynosa.* Pemex, July 1977.

Petroleos Mexicanos. *Informe del Director General de Petroleos Mexicanos.* Pemex, March 18. Annual.

Petroleos Mexicanos. *Memoria de Labores.* Pemex, annual.

Petroleos Mexicanos, *Potencial Actual y Futuro de la Industria Petrolera en Mexico.* Pemex, March 1978.

Petroleum Economist. Monthly.

Potential Gas Committee. *Potential Supply of Natural Gas in the United States.* Colorado School of Mines, Golden, Colo., 1976.

Smith, J. P. "Iraq Sits atop Huge Reserves." *The Washington Post* (August 7, 1978).

United States Central Intelligence Agency. *International Energy Statistical Review.* Monthly.

United States Department of Commerce, Bureau of the Census. *Annual Survey of Oil and Gas.* MA-13K(6)-1. Washington, D.C.: U.S. Government Printing Office, annual.

United States Department of Energy. *Monthly Energy Review.*

Wall Street Journal (November 14, 1977).

World Oil (February 15, various years).

PART III: OIL AND GAS: ESTIMATION OF UNDISCOVERED RESOURCES

Allais, M. "Method of Appraising Economic Prospects of Mining Exploration over Large Territories," *Management Science* 3 (1957): 285-347.

Alpert, M., and Raiffa, H. "A Progress Report on the Training of Probability Assessors." Unpublished manuscript, Harvard University, Cambridge, Mass., 1969.

Arps, J. J., Mortada, M., and Smith, A. E. "Relationships between Proved Reserves and Exploratory Effort." *Journal of Petroleum Geology* (June 1971): 671-75.

Arps, J. J., and Roberts, T. G. "Economics of Drilling for Cretaceous Oil Production on the East Flank of the Denver-Julesberg Basin." *AAPG Bulletin* 42, no. 11(1958): 2549-66.

Attanasi, E. D. "The Nature of Firm Expectations in Petroleum Exploration." *Land Economics* 55, no.3(1979): 299-312.

Attanasi, E. D., and Drew, L. J. "Field Expectations and the Determinants of Wildcat Drilling." *Southern Economic Journal,* 44. no. 1 (1977): 53-67.

Attanasi, E. D., Drew, L. J. and Schuenemeyer, J. H. "An Application to Supply Modeling." In *Petroleum-Resource Appraisal and Discovery Rate Forecasting in Partially Explored Regions,* U.S. Geological Survey Professional Paper 1138, Parts A, B, C, U.S. Department of the Interior, 1980.

Barouch, E., and Kaufman, G. M. "Oil and Gas Discovery Modelled as a Sampling Proportional to Random Size." MIT Sloan School of Management Working Paper 831-76, Massachusetts Institute of Technology, Cambridge, Mass., 1976.

Barouch, E., and Kaufman, G. M. "On Sums of Lognormal Random Variables." MIT Sloan Scgool of Management Working Paper 831-76, Cambridge, Mass. 1976.

Barouch, E. and G. M. Kaufman. "Estimation of Undiscovered Oil and Gas," In *Proceedings of the Symposia in Applied Mathematics,* vol. 21 (Providence, R.I.: American Mathematical Society, 1977). pp. 77-91.

Beall, A. D. "Dynamics of Petroleum Industry Investment in the North Sea." MIT Energy Laboratory Working Paper no. MIT-EL 76-007WP, Cambridge, Mass., June 1976.

Bernoulli. *Ars Conjectandi.* Basil, Switzerland, 1713.

Blondel, F., and Ventura, E. *Le Structure de la Distribution des Produits Mineraux dans le Monde.* Paris, 1956.

Bloomfield, P., Deffeyes, K. S., Watson, G.S., Benjamini, Y., and Stine, R. A. "Volume and Area of Oil Fields and Their Impact on the Order of Discovery." Resource Estimation and Validation Project, Department of Statistics and Geology, Princeton University, Princeton, N.J., February 1979.

Bromberg, L., and Hartigan, J. A. "Report to the Federal Energy Administration: United States Reserves of Oil and Gas." Department of Statistics, Yale University, New Haven, Conn., 1967.

Bromberg, L. and Hartigan, J. A. "U.S. Reserves of Oil and Gas." Report to the Federal Energy Administration. Department of Statistics, Yale University, New Haven, Conn., May 1975.

Brown, G., and J. W. Sanders. "Longnormal Genesis." *Journal of Applied Probability* 18 (1981): 542–47.

Bryant, T. A. "Policy Implications of Oil and Natural Gas Reserve Estimates." Unpublished M. S. thesis, Massachusetts Institute of Technology, Cambridge, Mass., May 1976.

Capen, E. C. "The Difficulty of Assessing Uncertainty." *Journal of Petroleum Technology* (August 1976): 843–50.

Cox, D. "Further Results on Tests of Separate Families of Hypotheses." *Journal of the Royal Statistical Society* Series B 24, no. 2 (1962): 406–23.

Cram, I., ed. *Future Petroleum Provinces of the United States – Their Geology and Potential.* Tulsa, Okla.: American Association of Petroleum Geologists, 1971.

Day, D. T. "The Petroleum Resources of the United States." In *Papers on the Conservation of Mineral Resources.* U.S. Geological Survey Bulletin 394, 1909, pp. 30–50.

Dempster, A. P., N. M. Laird, and D. R. Rubin. "Maximum Likelihood from Incomplete Data via the EM Algorithm." *Journal of the Royal Statistical Society* Series B 39, no. 1 (1977): 1–38.

Dietzman, William D. et al. "Nigeria – An Assessment of Crude Oil Potential." Analysis Memorandum prepared for Energy Information Administration, June 7, 1979.

Drew, L. J., Schuenemeyer, J. H., and Root, D. H. "An Application to the Denver Basin." In *Petroleum-Resource Appraisal and Discovery Rate Forecasting in Partially Explored Regions,* U.S. Geological Survey Professional Paper 1138 – Parts A, B, C, U.S. Department of the Interior, 1980.

Dunbar, C. O. *Historical Geology.* New York: Wiley, 1948.

Duncan, D. C., and McKelvey, V. E. "United States and World Resources of Energy." In *Symposium on Fuel and Energy Economics.* American Chemical Society, 9, no. 2, (1965): 1–17.

Eckbo, P. "Estimating Offshore Exploration Development and Production Costs." MIT Energy Laboratory Working Paper, Cambridge, Mass., September 1977.

Eckbo P., Jacoby, H., and Smith J. "Oil Supply Forecasting: A Disaggregated Approach." *Bell Journal of Economics and Management Science* 9, no. 1 (Spring 1978): 218–35.

Energy, Mines and Resources Bureau (Canada). *Oil and Gas Resources of Canada, 1976.* EMS Report EP77-1, 1977.

Erickson E., and Spann, R. "Supply Response in the Regulated Industry: The Case of Natural Gas." *Bell Journal of Economics and Management Science* 2, no. 1 (Spring 1971): 94–121.

Fan, P. Y. "Computational Problems in Modelling the Oil and Gas Discovery Process." Unpublished MS Thesis, Sloan School of Management, Massachusetts Institute of Technology, Cambridge, Mass. 1976.

Fisher, F. M. *Supply and Costs in the U.S. Petroleum Industry.* Baltimore: The Johns Hopkins University Press, 1964.

Freeman, D. A. "Statistics and the Scientific Method." In *Proceedings of the Social Science Research Council Conference on Analyzing Longitudinal Data for Age, Period and Cohorts Effects,* forthcoming.

Gompertz, B. "On the Nature of the Function Expressive of the Law of Human Mortality and on a New Mode of Determining the Value of Life Contingencies." *Philosophical Transactions of the Royal Society of London* 155 (1925): 513–85.

Grayson, C. J. *Decision under Uncertainty: Drilling Decisions by Oil and Gas Operators.* Harvard Business School Division of Research, Harvard University, Boston, Mass., 1960.

Haldane, J. B. S. *Possible Worlds.* New York: Harper and Brothers, 1928.

Hendricks, T. A. "Resources of Oil, Gas, and Natural Gas Liquids in the United States and in the World." In *U.S. Geological Survey Circular 522.* 1965.

Hogarth, R. M. "Cognitive Processes and the Assessment of Subjective Probability Distributions." *Journal of the American Statistical Association* 70 (1975): 271–94.

Hogarth, R. M. "Methods for Aggregating Opinions." In *Fifth Research Conference on Subjective Probability, Utility, and Decision Making.* European Institute of Business Administration and Centre European d'Education Permanente, 1976.

Holaday, B., and Houghton, J. "Private Sector Supply Forecasting and Decision Making." *Energy Modeling Forum* 5 (1979).

Hotelling, H. "The Economics of Exhaustible Resources." *Journal of Political Economy* 39, no. 2 (1931): 137–75.

Hubbert, M. K. "Nuclear Energy, and the Fossil Fuels." In *Drilling and Production Practice.* American Petroleum Institute, 1956, pp. 7–25.

Hubbert, M. K. Unpublished U.S. Geological Survey paper presented to National Resources Subcommittee of the Federal Council on Science and Technology, February 1961.

Hubbert, M. K. "Energy Resources: A Report to the Committee on Natural Resources: National Academy of Sciences." National Research Council Publication 1000–D, 1962.

Hubbert, M. K. "History of Petroleum Geology and Its Bearing upon Present and Future Exploration." *AAPG Bulletin* 50 (1966): 2,504–18.

Hubbert, M. K. "Degree of Advancement of Petroleum Exploration in the United States." *AAPG Bulletin* 51, no. 11 (1967): 2,207–2,227.

Hubbert, M. K. "U.S. Energy Resources: A Review as of 1972." In *A National Fuel and Energy Policy Study, Part I.* Serial No. 93–40. Committee on Interior and Insular Affairs, U.S. Senate, 1974, pp. 101–103.

Hubbert, M. K. "Ratio between Recoverable Oil per Unit Volume of Sediments for Future Exploratory Drilling to That of the Past for the Coterminous United States." Appendix to Section II, of Report of Panel on Estimate of Mineral Resources and Reserves, In *Mineral Resources and the Environment*, National Research Council, 1975.

Hudson. E. A., and Jorgenson, D. W. "U. S. Energy Policy and Economic Growth 1975-2000." *Bell Journal of Economics and Management Science* 5 no. 2 (August 1974): 461-514.

ICF Incorporated, "A Review of the Methodology of Selected U. S. Oil and Gas Supply Models." Alexandria, Va., 1979.

Jones, R. W. "A Quantitative Geologic Approach to Prediction of Petroleum Resources." In *Methods of Estimating the Volume of Undiscovered Oil and Gas Resources.* Edited by J. D. Haus. Tulsa, Okla.: American Association of Petroleum Geologists, 1975, pp. 186-95.

Kalecki, M. "On the Gibrat Distribution." *Econometrica* 13 (1945).

Kaufman, G. M. *Ststistical Decision and Related Techniques in Oil and Gas Exploration.* New York: Prentice-Hall, 1962.

Kaufman, G. M. "Issues Past and Present in Modelling Oil and Gas Supply." MIT Energy Laboratory Working Paper no. MIT-EL 80-032WP, Cambridge, Mass. September 1980.

Kaufman, G. M., Balcer, Y., and Kruyt, D. "A Probabilistic Model of Oil and Gas Discovery." In *Studies in Geology, No. 1 – Methods of Estimating the Volume of Undiscovered Oil and Gas Resources* (Tulsa, Okla.: American Association of Petroleum Geologists, 1975), pp. 113-42.

Kaufman, G. M., Runngaldier, W., and Livne Z. "Predicting the Time rate of Supply from a Petroleum Play." In *Economics of Exploration for Energy Resources, vol. 26 of Contemporary Studies in Economics and Financial Analysis.* Edited by J. Ramsey, pp. 69-102.

Kaufman, G. M. and Wang, J. "Model Mis-Specification and the Princeton Study of Volume and Area of Oil Fields and Their Impact on the Order of Discovery." MIT Energy Laboratory Working Paper no. MIT-EL 80-003WP Cambridge, Mass. January 1980.

Khazzoom, J. D. "The FPC Staff's Econometric Model of Natural Gas Supply in the United States." *Bell Journal of Economics and Management Science* 2, no. 1 (Spring, 1971): 51-93.

Krige, D. G. "A Statistical Approach to Some Basic Mine Valuation Problems on the Witwatersrand," *Journal of the Chemical, Metallurgical, and Mining Society of South Africa* 52, no. 6 (1951): 119-39.

LaPlace, P. S. *Essai Philosophique sur les Probabilities.* 5th ed. Paris: Bochelier, 1825.

Lindley, D. *Making Decisions.* New York: Wiley-Interscience, 1971.

MacAvoy, P. W., and Pindyck R. R. "Alternative Regulatory Policies for Dealing with the Natural Gas Shortage." *Bell Journal of Economics and Management Science* 4, no. 2 (Spring 1973): 454-98.

MacAvoy, P. W., and Pindyck R. S. *The Economics of the Natural Gas Shortage* (1960-1980). Amsterdam: North Holland, 1975.

MacAvoy, P. W., and Pindyck, R. S. *Price Controls and the Natural Gas Shortage.* Washington, D.C.: American Enterprise Institute, 1975.

Mallory, P. F. "Accelerated National Gas Resource Appraisal (ANOGRE)." In *Methods of Estimating the Volume of Undiscovered Oil and Gas Resources* Edited by J. D. Haun. Tulsa, Okla.: American Association of Petroleum Geologists, 1975, pp. 23–30.

Manne, A. S. "ETA: A Method for Energy Technology Assessment." *Bell Journal of Economics and Management Science* 7, no. 2 (August 1976): 379–406.

Mayer, L. S., Silverman, R., Zeger, S. L., and Bruce, A. G. "Modelling the Roles of Domestic Crude Oil Discovery and Production." Resource Estimation and Validation Project, Department of Statistics and Geology, Princeton University, Princeton, N.J. February 1979.

McCrossan, R. G. "An Analysis of Size Frequency Distribution of Oil and Gas Reserves of Western Canada." *Canadian Journal of Earth Sciences* 6, no. 201.

McKelvey, V. E. "Mineral Resources Estimates and Public Policy." *American Scientist* 60 (1972): 32–40.

McKelvey, V. E. "U.S. Energy Resources: A Review as of 1972." In *A National Fuel and Energy Policy Study, Part I.* Serial No. 93–40/38. Committee on Interior and Insular Affairs, U.S. Senate, 1974.

Meisner, J., and Demiren, F. "The Creaming Method: A Bayesian Procedure to Forecast Future Oil and Gas Discoveries in Mature Exploration Provinces." *Journal of the Royal Statistical Society* Series A 143 (1980).

Moore, C. L. "Projections of U.S. Petroleum Supply to 1980." Washington, D.C.: U.S. Department of the Interior, Office of Oil and Gas, 1966.

Moore, P. F. Appendix F. In *Future Petroleum Provinces of the United States.* Edited by Ira Cram. National Petroleum Council, 1970.

Moore, Peter F. "The Use of Geological Models in Prospecting for Stratigraphic Traps." Unpublished manuscript, 1974.

Moore, P. G. "The Managers' Struggles with Uncertainty." *Journal of the Royal Statistical Society* Series A, 140 (1977): 129–65.

Nairn, A. E. M. "Uniformitarianism and Environment." *Paleogeography, Paleoclimatology, Paleoecology* 1 (1965): 5–11.

Nelligan, J. D. "Petroleum Resources Analysis within Geologically Homogeneous Classes." PhD thesis, Department of Mathematics and Computer Sciences, Clarkson College of Technology, June 1980.

Oil and Gas Journal (various issues).

Pelto, C. R. "Forecasting Ultimate Oil Recovery." SPE Paper 4261. American Institute of Mining, Metallurgical, and Petroleum Engineers, 1973.

Petrov, A. A. "Testing Statistical Hypotheses on the Type of a Distribution on the Basis of Small Samples." *Theory of Probability Applications* 1 (1956): 223–45.

Pindyck, R. S. "The Regulatory Implications of Three Alternative Econometric Supply Models of Natural Gas." *Bell Journal of Economics and Management Science* 5, no. 2 (Autumn 1974): 633–45.

Pindyck, R. S. "Higher Energy Prices and the Supply of Natural Gas." *Energy Systems and Policy* 2, no. 2 (1978): 177–209.

Phillips, W. G. B. "Statistical Estimation of Global Mineral Resources." *Resources Policy* (December 1977): 168–80.

Pogue, S. E., and Hill, K. E. *Future Growth and Financial Requirements of the World Petroleum Industry*. The Chase Manhattan Bank, Petroleum Department, New York, 1956.

Pratt, W. E., Weeks, and Stebinger. Standard Oil of New Jersey Projections, 1942-1950.

Pratt, W. E. "The Impact of Peaceful Uses of Atomic Energy on the Petroleum Industry." In *Peaceful Uses of Atomic Energy*. Vol. 2. Joint Committee on Atomic Energy, U. S. Congress, 1956, pp. 89–105.

Press, S. J. "Qualitative Controlled Feedback for Forming Group Judgments and Making Decisions." Technical Report No. 35. University of California, Department of Statistics, Riverside, Calif., 1978.

Prokhorov, Y. V. "On the Lognormal Distribution in Geochemical Problems." *Theory of Probability and Its Applications* 7 (1967): 169–173.

Richards, F. J. "A Flexible Growth Curve for Empirical Use." *Journal of Experimental Botany* 10 (1959): 290–300.

Root, D. H. "Future Supply of Oil and Gas from the Permian Basin of West Texas and Southeastern New Mexico, U. S. Geological Survey Circular 828, U. S. Department of the Interior, 1980.

Root, D. H., and Schuenemeyer, J. H. "Mathematical Foundations." In *Petroleum-Resource Appraisal and Discovery Rate Forecasting in Partially Explored Regions,* U. S. Geological Survey Professional Paper 1138 —Parts A, B, C, U. S. Department of the Interior, 1980.

Rose, P. "Procedures for Assessing U. S. Petroleum Resources and Utilization of Results." In *First IIASA Conference on Energy Resources.* Edited by M. Grenon. Luxenburg, Austria: IIASA, 1975, pp. 291–310.

Ryan, J. M. "National Academy of Sciences Report on Energy Resources: Discussion of Limitations of Logistic Projections." *AAPG Bulletin* 49 (1965).

Ryan, J. M. "Limitations of Statistical Methods for Predicting Petroleum and Natural Gas Reserves and Availability." *Journal of Petroleum Technology* (1968): 281–86.

Ryan, J. M. "An Analysis of Crude-Oil Discovery Rate in Alberta." *Bulletin of Canadian Petroleum Geology* 21, no. 2 (June 1973): 219–35.

Savage, L. J. "Elicitation of Personal Probabilities and Expectations." *Journal of the American Statistical Association* 66 (1971): 783–801.

Scheunemeyer, J. H., and Root, D. H. "Resource Appraisal and Discovery Rate Forecasting in Partially Explored Regions: Part C." U. S. Geological Survey Professional Paper, forthcoming.

Singer, D. A., and Drew, L. J. "The Area of Influence of an Exploratory Hole." *Economic Geology* 71, no. 3 (1976): 643–47.

Smith, J. L .and Ward, G. L. "Maximum Likelihood Estimates of the Size Distribution of North Sea Oil Deposits." MIT Energy Laboratory Working Paper no. MIT-EL 80–027WP, Cambridge, Mass., 1980.

Steiner, I. D. "Models for Inferring Relationships Between Group Sizes and Potential Group Activity." *Behavioral Science* 11 (1966): 273–83.

U.S. Geological Survey, Oil and Gas Branch. "Geologic Estimates of Undiscovered and Recoverable Oil and Gas Resources in the United States." Circular 725, U.S. Department of the Interior, 1975.

U.S. Geological Survey. "Future Supply of Oil and Gas from the Permian Basin of West Texas and Southeastern New Mexico." U.S. Department of the Interior Geological Survey Circular 828, 1980.

Wang. J. "Adaptive Optimal Control Applied to Discovery and Supply of Petroleum." Unpublished PhD Thesis, Department of Aetonautics and Astronautics, Massachusetts Institute of Technology, Cambridge, Mass., 1980.

White, D.C.; Garrett, R.W. Jr.; Marsh, G.P.; Baker, R.H.; and Gohman, H.M. "Assessing Regional Oil and Gas Potential." In *AAPG Studies in Geology, No. 1.: Methods of Estimating the Volume of Undiscovered Oil and Gas Reserves.* Tulsa, Okla.: American Association of Petroleum Geologists. 1975.

Winkler, R.L. *The American Statistician* 32, no. 2 (May 1978).

Wiorkowski, J.J. "Estimating Volumes of Remaining Fossil Fuel Resources: A Critical Review." *Journal of the American Statistical Association* 76, no. 375 (September 1981): 534–47.

Wood, D. "Model Assessment and the Policy Research Process: Current Practice and Future Promise." Workshop on Validation and Assessment of Energy Models, NBS, Gaithersburg, Md., January 10–11, 1979.

Zapp, A.D. *Future Petroleum Producing Capacity of the United States."* U.S. Geological Survey Bulletin 1142H. Washington, D.C.: U.S. Government Printing Office, 1962.

PART IV: COAL: AN ECONOMIC INTERPRETATION OF RESERVE ESTIMATES

American Institute of Mining Engineers. *Mining Engineering Handbook.* Summer 1973.

Averitt, P. *Coal Resources of the United States.* U.S. Geological Survey Bulletin 1412, January 1, 1974. Washington, D.C.: U.S. Government Printing Office, 1975.

Boulter, G. "Cyclical Methods–Draglines and Clamshells." In *Surface Mining.* Edited by E.P. Pfleider. New York: American Institute of Mining, Metallurgical, and Petroleum Engineers, 1968.

ICF Incorporated *The National Coal Model: Description and Documentation.* Contract No. C0–05–50198–00, submitted to Federal Energy Administration, Washington, D.C., 1976.

Keystone Coal Industry Manual. New York: McGraw-Hill, various years.

Lowrie, R. *Recovery Percentage of Bituminous Coal Deposits in the United States,* Part I, Underground Mines. U.S. Bureau of Mines Report of Investigations 7109. Washington, D.C.: U.S. Government Printing Office, 1968.

Matson, R.E., and Blume, J.W. *Quality and Reserves of Strippable Coal, Selected Deposits, Southeastern Montana.* Bulletin 91. Butte: Montana Bureau of Mines and Geology, 1974.

Paul Weir Company. *Economic Study of Coal Reserves in Pike County Kentucky and Belleville, Illinois.* Job No. 1555, January 1972.

Peters, W. C. *Exploration and Mining Geology.* New York: Wiley and Sons, 1978.

Rumfelt, H. "Computer Method for Estimating Proper Machinery Mass for Stripping Overburden." *Mining Engineering* 13 (1961): 480-87.

Simon, J. A., and Smith, W. H. *An Evaluation of Illinois Coal Reserve Estimates.* Proceedings of the Illinois Mining Institute, 1968.

U. S. Bureau of Mines. *Basic Capital Investment and Operating Costs for Coal. Strip Mines.* Information Circular 8535. Washington, D. C.: U. S. Government Printing Office, 1972.

U. S. Bureau of Mines. *Projects to Expand Fuel Sources in Western States.* Information Circular 8719. Washington, D.C.: U. S. Government Printing Office, 1976.

U. S. Bureau of Mines. *Strippable Reserves of Bituminous Coal and Lignite in the United States.* Information Circular 8531. Washington, D.C.: U. S. Government Printing Office, 1971.

U. S. Bureau of Mines. *The Reserve Base of U. S. Coals by Sulfur Content.* Information Circulars 8680 and 8693. Washington, D.C.: U. S. Government Printing Office, 1975.

U. S. Geological Survey. *Coal Resource Classification System of the U.S. Bureau of Mines and the U.S. Geological Survey.* U.S. Geological Survey Bulletin 1450-B. Washington, D.C.: U. S. Government Printing Office, 1975.

U. S. Geological Survey. *Principles of the Mineral Resources Classification System of the U.S. Bureau of Mines and U.S. Bureau of Mines and U.S. Geological Survey.* U.S. Geological Survey Bulletin 1540-A. Washington, D.C.: U. S. Government Printing Office, 1976.

Woodruff, S. *Methods of Working Coal Mines,* vol. 3. Elmsford, N.Y.: Pergamon Press, 1966.

Zimmerman, M. B. "Estimating a Policy Model of U. S. Coal Supply." *Materials and Society* 2 (1978): 67-83.

PART V: URANIUM: ESTIMATES OF RESERVES AND RESOURCES

Agterberg. F. P., and S. R. Divi. "A Statistical Model for the Distribution of Copper, Lead, and Zinc in the Canadian Appalachian Region." *Economic Geology* 73, no. 2 (March-April 1978): 230-45.

Ahmed, S. B. "Economics of Nuclear Fuels." Unpublished manuscript, Western Kentucky University, Bowling Green, 1978.

Appelin, C. W. "The Meaning and Distribution of Uranium Ore Reserves." Grand Junction Office Memorandum, U.S. Atomic Energy Commission, January 1, 1972.

Battelle-Northwest. *Assessment of Uranium and Thorium Resources in the United States and the Effect of Policy Alternatives.* Richland, Wash.: December 1974.

Blanc, R. P. "U.S. Uranium Potential Resources." In COMRATE, *Reserves and Resources of Uranium in the United States*. Washington, D.S.: National Academy of Sciences, 1975.

Brinck, J. W. "Calculating the World's Uranium Resources." *Euratom Bulletin* 6, no. 4 (1967).

Brinck, J. W. *The Prediction of Mineral Resources and Long-Life Price Trends in the Non-Ferrous Metal Mining Industry*. Section 4, 14th IGC, 1972.

Charles River Associates. *Uranium Price Formation*. EPRI–EA 498. Palo Alto, Calif.: Electric Power Research Institute, October 1977.

Committee on Mineral Resources and Environment. (COMRATE). *Reserves and Resources of Uranium in the United States*. Washington, D.C.: National Academy of Sciences, 1975.

Committee on Nuclear and Alternative Energy Systems (CONAES), "Problems of U.S. Uranium Resources and Supply to the Year 1010." Report of the Uranium Resource Group. Washington, D. C.: National Academy of Sciences, 1978.

Curry, D. L. "Estimation of Potential Uranium Resources." In *Concepts of Uranium Resources and Producibility Workshops*. Washington, D.C.: National Academy of Sciences, September 20, 1977.

deHalas, D. R. *EPRI–A Program to Investigate Uranium Supply and Price Dynamics*. Contract SOA-77-401. Prepared for the Electric Power Research Institute, Palo Alto, Calif., September 1977.

Denver Research Institute. *Development of a Conceptual U_3O_8 Production Forecasting Model*. Prepared for Grand Junction Office, U. S. Energy Research and Development Administration, Denver, Colo., May 1977.

Drew, M. W. "U.S. Uranium Deposits: A Geostatistical Model," *Resources Policy* (March 1977).

Erickson, R. L. "Crustal Abundance of Elements, and Mineral Reserves, and Resources." In *United States Mineral Resources* U.S. Geological Survey Professional Paper 820, 1973, pp. 21–25.

Finch, W. I. et al. "Nuclear Fuels." In *United States Mineral Resources,* U.S. Geological Survey Professional Paper 820, 1973, pp. 455–68.

Gaskins, D. W., Jr., and Haring, J. R. Letter to *Science* in response to "United States Uranium Resources—An Analysis of Historical Data" by M A. Lieberman, 196, May 6, 1977.

Grand Junction Office of the U.S. Atomic Energy Commission. "Nuclear Fuel Resource Evaluation Concepts, Uses, Limitations." GJO-105. Washington, D.C., May 1973.

Grand Junction Office of the U. S. Department of Energy. *Uranium Industry Seminar*. GJO-108(77), 1977.

Grand Junction Office of the U. S. Department of Energy. *A System for Deriving Favorability Factors in Estimating Potential Resources of Uranium in the U.S.* For internal use only, date unknown.

Harris, D. P. *A Subjective Probability Appraisal of Uranium Resources in the State of New Mexico*. GJO-110(76). Grand Junction Office, U.S. Energy Research and Development Administration, December 1975.

Harris, D. P. *A Critique of the NURE Appraisal Procedure as a Basis for a Probabilistic Description of Potential Resources, and Guides to Preferred Practice.* Unpublished manuscript prepared at the request of U. S. Energy Research and Development Administration, 1976.

Harris, D. P. *The Estimation of Uranium Resources by Life-Cycle or Discovery-Rate Models: A Critique.* GJO-112(76). Grand Junction Office, U. S. Energy Research and Development Administration, October 1976.

Harris, D. P. *Information and Conceptual Issues of Uranium Resources and Potential Supply* Presented at a Workshop of Energy Information for the U. S. Energy Information Administration, sponsored by The Institute of Energy Studies, Stanford University, December 1977.

Harris, D. P. *Quantitative Methods for the Appraisal of Mineral Resources.* Research Report GJO-6344. U. S. Energy Research and Development Administration Contracts AT-05-1-16344 and E(05-1)-1665, January 1, 1977.

Harris, D. P. "Undiscovered Uranium Resources and Potential Supply: A Nontechnical Description of Methods for Estimation and Comment on Estimates Made by U. S. ERDA, Lieberman, and the European School (Brinck and PAU)." Part VI of *Mineral Endowment, Resources, and Potential Supply: Theory, Methods for Appraisal, and Case Studies.* MINRESCO, Tucson, Ariz., January 1, 1977.

Hetland, D. L. "Potential Resources." In *Uranium Industry Seminar,* GJO-108 (75). Grand Junction Office, U. S. Energy Research and Development Administration, October 1975.

Hetland, D. L "Discussion of the Preliminary NURE Report and Potential Resources." In *Uranium Industry Seminar,* GJO-108(76). Grand Junction Office, U. S. Energy Research and Development Administration, October 1976.

Hetland, D. L., and W. D. Grundy. "Potential Resources." In *Uranium Industry Seminar,* GJO-108(77). Grand Junction Office, U. S. Department of Energy, 1977.

Hetland, D. L., and W. D. Grundy. "Potential Uranium Resources." In *Uranium Industry Seminar.* GJO-108(80), Grand Junction Office, U. S. Department of Energy, 1980.

Klemenic, J. "Uranium Supply and Associated Economics." In *Uranium Industry Seminar.* GJO-108(75). Grand Junction Office, U. S. Energy Research and Development Administration, October 1975.

Klemenic, J. "Analysis and Trends in Uranium Supply." In *Uranium Industry Seminar.* GJO-108(76). Grand Junction Office, U. S. Energy Research and Development Administration, October 1975.

Klemenic, J. *Examples of Overall Economies in a Future Cycle of Uranium Concentrate Production for Assumed Open Pit and Underground Mining Operations.* Grand Junction Office, U. S. Energy Research and Development Administration, October 19, 1976.

Klemenic, J. "Production Forecasting Methodology." In *Concepts of Uranium Resources and Producibility Workshop.* Washington D. C.: National Academy of Sciences, September 20, 1977.

Klemenic, J. "Production Capability and Supply." In *Uranium Industry Seminar,* GJO-108(77). Grand Junction Office, U. S. Department of Energy, October 1977.

Klemenic. J., and L. R. Sanders. "Discovery Rates and Costs for Additions to Constant-Dollar Uranium Reserves." In *Uranium Industry Seminar,* GJO–108(76). Grand Junction Office, U.S. Energy Research and Development Administration, 1976.

Lieberman, M. A. "United States Uranium Resources—An Analysis of Historical Data." *Science* 192, no. 4238 (1976): 431–436.

McKelvey, V. E. "Mineral Resource Estimates and Public Policy." In *United States Mineral Resources,* U.S. Geological Survey Professional Paper 820, 1973, pp. 9–19.

McLuhan, M. *The Medium is the Message.* New York: Random House, 1967.

Meehan, R. J. "Uranium Reserves and Exploration Activity." In *Uranium Industry Seminar,* GJO–108(75). Grand Junction Office, U.S. Energy Research and Development Administration, October 1975.

Meehan, R. J. "Uranium Ore Reserves." In *Uranium Industry Seminar,* GJO–108(76). Grand Junction Office, U.S. Energy Research and Development Administration, October 1976.

Meehan, R. J. "Ore Reserves." In *Uranium Industry Seminar,* GJO–108(77). Grand Junction Office, U.S. Department of Energy, October 1977.

Meehan, R. J., and W. D. Grundy. "Estimation of Uranium Ore Reserves by Statistical Methods and a Digital Computer." Grand Junction Office Memorandum, U.S. Atomic Energy Commission, 1972,

Nuclear Energy Policy Study Group. *Nuclear Issues and Choices.* MITRE Corporation for the Ford Foundation. Cambridge, Mass.: Ballinger, 1977.

NUS Corporation. *Development of U_3O_8 Mining/Milling Cost and Materials Models: Phase I Progress Report.* Vol.1. Prepared for Grand Junction Office, U.S. Energy Research and Development Administration, Denver, Colo. May 1977.

Perl, Lewis, J. "Testimony of Dr. Lewis J. Perl on Behalf of the GESMO Utility Group on U_3O_8 Prices," in the matter of Generic and Environmental Statement on Mixed Oxide Fuel. National Economic Research Association, New York, April 14, 1977.

Rackley, R. I. "Problems of Converting Potential Uranium Resources into Minable Reserves." In *Reserves and Resources of Uranium in the United States.* Washington, D.C.: COMRATE, National Academy of Sciences, 1975.

Schantz, J. J., Jr. "Oil and Gas Resources—Welcome to Uncertainty." *Resources* 58, Resources for the Future. Washington, D.C., March 1978.

Schantz, J. J., Jr. *Resources Terminology: An Examination of Concepts and Terms and Recommendations for Improvements.* EPRI–336. Palo Alto, Calif.: Electric Power Research Institute, August 1975.

Searl, M. F., et al. *Uranium Resources to Meet Long-Term Uranium Requirements.* EPRI SR–5. Palo Alto, Calif.: Electric Power Research Institute; 1974.

Searl, M. F., and Platt, J. "Views on Uranium and Thorium Resources." *Annals of Nuclear Energy* 2 (1975): 751–62.

Searl, M. F., and Platt, J. Letter to *Science* in response to "United States Uranium Resources—An Analysis of Historical Data" by M. A. Lieberman, *Science* vol. 196, May 6, 1977.

Singer, D. A. "Long-Term Adequacy of Metal Resources." *Resources Policy* (June 1977): 127–133.

Skinner, B. F. *Earth Resources*. 2nd ed. New York: Prentice-Hall, 1976.

Skinner, B. F. "A Second Iron Age Ahead?" *American Scientist* 64, no. 3 (1976): 258–69.

Stoll, M. G. "Field Methods of Determining Uranium Ore Reserves." Grand Junction Office Memorandum, Atomic Energy Commission, 1972.

S. M. Stoller Corporation. "Report of Uranium Supply: Task III of EEI Nuclear Fuels Supply Study Program." Appendix II in *Nuclear Fuels Supply*. New York: Edison Electric Institute, 1976.

S. M. Stoller Corporation. *Uranium Data*. EPRI EA–400. Palo Alto, Calif.: Electric Power Research Institute, June 1977.

S. M. Stoller Corporation. *Uranium Exploration Activities in the United States*. EPRI EA–401. Palo Alto, Calif.: Electric Power Research Institute, June 1977.

Taylor, V. "How the U. S. Government Created the Uranium Crisis (and the Coming Uranium Bust)." Los Angeles: Pan Heuristics, June 1977.

Taylor, V. "The Myth of Uranium Scarcity." Unpublished Manuscript, Pan Heuristics, Los Angeles, April 25, 1977.

Thiel, H. *Principles of Econometrics*. New York: John Wiley and Sons, 1971.

INDEX

ABOUT THE AUTHORS

M. A. Adelman is professor of economics at M. I. T. He received his Ph. D from Harvard University and since then has served in numerous professional capacities, most recently as president of the International Association of Energy Economists. Aside from his extensive publications in energy journals, Professor Adelman has written a number of books, including *The Supply & Price of Natural Gas* and *The World Petroleum Market*.

John C. Houghton received his Ph. D. in engineering from Harvard University and currently serves as senior geologist with the U.S. Geological Survey, where he is developing new methods for assessing future reserves of oil, gas, and minerals. In 1979–81 he served as senior policy analyst, Executive Office of the President, where he administered natural resource issues for the president's science advisor in the Office of Science and Technology.

Gordon Kaufman is professor of operations research and management at the Sloan School of Management, M. I. T. His principal research interests involve the applications of mathematics and statistics to problems arising in exploring for and producing oil, gas, and uranium. He is the author of twenty-five papers and a book treating these problems, and has served for nearly twenty years as a consultant to both government and industry on energy related issues.

Martin Zimmerman is associate professor of business economics in the Graduate School of Business at the University of Michigan. He studied at Dartmouth College and received a Ph. D. in economics from M. I. T., where he was on the faculty of the Sloan School of Management. He has written extensively about energy economics, and coal industry, and the economics of nuclear power. He is the author of a recent book, *The U. S. Coal Industry: The Economics of Policy Choice*.